GEOGRAPHY
MARK-UP
LANGUAGE (GML)

GEOGRAPHY MARK-UP LANGUAGE (GML)

Ron Lake
David S. Burggraf
Milan Trninić
Laurie Rae

Galdos Systems Inc.

John Wiley & Sons, Ltd

Email (for orders and customer service enquiries): cs-books@wiley.co.uk
Visit our Home Page on www.wileyeurope.com or www.wiley.com

Other Wiley Editorial Offices

John Wiley & Sons Inc., 111 River Street, Hoboken, NJ 07030, USA

Jossey-Bass, 989 Market Street, San Francisco, CA 94103-1741, USA

Wiley-VCH Verlag GmbH, Boschstr. 12, D-69469 Weinheim, Germany

John Wiley & Sons Australia Ltd, 33 Park Road, Milton, Queensland 4064, Australia

John Wiley & Sons (Asia) Pte Ltd, 2 Clementi Loop #02-01, Jin Xing Distripark, Singapore 129809

John Wiley & Sons Canada Ltd, 22 Worcester Road, Etobicoke, Ontario, Canada M9W 1L1

Wiley also publishes its books in a variety of electronic formats. Some content that appears
in print may not be available in electronic books.

Library of Congress Cataloging-in-Publication Data

Geography mark-up language: foundation for the geo-web / Ron Lake ... [et al.].
      p. cm.
Includes bibliographical references and index.
ISBN 0-470-87153-9 (cloth : alk. paper) – ISBN 0-470-87154-7 (pbk. : alk.
paper)
1. Geography – Data processing. 2. Geography – Computer-aided design.
3. Geographic information systems. I. Lake, Ron, 1949–
G70.2.G4775 2004
910′.285′674 – dc22

2004005078

British Library Cataloguing in Publication Data

A catalogue record for this book is available from the British Library

ISBN 0-470-87153-9 (Cloth)
ISBN 0-470-87154-7 (Paper)

Typeset in 10/12pt TimesTen by Laserwords Private Limited, Chennai, India
Printed and bound in Great Britain by Antony Rowe Ltd, Chippenham, Wiltshire
This book is printed on acid-free paper responsibly manufactured from sustainable forestry
in which at least two trees are planted for each one used for paper production.

Contents

About Ron Lake

Ron Lake is the founder and CEO of Galdos Systems Inc., a world leader in the application of XML and web-service technologies to distributed Geographic Information Systems (GIS). He was the original creator of GML, and a founding member of the OpenGIS Consortium. He has over 30 years of experience in advanced information technology, including spatial information systems, distributed database systems, real-time control systems, embedded computing and systems for simulation and mathematical analysis. He holds M.Sc. (Mathematics) and M.A.Sc. (Aerospace) degrees from the University of Toronto.

Introduction

The Geography Mark-Up Language (GML) and the associated OGC specifications represent a major development in the evolution of Geographic Information Systems (GIS). GML holds the key to integrating these systems into the wider world of electronic-information processing. More profoundly, GML is expected to lead to the development of the Geo-Web, a seamless fabric of geographic data and services that may form the basic infrastructure for managing our impact upon the world around us.

The *Geography Mark-Up Language (GML) 3.0 – Foundation for the Geo-Web* has been organized into the following two separate sections: *Part I GML: An Introduction* and *Part II GML: A Technical Reference Guide*. Part I provides a relatively non-technical overview of the key concepts of GML, and Part II is a technical supplement to the *GML Version 3.00 OpenGIS® Implementation Specification*.

About Part I GML: An Introduction

The object of Part I is to introduce readers to the new world of GML and to provide them with the tools to develop their own view of the impact of GML on business, government and everyday life. Every effort has been made to provide the ideas in as simple a manner as possible. Wherever possible, background information is also provided for supporting concepts and technologies.

Part I should be of interest to anyone new to GML, but it will be of particular interest to managers of software and data-management systems and projects. This part will get you acquainted with the basic concepts of XML and GML and enable you to make decisions on the utility of GML in your projects and software acquisitions.

About Part II GML: A Technical Reference Guide

Part II provides software developers, data modellers and database administrators with detailed information about key constructs and components in GML 3.0. This part is intended to answer questions respecting the meaning and structure of GML schema components, the development of GML application schemas and the use of GML in connection with web services, legacy GIS and relational databases. Readers of Part II should be familiar with XML, XML Schema and the fundamental concepts of software development.

Organization of the Book

The book is organized into two different sections, *Part I GML: An Introduction* and *Part II GML: A Technical Reference*. Part I contains seven chapters and Part II contains 13 chapters, all of which are described below:

PART I GML: AN INTRODUCTION

- **Chapter 1 Once over lightly**. Introduces the key ideas of GML and geospatial web services in non-technical form and provides a brief history of GML.

- **Chapter 2 XML and GML**. Discusses the differences between binary and text-based encodings as an introduction to the basics of XML. Outlines the XML technologies, including XML Schema, that are particularly relevant to GML. Because this material is introductory in nature, you might prefer to skip this chapter if you are familiar with XML and related Internet technologies.

- **Chapter 3 Basic concepts of GML**. Introduces the basic concepts of GML, including the GML object-property model, features, feature relationships and geometries. Some basic encoding examples are provided.

- **Chapter 4 GML core and application schemas**. Provides an overview of the GML core schemas and discusses the key issues concerning GML application schemas that software development managers should be aware of.

- **Chapter 5 Technical issues for deploying GML**. Examines some of the technical issues – such as performance, data management and data volumes – that are associated with deploying GML in real-world applications. Provides examples of GML solutions for real-world deployment.

- **Chapter 6 GML and geospatial web services**. Provides a more in-depth discussion of GML and geospatial web services, particularly in relation to OGC geospatial web services, such as the Web Feature Service (WFS) and the Web Coverage Service (WCS). Discusses some of the technologies that are used for web-service description.

- **Chapter 7 Real-world deployment examples**. Outlines the application of GML in a few fields of interest, including local government, utilities, natural resource management, disaster management and location-based services.

PART II GML 3.0: A TECHNICAL REFERENCE GUIDE

- **Chapter 8 Basic concepts**. Discusses the role of GML in relation to the Geo-Web, the basic concepts of GML instances and schemas, the changes since GML 2 and the background concepts from XML Schema that the reader should be familiar with.

- **Chapter 9 Introducing the GML model and GML features**. Introduces the GML object-property model and discusses the rules for encoding GML features and properties. Also discusses feature collections, feature relationships and other concepts relating to features.

- **Chapter 10 GML core schemas overview**. Provides an overview of the 28 core GML 3.0 schemas and covers the key components defined in the gmlBase.xsd schema.

- **Chapter 11 Developing and managing GML application schemas**. Covers the rules for creating and managing GML application schemas, including specific examples of user-defined feature types. Also discusses schema dependencies in GML 3.0 and the creation of application-specific metadata schemas.
- **Chapter 12 GML geometry**. Discusses the most commonly used constructs from the five geometry schemas provided by GML 3.0, including `LineString`, `Curve`, `Surface` and `Solid`.
- **Chapter 13 GML topology**. Introduces the GML 3.0 topology model and provides encoding examples of the topology primitives, including `Node` and `Edge`.
- **Chapter 14 GML temporal elements and dynamic features**. Covers the constructs for modelling temporal properties in GML 3.0, in particular, temporal objects and dynamic features with time-varying properties.
- **Chapter 15 GML coordinate reference systems**. Discusses the dictionary model for encoding CRS dictionaries, provides an overview of the CRS and support-component definitions and includes examples for various applications that use CRS.
- **Chapter 16 Units of measure, values and observations**. Discusses the units of measure types and the GML model for encoding units of measure dictionaries. Also provides an overview of the different kinds of value objects in GML and includes the rules for encoding observations and observation collections.
- **Chapter 17 GML Coverages**. Introduces the constructs for encoding Coverages in GML and covers the different kinds of Coverages that can be encoded, including multipoint and grid Coverages.
- **Chapter 18 GML default styling**. Discusses the GML styling description mechanism that can be used to apply a visual presentation style to a GML data set.
- **Chapter 19 GML and geospatial web services revisited**. Provides a more technical discussion of GML in relation to geospatial web services, focusing primarily on WSDL definitions and Web Features Service (WFS) request and response messages.
- **Chapter 20 GML, relational databases and legacy GIS**. Provides an introductory reference guide to mapping GML features and geometries to relational databases and legacy Geographic Information Systems (GIS).

Each chapter contains a chapter summary that provides an overview of the concepts covered in the chapter. If you are already familiar with the concepts discussed in the chapter, you might prefer to read only the chapter summaries. This book also has the following four appendices:

- **Appendix A GML core schemas**. Lists all of the core schemas in GML 3.0 in tabular format, describing the purpose and typical use of each schema.
- **Appendix B Resources**. Lists the sources cited throughout the guide, plus additional sources that can provide further information about GML 3.0 and related technologies.

- **Appendix C Glossary of terms**. Contains a glossary of important terms covered in the book.
- **Appendix D XMLSpy tutorial**. Covers the basic steps for using XMLSpy to create GML instance documents and application schemas.

Document Conventions

All of the body text is written in Times Ten.

`Lucida Typewriter` is used for

- all code examples;
- anything that appears in a GML or XML instance or schema, including element names, types, features, properties and attributes;
- anything that appears in a program.

Note that most of the examples in this guide are simple examples, some of which are extracts of complete schemas and instances contained on the *Worked Examples CD*. The following URI is used in most of the references that appear in the code fragments: http://www.ukusa.org. The files 'located' at this URI are not real documents, and should be treated as illustrative examples.

Namespace prefixes are used in all of the examples in this book. The `app` prefix is used for user-defined elements and types (those are not defined in GML or one of the supporting technologies, but in GML application schemas) that 'reside' in the example namespace, http://www.ukusa.org. The `exp` namespace prefix is used for user-defined elements and types that 'reside' in another example namespace, http://www.examples.org. All namespace prefixes are included in the example fragments to illustrate the namespace of each type. Note that namespace prefixes are not included in the text, except in cases in which the context is not clear.

PART I

GML: AN INTRODUCTION

Chapter 1

Once over lightly

This chapter provides a very high-level overview of the Geography Mark-up Language (GML) by answering the following questions:

- What is GML?
- What is a web service?
- What is a geospatial web service?
- What is the role of GML in relation to geospatial web services?
- What is the Geo-Web?
- What is the relationship between GML and G-XML?
- What is the relationship between GML and ISO/TC 211?

If you are already familiar with these topics, you might prefer to skip this chapter or simply review the chapter summary.

1.1 What is GML?

GML is a mark-up language that is used to describe geographic objects in the world around us. By building on broader Internet standards from the World Wide Web Consortium (W3C), GML expresses geographic information in a manner that can be readily shared on the Internet. In particular, GML builds on the eXtensible Mark-up Language (XML), which is discussed in further detail in Chapter 2.

In GML, real-world objects are called features, which are categorized into particular types. GML features can be concrete and tangible, such as rivers, buildings, streets or fire hydrants; or abstract and conceptual, such as political boundaries or health districts.

A feature is described in terms of its properties, which can be geometry properties, such as location, form and extent; or non-geometry properties, such as colour, height, speed and density. For example, a GolfCourse feature can have a property that describes its physical extent and another property that describes its name. Chapter 3 discusses features in more detail.

Note that specific feature types like rivers or roads are not defined in GML itself. These feature types are defined in application schemas, which are typically created by database administrators. Application schemas are covered in Chapter 4.

Geography Mark-up Language (GML).   R. Lake, D. S. Burggraf, M. Trninić, L. Rae © 2004 Galdos Systems Inc.
Published by John Wiley & Sons, Ltd   ISBNs: 0-470-87153-9 (HB); 0-470-87154-7 (PB)

In addition to being a mark-up language that describes objects in the world around us, GML also uses XML to transport these descriptions over the Internet, essentially as text data. GML can be used to transport descriptions of features, such as a Road, and to transport the actual feature data, such as data for a particular Road. This allows for the exchange of descriptions and instances of features, such as rivers or roads, from one person to another, or from one application to another, over the wired or wireless Internet.

1.2   What is a web service?

In the simplest sense, a web service is an application that accepts and processes requests from other applications across a network, such as the Internet. An application on one machine (the 'calling' application) invokes a web service (the 'receiving' application) on another machine to process a request. Typically, the web service is asked to compute something, look something up or return values from a database. Table 1 lists examples of requests that applications might send to a web service.

Table 1   Sample web service requests

Application	Sample Web Service Request
Cut Block Planning Application	A user in the forestry industry wants to get an estimate of the cost of a cut block plan. He provides the location and extent of the cut block.
Cable Location Application	Before digging or excavating, a contractor wants to avoid hitting a cable from the phone company or a gas pipe from the gas company. He supplies the location of the planned excavation.

A web service can also be described as an exchange of messages between the service and its consumer clients. The client sends a request message to the service and the service sends a response message back to the client. Figure 1 illustrates this exchange of messages with a Corn Price Service example. In this example, a client sends a request message to get the current price of corn, and a Corn Price Service obtains the price and sends the client a response message with the current price.

Note that in Figure 1, 'client' refers to the 'calling' application that sends and receives these messages and not to the end-users who might use this information. In other words, the concept of a web service is quite different from that of a portal or web site, which directed at humans and provides visual information in the form of text or images. With a web service, the communication is between programs, that is, the client and the service. Although a web service can present

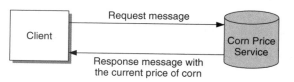

Figure 1   Corn Price Service example.

visual information through a portal or web site, this is not a fundamental aspect of a web service.

Web services rely on standardized Web protocols, such as HyperText Transfer Protocol (HTTP) or Simple Mail Text Protocol (SMTP), to send and receive information over the Internet. It should also be noted that all requests that are sent to a web service are typically formatted in XML.

To describe a web service, you need to know the content and structure of the messages exchanged between the client and the service. Consider again the Corn Price Service example in Figure 1. A client needs to know how to make a request and how to interpret the response. For example, how does the client ask for the price of corn? How does the client interpret the price or the type of currency used? If you want to create your own Corn Price Service, you need to know the answers to these questions.

1.3   What is a geospatial web service?

Geospatial web services are services that deal with geographic information. These web services can do the following:

- Provide access to geographic information stored in a database.
- Perform geographic computations, for example, find the area of a land parcel.
- Perform complex computations that depend on the geometry of a set of geographic objects and their distribution in time and space.
- Return messages that contain geographic information, which can be delivered as text, numeric data or geographic features.

Different industries can use geospatial web services to acquire information. Consider the following simple example, as shown in Figure 2. A trucking company

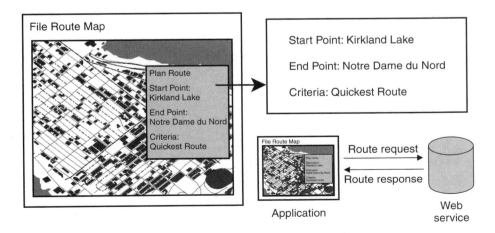

Figure 2 Quickest route example.

wishes to find the quickest route between two cities, taking into account the speed limits along various roads and highways. The client provides the web service with the name and location of the selected start and end points, and the web service returns the quickest route as a collection of road segments.

Now consider another example of a utility company that wants to find a suitable corridor in which to run a new high voltage transmission line. Candidate corridors must meet specific restrictions such as terrain type, absence of major roads, public land ownership and freedom from environmental problems. The utility company's client provides the web service with the corridor end points and associated criteria, and the web service responds with a set of candidate corridor geographic features. Note that some types of geospatial web services may require access to larger amounts of geospatial information than can be practically transmitted in the request message.

1.4   What is the relationship between GML and web services?

GML can play several roles in relation to geospatial web services. For example, GML data can be in the request and response messages that are sent to and from the service. In the routing service example (Figure 2), GML can be used to communicate both the start and end points in the request message and the route determined by the routing service. In the utility example, the client can use GML to communicate the approximate location of the corridor, including its start and end points. The planning service can also use GML to describe the candidate corridors in its response message.

Web services can provide GML descriptions of the geographic content in requests and responses. For example, the following statement, 'this service provides routes between cities', can be provided in GML. Registries that describe geospatial web services can make use of GML descriptions of the geographic content of requests and responses. This makes it easier to find and use web services.

The Open GIS Consortium (OGC) is developing a set of standard geospatial web services, including the following:

- Web Map Service (WMS)
- Web Feature Service (WFS)
- Web Coverage Service (WCS).

GML is used in both the request and response messages of the WFS, which is a standard service for accessing geographic feature data. Chapter 6 covers the relationship between GML and different OGC geospatial web services.

1.5   What is the Open GIS Consortium (OGC)?

The OGC is an international industry consortium whose mission is to 'deliver spatial interface specifications that are openly available for global use' (http://www.opengis.org/about/?page=vision). In the OGC, geoprocessing technology users, vendors and other technology providers work together to establish open-interface standards for different Geographic Information Systems (GISs).

The OGC was established on September 25, 1994, to address the need for an industry consortium that could develop open-interface specifications for geographic information systems. At the end of 1994, the OGC had 20 members. As of September 2003, the OGC had 255 members, including North American, European, Asian and South American corporations, government agencies and universities (http://www.opengis.org/about/?page=members&view=Name).

The OGC maintains liaisons with various members of the world standards community, including the International Organization for Standardization (ISO), the Location Interoperability Forum (LIF), the Object Management Group (OMG), OASIS and the W3C. Many geospatial open-interface specifications have been developed by the OGC, including GML, WFS, WCS and WMS.

1.6 What is the Geo-Web?

The Geo-Web is a distributed network of interconnected geographic information sources and processing services that are

- globally accessible, that is, they live on the Internet and are accessed through standard OGC and W3C interfaces,
- globally integrated data sources that make use of the GML data representation, and where appropriate, explicitly refer to one another.

The Geo-Web is being developed to address the need for access to current and accurate geographic information from diverse sources around the world. Because the diversity of information is often very great, it is difficult to predict the exact information that is required. For example, consider a common urban emergency such as an earthquake. In this situation, you need information about many things, including

- location of water mains and electric utilities,
- location of roads and bridges,
- hazardous material depots,
- distribution of soil types.

The same problem in a different jurisdiction might require information about debris flows and the location of port facilities, tunnels and military installations. How is it possible to accommodate such a diversity of information sources? How do you ensure that this information is current and accurate?

With the Geo-Web, data is locally maintained – that is, it is created and updated by the organization responsible for the data in a particular jurisdiction. For example:

- Railway authorities maintain information about railway lines, signals and overpasses.
- Highway authorities maintain information about highways, highway bridges and signage.

- Municipal or regional governments maintain information about land ownership and local facilities.

Drawing on information sources across the Geo-Web, users can create dynamic 'sub-nets' of geographic information that apply to their problem domain (for example, earthquake response).

The information sources in the Geo-Web include geospatial web services and linked databases, some of which are located within geospatial web services. GML provides the mechanisms for linking geographic information across the databases and web services on the Geo-Web. With these mechanisms, features in one jurisdiction's database can be related to, or can refer to, features in the database of another jurisdiction. With GML, the Geo-Web can provide global access to unified geographic information.

1.7   What is the connection between GML and G-XML?

G-XML is an encoding protocol that is being developed in Japan through a consortium effort funded by the Japanese Ministry of Economy, Trade and Industry, with development and consortium guidance provided by the Database Promotion Center (DPC). There is a shared history between GML and G-XML. At an OGC meeting in Atlanta in February 1999, discussion began between NTT Data, a member of the DPC, and Galdos Systems Inc. of Vancouver, Canada. At the time, Galdos was developing a language called X-GML, and NTT Data had started working on G-XML.

Following discussions between Mr Akifumi Nakai and Mr Takaaki Kami-higashi of NTT Data and Mr Ron Lake of Galdos Systems Inc., NTT Data participated in the OGC WMT I Testbed project as part of the Xbed team. This led to the creation of a prototype XML grammar called Simple Feature XML (SFXML).

Galdos eventually changed the name of SFXML to GML. In October 1999, Galdos submitted a Request for Comments (RFC) about GML to the OGC, based on SFXML and their own research on W3C's Resource Description Framework (RDF). The RFC was released for public comment in December 1999. After considerable discussion, the GML 1.0 specification became an OGC Recommendation Paper in May 2000.

During this period, work on G-XML continued in Japan. Technical discussions also continued between Galdos and NTT Data and – at an organizational level – between the DPC and the OGC to further enhance the development cooperation of the two consortia. In January 2001, an agreement was made to incorporate several key concepts from G-XML into GML 3.0, including an expanded metadata mechanism and support for the following G-XML functionalities:

- Temporal properties
- Default styling
- Observations
- Units of measure

- A network topology model
- The state of moving objects.

As work continued on GML 3.0, G-XML Version 2.0 was adopted as a Japanese Industrial Standard (JIS) in August 2001. The G-XML Working Group and DPC then developed G-XML 3.0, building on GML 3.0. GML 3.0 was approved by the OGC in January 2003, and G-XML 3.0 was subsequently released at the end of May 2003 as a GML application schema for location-based services. Additional information about the DPC and G-XML is available at http://gisclh.dpc.or.jp/gxml/contents-e/.

1.8   What is the connection between ISO TC/211 and GML?

The ISO is one of the key international bodies for all forms of standardization. A technical committee of the ISO, TC/211, has been working on a unified set of standards for geographic information since 1994. TC/211 has published a collection of standards in the 19XXX series, most of which provide an abstract framework for the description of geographic objects, including their relationships and coordinate reference systems. For more information about TC/211, please see http://www.isotc211.org/.

In 2002, the OGC approached the ISO about making GML an ISO standard, and a New Work Item Proposal was created for this purpose. A joint ISO/OGC project team (ISO 19136) is now working on making GML an ISO standard. It is anticipated that the team will add some additional components to GML but will not significantly alter the GML 3.0 specification passed in January 2003 at the OGC. It is anticipated that GML will be an ISO TC/211 draft specification sometime in 2004. In the development of GML 3.0, many GML schemas were based, wholly or partially, on ISO TC/211 specifications, including 19109, 19107, 19108, 19111, 19117 and 19123.

1.9   Chapter summary

GML is an XML-based mark-up language that is used to encode information about real-world objects. In GML, these real-world objects are called features, and they have geometry and non-geometry properties. As an XML-based language, GML-encoded information can be easily transported over the Internet.

Web services are essentially applications that respond to requests sent to them from remote clients over the Internet. Geospatial web services – which are web services that provide access to geographic data and perform data processing – are becoming available.

GML is well suited for encoding the geographic information sent to and from these new geospatial web services. As more geospatial web services emerge, the Geo-Web – a globally integrated web of geographic information – will come into being. GML provides the mechanisms for linking information in the Geo-Web.

GML is not the only XML-based language that has been developed for describing geographic objects. The DPC in Japan has also developed a mark-up language called G-XML. Through a series of discussions, the DPC and the OGC have worked together to converge these two languages. As a result, GML 3.0 has added new sections based on G-XML, and the DPC has issued a new version of G-XML based on the new version of GML.

GML is currently progressing to become an ISO Standard under the TC/211 19XXX series of standards. This is a joint project of the ISO TC/211 Technical Committee and the OGC. It is anticipated that GML will be a Draft International Standard (DIS) sometime in 2004.

References

http://www.opengis.org/about/?page=vision (October 5, 2003).

http://www.opengis.org/about/?page=members&view=Name (September 20, 2003).

http://gisclh.dpc.or.jp/gxml/contents-e/ (September 20, 2003).

http://www.isotc211.org/ (October 5, 2003).

Chapter 2

XML and GML

This chapter discusses the basic ideas of an eXtensible Mark-up Language (XML) encoding and the different members of the XML Technology Family that are particularly relevant to GML. This chapter does not provide a comprehensive discussion of XML. If you feel that you need to read additional material about XML, please refer to *XML: A Manager's Guide* by Kevin Dick (Dick, 2000).

2.1 What is an XML encoding?

In software applications, all information must be encoded. This section examines the differences between two kinds of information encoding: binary and text-based. Then it compares traditional text-based encoding with XML encoding and explains some of the basic concepts and advantages of XML.

With binary encodings, information is encoded in sequences of binary numbers, that is, 0s and 1s. All electronic information systems make use of binary encoding mechanisms, which are the most basic form of information encoding. A text-based encoding is, as the name implies, information encoded in text form. Table 1 compares some of the advantages and disadvantages of binary and text-based encodings.

Text-based encodings have been deployed successfully in a variety of applications, because it is easy for human developers and maintenance personnel to understand these encodings. Text-based encodings have been used for many different purposes, including the following:

- Transmitting data between a sensor device and a receiving computer in measurement and sensor systems.
- Defining the appearance and content of web pages on the Internet.
- Storing and exchanging geographic information. Text-based encoding systems were developed successfully for geographic information more than 15 years ago. For example, 7000 TRIM (1:20,000) map sheets were encoded for the Canadian province of British Columbia in the Spatial Archive and Interchange Format (SAIF) text-based format (http://srmwww.gov.bc.ca/bmgs/trim/trim/trim_overview/trim_program.htm).

XML encodings are a special case of text-based encoding that address many of the limitations of arbitrary text-based encodings. Table 2 shows how XML handles some of these limitations.

Geography Mark-up Language (GML). R. Lake, D. S. Burggraf, M. Trninić, L. Rae © 2004 Galdos Systems Inc.
Published by John Wiley & Sons, Ltd ISBNs: 0-470-87153-9 (HB); 0-470-87154-7 (PB)

Table 1   Binary versus text-based encodings of data content

Text-Based Encodings	Binary Encodings
Are typically much larger than equivalent binary encodings.	Are typically quite compact and can be processed very efficiently.
Are flexible, extensible and easy to understand and require minimal external documentation.	Are not typically flexible or readily extensible. Binary encodings require additional external documentation to be understood by human readers.
Typically require you to develop custom parsing and querying tools.	Typically require a special program to extract the content from the encoding.
Are more appropriate for describing complex real world objects and their interactions.	Are far removed from the description of complex real world objects and their interactions. Although it is possible to develop a direct binary encoding for more complex cases, it is more difficult to develop and manage the evolution of the encoding over time.

Table 2   Traditional text-based and XML encoding

Traditional Text-Based Encoding	XML Encoding
Often did not have a formally defined grammar, which made it difficult to write tools to check the data.	Has a well-defined grammar, which means that general parsing and querying tools can be used to process XML data.
Was not easy to process, because of the lack of standard tools for extracting text-encoded data.	Is a widely used standard that provides many useful tools for editing, querying, transforming and presenting data.

In addition to the above-mentioned advantages, XML also

- ensures data integrity because XML editors can check that documents satisfy defined constraints on content and structure;
- provides extensibility, which means it can also be used to write other mark-up languages, such as GML;
- provides a formal schema language that is more or less human readable. This means that XML encodings are to a large degree self-documenting. In other words, programmers do not need to provide extensive external documentation comments to explain their data coding. This is especially important for data archival applications.

In XML, special words called tags are enclosed in angle brackets, for example <Book>. Content is placed between an opening tag (<Book>) and its corresponding closing tag (</Book>). This content, together with containing tags, is called an element. XML elements may contain other elements that contain elements or text, giving XML a natural hierarchical structure. The following example shows a simple book encoding, which is also called an instance document.

```
<Book>
    <Title>Hunger</Title>
    <Author>
        <First>Knut</First>
        <Last>Hamsun</Last>
    </Author>
    <Chapter>It was a long winter. There seemed no end of
        ⌣it, and the cold burned into the heart of his
        ⌣life.</Chapter>
</Book>
```

In this Book instance document, a single XML element, Book, contains the 'child' elements, Title, Author and Chapter. The Author element has two element children for the author's first (First) and last (Last) names. The Chapter element contains the chapter text.

2.2 The XML Technology Family

One can view XML as a family of technologies that includes the following:

- XML 1.0 and Document Type Definitions (DTDs)
- XML Schema
- XLinking Language (XLink)
- XPointer Language (XPointer)
- Resource Description Framework (RDF) and RDF Schema (RDFS)
- eXtensible Stylesheet Language Transformations (XSLT)
- Scalable Vector Graphics (SVG)
- Web Services Description Language (WSDL)
- Simple Object Access Protocol (SOAP).

Although this list does not reflect the complete set of XML languages, the listed items were selected based on their relevance to GML and the maturity of the specifications. Other important languages include XML Signatures, XML Encryption and the Business Process Execution Language (BPEL). Many new members will be added to the family in the future.

The XML family includes, in particular, languages that allow you to specify the structure of XML documents. These languages include XML 1.0 DTD, XML Schema and RDF (although there are many other languages, only these three are discussed in this book).

2.2.1 XML 1.0 and Document Type Definitions (DTDs)

The original XML specification, XML 1.0, defined the basic grammar of XML. In particular, it defined the concept of a well-formed XML document, which is a document that satisfies well-formedness rules, ensuring that all XML documents retain a certain structural integrity. All XML documents must be well formed. The well-formedness rules include the following:

- All tags that are opened must be closed in order to form complete elements.

- Child tags must be closed before their parent tags so that a child tag does not span the closing of its parent tag.

XML 1.0 also introduced the notion of a DTD, which is a non-XML mechanism for defining and controlling the structure of an XML document. In other words, DTDs are metadata files. The following example shows a DTD for the Book document that was discussed earlier in this chapter.

```
<!element book(title, author, chapter*)>
<!element author(first last)>
<!element first #PCDATA>
<!element last  #PCDATA>
<!element title #PCDATA>
<!element chapter #PCDATA>
```

In the above example, #PCDATA means that the title or chapter elements can contain simple strings of text. The rest of the listing provides the hierarchical structure of the Book element. The Book DTD is essentially a metadata file that specifies how a Book instance document should be structured.

In XML, if an instance document conforms to a specific DTD, it is considered valid with respect to the DTD. Note that while all XML documents must be well formed, they do not have to conform to a DTD. In GML 1.0, DTDs were used as metadata files to specify the content of GML instance documents.

2.2.2   XML Schema

DTDs have many limitations as a data description mechanism for XML documents. For example, DTDs are not written in XML and require a special non-XML parser. In contrast, XML Schemas are written in XML and do not require a special parser. Also, with XML Schemas you do not need to learn a completely different grammar simply for data description.

XML Schemas, however, are much more complex than DTDs and can seem quite daunting to inexperienced programmers. A DTD is also much more compact than the equivalent XML Schema. However, the expressive power and precision of XML Schema far outweighs these disadvantages, and almost all current XML languages, including GML, are written using XML Schema. While XML Schema has not completely replaced DTDs, this is likely to happen in the near future.

XML Schema provides elements that define other elements. The following example shows how to use XML Schema to define the Author element from the Book instance document discussed earlier in this chapter. In this example, the Author element comprises a sequence of two elements, First and Last, which are both strings of text.

```
<element name="Author">
  <complexType>
```

```
      <sequence>
         <element name="First" type="string"/>
         <element name="Last" type="string"/>
      </sequence>
   <complexType>
</element>
```

A powerful language for data description, XML Schema provides an extensive set of built-in data types. XML Schema also provides mechanisms that enable database administrators to build additional data types from the existing built-in ones. In other words, XML Schema is flexible and extensible.

XML Schema has been used to create GML schemas in all versions of GML since version 2.0, which contained three core GML schemas. Many more schemas have been added to GML 3.0. Database administrators can create additional domain-specific GML application schemas by building on the GML 3.0 core schemas. Chapter 4 discusses GML application schemas.

2.2.3   XML Linking Language (XLink) and XML Pointer Language (XPointer)

Most people are familiar with clicking on hyperlinks in web pages to access information on the Internet. The World Wide Web can be viewed as a giant collection of web pages connected by hyperlinks. This makes the web very flexible and allows everyone to reuse information just by linking to it.

Although they are a very powerful mechanism, web-page hyperlinks do not allow you to create associations or relationships

- between multiple resources,
- that are traversable in more than one direction,
- that are simply an expression of the relationship between two kinds of objects,
- that are not related to a visual presentation on a web page.

XLink was introduced to accommodate these more abstract notions of association. With XLink, arbitrary resources – especially individual XML elements – can be associated with one another. XPointer provides XLink with a means of pointing at XML resources. XPointer is based on another language in the XML family called XPath that is also part of XSLT.

By combining XLink and XPointer, it is possible to create links that are far more complex and abstract than conventional web-page hyperlinks. These abstract links are used in GML to express associations between geographic features – such as `Bridge`, `River` and `Road` – and between geographic features and other GML objects, such as geometries and topologies.

2.2.4   Resource Description Framework (RDF) and RDF Schema (RDFS)

The W3C developed the RDF and RDFS specifications to provide a framework for describing different resources on the Internet, such as web pages, persons or places. GML 1.0 was based in part on RDF, in that RDF supported namespaces and a form of inheritance. Although RDF has not been used in GML since Version 2.0, GML has borrowed many useful ideas from RDF, including the following:

- The class/property model, which is used in GML as the basis of the object-property model. Chapter 3 discusses the object-property model in more detail.
- The `rdf:resource` attribute, which points to the values of the property to which the `rdf:resource` attribute is attached. In GML, the `xlink:href` attribute is used instead of `rdf:resource`.
- The notion that a property, encoded as an XML element, can be both an attribute and an association.
- The `about` attribute on the `metaDataProperty` that points to the resource that the associated metadata describes.

2.2.5   eXtensible Stylesheet Language Transformations (XSLT)

XSLT is a language for transforming XML from one XML grammar to another or even to non-XML languages like the HyperText Mark-up Language (HTML) or plain text. Although you can also use other text manipulation languages such as AWK or PERL, XSLT is a popular option for XML. XSLT is a declarative language, which means that you state what you want the language to do and not how to do it.

XSLT specifies a set of transformation rules, such as:

```
Find book titles
and then
take this action
```

In XSLT, this rule can be written as shown below:

```
<xsl:template match="//Book/Title">
    ...Here we put the desired action...
</xsl:template>
```

This XSLT transformation rule searches XML data for Book instances with Title child elements and then executes the desired action. This desired action can be quite simple, such as changing the font of all Title elements, or it can be more complex, such as generating a Table of Contents from the list of titles or invoking further XSLT searches and actions. Note that XSLT is written in XML, and XPath – which is also part of XPointer – is used to write the match part of the listing.

XSLT processors are everywhere. They can be readily inserted in many web browsers and are built into Microsoft Internet Explorer 6.0 and above. XSLT processors are also becoming available in many portable devices, including PDAs, Internet appliances and cellular telephones.

You do not need to know XSLT to read or understand GML. However, XSLT can perform many useful geospatial tasks with GML data including the following:

- Mapping schemas
- Transforming coordinates
- Generating graphical maps (see SVG in the next section).

XSLT is a very useful tool for anyone who works with XML and GML.

2.2.6 Scalable Vector Graphic (SVG)

SVG is a powerful XML language for encoding two-dimensional graphical drawings. With a few lines of SVG, you can create a wide range of drawings, from simple geometric shapes such as rectangles and circles to complex maps with annotation and animation.

In addition to vector graphics, SVG also supports raster graphics. Not only can you display raster image fragments such as Graphics Image Format (GIF) or Joint Photographic Expert Group (JPEG) images, but you can also produce fancy raster effects. Figure 1 shows an example of an SVG graphic with shadow and frosting effects, and Figure 2 demonstrates some of the different text and line styles that can be encoded in SVG. For example, with SVG, you can use many different line styles, generate text that follows a line or curve, and control the different fonts and colours of text styles.

SVG is widely available. Adobe provides a free plug-in for most browsers and on most platforms, and many graphics-processing applications, such as Adobe Illustrator and Corel Draw, have SVG processing tools. SVG Mobile – a special profile of SVG specifically directed at mobile devices – is being developed for PDAs and cellular phones.

SVG is an ideal companion to GML. While GML describes the content of geographic features, SVG provides the means for providing a graphical representation of this content. For example, you can generate maps. Figure 3 shows an example of a simple SVG map of Japan.

2.2.7 Web Services Description Language (WSDL)

WSDL provides a description, written in XML Schema, of the messages exchanged between a web service and its clients. A WSDL web service description:

- Describes the interfaces to the service in an abstract manner. These interfaces are described as sets of operations, each with specific input, output or fault messages.

Figure 1 SVG sample with shadow effects.

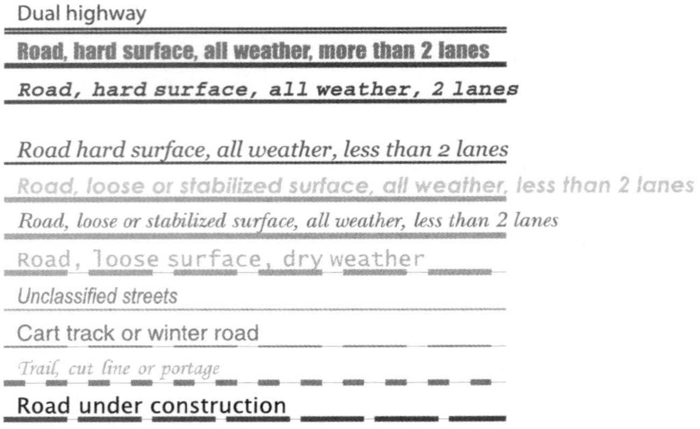

Dual highway
Road, hard surface, all weather, more than 2 lanes
Road, hard surface, all weather, 2 lanes
Road hard surface, all weather, less than 2 lanes
Road, loose or stabilized surface, all weather, less than 2 lanes
Road, loose or stabilized surface, all weather, less than 2 lanes
Road, loose surface, dry weather
Unclassified streets
Cart track or winter road
Trail, cut line or portage
Road under construction

Figure 2 SVG sample with various text and line styles.

Sapporo
Sea of Japan
JAPAN
Tokyo
Nagoya
Yokohama
Osaka
Fukuoka
Ryukuyu Islands

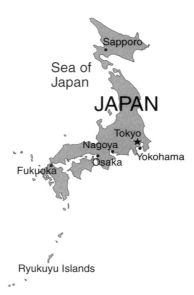

Figure 3 SVG sample map of Japan.

- Provides information about the actual implementation of the service, and in particular describes the binding between the abstract message components and a specific transport protocol.
- Provides a set of types that support the abstract message definitions.

You can use GML to create the geographic components of the messages described in WSDL. This is particularly useful for geospatial web services. Chapter 6 discusses WSDL in relation to GML and geospatial web services.

2.2.8 Simple Object Access Protocol (SOAP)

Although SOAP has several forms, its most significant function in relation to XML is that it provides a standard means for sending XML messages over the

Internet. SOAP functions as an 'envelope' that contains the information that is being transmitted. This envelope provides a framework for describing the contents of the message and how to process those contents. A SOAP message can contain additional information about the XML content and can also transport associated non-XML content, such as a binary data file. SOAP is written in XML Schema and uses XLink.

SOAP is already being used for geospatial web services. For example, the Microsoft MapPoint service is a SOAP-based geospatial service. Many other SOAP-based geospatial services and server products will also be available soon. Given that GML is an XML-based language, SOAP is the ideal means for sending GML-based geographic data requests, especially in relation to geospatial web services. You can find examples of web services that use SOAP at http://www.xmethods.com/.

2.3 Chapter summary

In software applications, there are many ways to encode information, such as binary and text-based encodings. Although text-based encodings are less compact than their binary counterparts, they are more flexible, extensible and easier to understand. However, text-based encodings are larger than binary encodings, and to extract content from them, you need to create custom parsing and querying tools.

XML eliminates many of the restrictions of traditional text-based encodings. In particular, XML provides

- a standard grammar that ensures that all XML documents have a certain level of structural integrity;
- standard tools for parsing the content of XML documents and ensuring that they comply with an external structure description.

The XML Technology Family comprises many languages. Table 3 lists those that are relevant to GML.

Table 3   XML Technology Family and GML

Technology	Relevance to GML
XML 1.0 and Document Type Definitions (DTD)	GML 1.0 was based on XML 1.0, and DTDs were used to describe the contents of instance documents in both GML 1.0.
XML Schema	XML Schema has been used to create GML schemas in all versions of GML since version 2.0, which contains three core GML schemas. Many new core GML schemas have been added in GML 3.0.
XML Linking Language (XLink) and XML Pointer Language (XPointer)	XLink and XPointer are used in GML to create links that express associations between geographic features and with other GML elements.

(continued overleaf)

Table 3   (*continued*)

Technology	Relevance to GML
Resource Description Framework (RDF) and RDF Schema (RDFS)	GML 1.0 also used RDF to describe the contents of instance documents because it supported the concepts of namespaces and inheritance. RDF has not been used in any of the newer versions of GML. However, GML continues to borrow ideas from RDF.
eXtensible Stylesheet Language Transformations (XSLT)	You can use XSLT to perform many useful tasks with GML data, such as translating schemas or generating graphical maps.
Scalable Vector Graphic (SVG)	SVG provides a means of generating graphical representations of GML descriptions of geographic features. For example, SVG can be used to create vector maps based on GML data.
Web Services Description Language (WSDL)	WSDL is useful for describing geospatial web services. GML can be used as part of the WSDL service description.
Simple Object Access Protocol (SOAP)	SOAP can be used for sending GML-based geographic requests and responses, especially in relation to geospatial web services.

References

DICK, K. (2000) *XML: A Manager's Guide.* Addison-Wesley Longman, Inc., Reading, Massachusetts.

http://srmwww.gov.bc.ca/bmgs/trim/trim/trim_overview/trim_program.htm (September 23, 2003).

Additional references

http://www.w3.org/TR/REC-xml (September 20, 2003).
http://www.w3.org/TR/xmlschema-0/ (September 20, 2003).
http://www.w3.org./TR/xptr/ (September 20, 2003).
http://www.w3.org/XML/Linking (September 20, 2003).
http://www.w3.org/TR/rdf-schema/ (September 20, 2003).

http://www.w3.org/TR/REC-rdf-syntax/ (September 20, 2003).
http://www.w3.org/TR/xslt (September 20, 2003).
http://www.w3.org/TR/SVG11/ (September 20, 2003).
http://www.w3.org/TR/wsdl (September 20, 2003).
http://www.w3.org/TR/SOAP/ (September 20, 2003).

Chapter 3

Basic concepts of GML

Throughout this book, two kinds of GML encodings are discussed: schemas and instances. GML schemas, which are based on XML Schema, describe the structure of GML data and define elements and attributes that are used in data instances. GML instances are files or parts of files that contain the actual geographic data – for example, particular roads or rivers – encoded in GML.

In every domain that uses GML to encode geospatial data, database administrators work closely with system and business analysts to create and manage domain-specific GML schemas and data instances. For example, if you are in the transportation industry, you might want to use GML to encode data about a bridge, such as the Golden Gate Bridge in San Francisco or Hijiribashi in Japan. To do this, you need to create a `Bridge` instance.

In order for GML-aware software to understand the content of this `Bridge` instance, there needs to be an existing schema that defines how that `Bridge` instance should be structured. In other words, you need to have a GML application schema that defines a `Bridge` element, which, in turn, has a content model that describes how a `Bridge` instance should be structured.

Note that the `Bridge` element is not defined in the GML core schemas. Instead, a database administrator in the transportation industry can define the `Bridge` element in a transportation-specific GML application schema. To define the `Bridge` element, the database administrator must use the GML core schemas. These core schemas provide the framework that database administrators use to create GML application schemas for their domain. GML core and application schemas are discussed in further detail in Chapter 4. These ideas are illustrated in Figure 1.

Before discussing the different schemas in GML, it is necessary to understand the basic concepts of GML instances and schemas. To begin, you need to understand and be familiar with features and properties, and understand how they are encoded in GML data instances.

3.1 Features and properties

3.1.1 Features

As discussed in Chapter 1, GML is a mark-up language that is used to describe objects in the world around us. Many of these objects are geographic features,

Geography Mark-up Language (GML). R. Lake, D. S. Burggraf, M. Trninić, L. Rae © 2004 Galdos Systems Inc. Published by John Wiley & Sons, Ltd ISBNs: 0-470-87153-9 (HB); 0-470-87154-7 (PB)

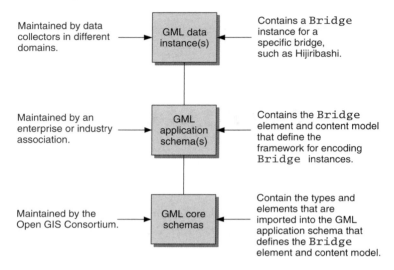

Maintained by data collectors in different domains. ⟶ **GML data instance(s)** ⟵ Contains a `Bridge` instance for a specific bridge, such as Hijiribashi.

Maintained by an enterprise or industry association. ⟶ **GML application schema(s)** ⟵ Contains the `Bridge` element and content model that define the framework for encoding `Bridge` instances.

Maintained by the Open GIS Consortium. ⟶ **GML core schemas** ⟵ Contain the types and elements that are imported into the GML application schema that defines the `Bridge` element and content model.

Figure 1   GML data, core and application schemas.

which can be concrete physical objects, such as roads or bridges, or abstract objects, such as political boundaries or health districts.

To encode features in GML, you need to create feature instances. Consider the example of an instance of a very simple `Bridge` feature, as shown below.

```
<Bridge gml:id="b1">...</Bridge>
```

This `Bridge` instance has the unique identifier of b1. Every instance must have its own unique identifier. For example, if you want to create a `Bridge` instance for the Hijiribashi bridge in Japan, you must include an appropriate identifier, such as `Hijiribashi`. The identifier that is used is typically based on existing standards in your domain. Note also that these identifiers only need to be unique within a GML document because they are XML identifiers.

3.1.2   Feature properties

In GML, properties are elements that are used to describe features. For a `Bridge` feature, properties can include the span, height or material of the bridge, as shown in Figure 2. Features in GML can also have geometry properties, which are discussed in detail later in this chapter.

Note that in Figure 2, the first letter is upper case for the `Bridge` feature and lower case for all of the properties. This reflects the naming conventions for properties and features used throughout GML. For both features and properties, all subsequent words start with an uppercase letter, for example, `heightAt-CenterOfSpan`.

The following example shows how properties can be included in the b1 `Bridge` instance. The `span` is `400`, the `height` is `50` and the `material` is `wood`.

```
<Bridge gml:id="b1">
  <span>400</span>
```

```
    <height>50</height>
    <material>wood</material>
</Bridge>
```

This instance does not specify the unit of measure of the span and height of the Bridge. For example, in this instance the Bridge could have a height of 50 feet or 50 metres. Support for units of measure has been introduced in GML 3.0, and is discussed in Chapter 16.

3.1.3 Remote properties

In the previous example, all of the property values are declared inside of the property tags. In GML, these are called in-line properties. You can also have remote properties, which are properties that refer to values that are located elsewhere. The following example shows a Bridge instance with a remote property. Instead of declaring a value inside of the tags, the property includes the xlink:href attribute, which points to a value that is located at http://www.ukusa.org/.

```
<Bridge gml:id="b1">
    <span>400</span>
    <height>50</height>
    <material>wood</material>
    <bridgeType
        ⌣xlink:href="http://www.ukusa.org/transportation.xml
        ⌣#bt1"/>
</Bridge>
```

Note that the bridgeType property does not have separate opening and closing tags, because the value is indicated by an attribute, and there is no value to place between the property tags. The '/' at the end of the tag indicates that the tag is closed.

3.1.4 Feature elements and type definitions

The Feature schema – which is one of the core GML schemas – provides the framework for defining feature types for different applications. Table 1 shows a list of some of the feature types that might be created by a database administrator in the transportation industry.

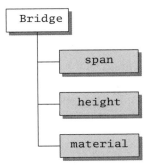

Figure 2 A Bridge feature and its properties.

Table 1   Examples of features in the transportation industry

Domain	Features
Transportation	Road, Street, Highway, ServiceStation, Bridge, Intersection, Route, Tunnel, Event, Airport, Aqueduct, Cableway, Lock, ParkingSite, Railroad, Trail, OnRamp, OffRamp, Highway, RunawayRamp, Causeway, Cut, Embankment, Barrier, Gate, SnowShed

The following schema fragment shows a `Bridge` element and content model that a database administrator might create to define the content and structure of all `Bridge` instances for a Transportation application.

```
<element name="Bridge" type="app:BridgeType"
   ⌐substitutionGroup="gml:_Feature"/>

<complexType name="BridgeType">
   <complexContent>
      <extension base="gml:AbstractFeatureType">
         <sequence>
            <element name="span" type="integer"/>
            <element name="height" type="integer"/>
            <element name="material" type="string"/>
         </sequence>
      </extension>
   </complexContent>
</complexType>
```

The above schema fragment has two parts: an element declaration of `Bridge` that declares the name of the feature and its type, and a content model (also called a type definition) that defines the content and structure of the feature. Element declarations and content models are concepts from XML Schema that are discussed in more detail in Chapter 11.

Note: In this book, the term 'feature type' is often used to refer to a kind (or class) of feature, such as `Bridge`. Throughout this book, the term 'feature type' refers to the element name (`Bridge`), not the XML Schema structural `type` (`BridgeType`). To avoid confusion, 'element' is often used to refer to the feature type name and 'content model' to refer to its structural type.

Note that this schema fragment also defines the `span`, `height` and `material` properties for the `Bridge` element and specifies the data type of each value. For example, the content model specifies that all values for the `span` of a `Bridge` must be integers, while all values for the `material` property must be a string (that is, a fragment of text).

3.1.5 Abstract types

How does GML software recognize that something is a feature, especially since feature elements are only defined in GML application schemas and not in GML itself? The answer is by using abstract types. Software can read the GML application schemas and figure out which elements are features, and then use this information when inspecting data instances that conform to those application schemas.

In the schema fragment discussed above, the `BridgeType` extends `AbstractFeatureType`, which is an abstract type for creating new features in GML. Abstract elements and types are only used to define new elements and types in GML core and application schemas; they cannot be included in GML instances.

In GML 3.0, there are many different abstract types. To create a new feature element in GML, a database administrator designates an object as a feature by having its content model inherit directly or indirectly from `AbstractFeature-Type`, as shown in Figure 3.

As the arrows indicate, an `AbstractPointOfInterest` and a `Bridge` are features whose content models inherit directly from the GML `Abstract-FeatureType`, and a `Museum` feature's content model inherits directly from the `AbstractPointOfInterest` feature's content model (and, therefore, indirectly from `AbstractFeatureType`). All GML-aware software can read this inheritance tree and understand how the different feature elements and types are classified in the GML application schema. The most important thing to remember is that, whenever you create a new feature, its content model must inherit (directly or indirectly) from `AbstractFeatureType`. Inheritance is covered in Chapters 10 and 11.

3.1.6 Namespaces in GML

In Figure 3, the `app`, `exp` and `gml` prefixes indicate the namespaces that the different elements and types belong to. A namespace is a concept from XML that is used to assign unique identities to sets of elements and attributes. For example, the transportation industry is not the only domain that might define a `Bridge` element in a GML application schema. By placing a transportation namespace

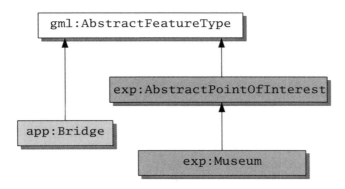

Figure 3 Type inheritance hierarchy example.

prefix in front of the `Bridge` element, this indicates that the `Bridge` element belongs to the transportation namespace and distinguishes it from other `Bridge` elements.

In this book, the `app` and `exp` prefixes refer to generic namespaces and are simply used to indicate that different parts of GML instances and schemas belong to different namespaces. The `gml` prefix refers to the GML namespace that is http://www.opengis.net/gml. Namespaces are discussed in more detail in Chapter 4.

3.2   Feature relationships

In GML, as in the real world, features are objects that relate to other objects. GML can express relationships between features such as

- `Bridge` spans `Gorge`
- `Room` is inside of a `Building`
- `Building` fronts on `Street`
- `Park` is enclosed by a `Hedge`.

These are all simple relationships that involve two participants. In simple relationships, GML can express the roles of either participant in the relationship. For example, you can say 'Bridge spans Gorge', or 'Gorge is spannedBy Bridge'. GML 3.0 also supports complex relationships between multiple participants and provides a topology model that can be used to express certain kinds of spatial relationships in a formal mathematical way. These topics are outside the scope of this volume and are covered in *Volume II: GML. A Technical Reference Guide*.

Simple relationships are expressed with GML properties where the property name designates the role of the target participant with respect to the source in the relationship. Consider the 'Bridge spans Gorge' example, as shown in Figure 4. The `spans` property is used to indicate the relationship between the `Bridge` and the `Gorge`, where `Bridge` is the source and `Gorge` is the target.

This relationship can also be expressed schematically in GML, as shown in Figure 5. In addition to the `spans` property, which expresses the relationship between the `Bridge` and the `Gorge`, the `Bridge` feature also has `height` and `material` properties, which are not part of the relationship with the `Gorge`.

The following example shows how this simple relationship can be written in GML. Note that a `Gorge` instance is the value of the `spans` property and that the value is declared inside of the property. In other words, it is an in-line property.

```
<app:Bridge gml:id="b1">
    <app:spans>
        <app:Gorge gml:id="h1"/>
    </app:spans>
    <app:height>50</app:height>
</app:Bridge>
```

Figure 4 A simple relationship with two participants.

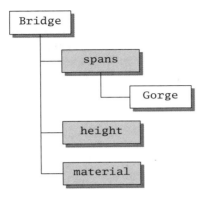

Figure 5 Schematic view of GML for a simple relationship.

Remote properties are often used to write relationships between features, as shown below.

```
<app:Bridge gml:id="b1">
    <app:spans xlink:href="#h1">
    <app:height>50</app:height>
</app:Bridge>
<app:Gorge gml:id="h1">
    <app:depth>50</app:depth>
    <app:spannedBy xlink:href="#b1"/>
</app:Gorge>
```

Note that, in this example, there is a separate Gorge instance, which is independent of the Bridge instance. While the Bridge instance has a feature relationship where the Bridge spans the Gorge, the Gorge is spannedBy the Bridge in the Gorge instance, and Gorge becomes the source participant, while Bridge becomes the target.

3.3   Geometries in GML

GML provides a number of geometry elements (also called geometries) that can be used to describe the geometric aspects of a feature, such as the position of a Bridge, the centre line of a Road or the extent of a Lake. Some features only have one geometry-valued property, while others have several geometry-valued properties that describe different aspects of the feature. For example, a Bridge can have a position property that describes its location on the earth's surface and a shape property that describes the actual structure of the bridge.

> *Note:* Instead of discussing the geometry of a feature, it is more accurate to consider the geometry-valued properties of a feature. In other words, a Building is not actually a Point, but it can have a position property whose value is a Point geometry.

Geometries can be stand-alone objects that are referenced through the properties of features and other GML objects. In this way, GML allows geometry objects to be shared between GML features.

Table 2   Geometries in GML 2

Geometries	Example	Might Be Used to Define...
Point		An exact point of a feature, such as its position.
LineString		A part of a Route.
Polygon		The physical extent of a feature, such as a Lake or a Field.
MultiPolygon		The physical extent of a group of islands that might comprise a Park.

3.3.1   Geometry types in GML 2

In GML 2, there are only a few geometries, all of which are linear geometries; that is, they are composed of straight-line segments. Table 2 lists the main geometries in GML 2.

3.3.2   New geometries in GML 3.0

Most fields of interest should not require more geometries than those offered in GML 2. However, users can add their own geometries to GML, and many new geometries have been added to the Geometry schemas in GML 3.0, including the following:

- Curve
- Surface
- Solid.

All of the GML 3.0 geometry types are discussed in detail in *Volume II: GML. A Technical Reference Guide*, and in the *GML Version 3.00 OpenGIS® Implementation Specification* (http://www.opengis.org/docs/02-023r4.pdf).

3.3.3   Convenience geometry-valued properties

GML geometry-valued properties are used to describe the role of geometries in relation to a particular feature. GML provides several convenience geometry-valued properties, including the following:

- position
- location

- centerOf
- edgeOf
- centerLineOf
- extentOf.

These convenience properties do not have to be used to express the geometry of a feature. Database administrators can create their own geometry-valued properties when they define their GML application schemas. However, the convenience properties have been provided to allow you to express a feature's geometry without defining new geometry-valued properties.

3.3.4   `Bridge` feature with geometry example

The following example shows a `Bridge` feature instance with a `centerLineOf` geometry-valued property, which is used to describe the path of the `Bridge`.

```
<app:Bridge gml:id="b1">
   <app:spans>
      <app:Gorge gml:id="h1"/>
   </app:spans>
   <app:height>50</app:height>
   <gml:centerLineOf>
      <gml:LineString srsName="#myRefSystem">
         <gml:posList>...</gml:posList>
      </gml:LineString>
   </gml:centerLineOf>
</app:Bridge>
```

Note that the `centerLineOf` property contains an instance of the `LineString` geometry, which has the `posList` property that provides the coordinate values for the geometry of the `Bridge`. This is the standard pattern that is used to express a feature's geometry. A feature has a geometry-valued property that contains a geometry element that has a property, and so on.

In GML 2 and 3.0, coordinates are typically expressed as coordinate strings, for example, `<coordinates>100.1, 23.2</coordinates>`. These coordinates are linked to a Coordinate Reference System (CRS) that provides the real-world context for the coordinates. GML 3 provides the following: core schemas that you can use to create CRS components, some support for three-dimensional geometry and additional mechanisms for defining coordinates. These are called `pos` and `posList`, and they are described in detail in *Volume II: GML. A Technical Reference Guide*. Note that `posList` is being introduced in GML version 3.1, which is currently an OGC Recommendation Paper.

3.4   Other objects in GML

Features are not the only kinds of objects that can be encoded in GML. In GML 2, only feature and geometry objects can be defined, but many additional kinds of objects can be defined in GML 3.0. These objects include geometries, topologies, units of measure and coordinate reference systems. Properties for objects are encoded in the same way as they are for features. In Volume I of this book, the discussion focuses primarily on features. Once you understand the role of features in GML, it should be quite easy to understand how other objects fit into GML. These other objects are described in detail in *Volume II: GML. A Technical Reference Guide*.

3.5 Chapter summary

Database administrators and data modellers are typically responsible for authoring GML schemas to define the structure of geographic data in their domain. These schemas are called GML application schemas and are based on the core GML schemas, all of which are discussed in Chapter 4. Schemas define elements and content models that describe the content of instances. For example, a `Bridge` element declaration and its content model describe the content of a `Bridge` feature.

In GML, geographic features are encoded in feature instances with unique identifiers. Properties are elements that are used to describe features. For example, a `Bridge` might have `span` and `height` properties. Property values can be declared in-line, within a data instance or remotely.

All feature elements must have content models that inherit directly or indirectly from `AbstractFeatureType`, which designates an object as a feature. `AbstractFeatureType` is a GML abstract type that is defined in the GML Feature schema. The `gml` prefix indicates the namespace that `AbstractFeatureType` belongs to. GML software uses these abstract types to determine the nature of elements in the data stream. For example, by looking at the type hierarchy, GML software can identify if that object is a feature or some other kind of object.

Features are usually related to other features, and GML can be used to express these relationships. For example, a `Bridge` can span a `Gorge`. In this relationship, the `span` property connects the two features. Simple feature relationships involve two features, while complex feature relationships can involve multiple features.

Geometry elements, which are also called geometries, are used to describe a feature's geometric aspects. For example, a geometry type can describe the `position` of a `Bridge`. A feature's geometry is expressed through its geometry-valued properties. A geometry property describes the role of the associated geometric object in relation to the feature (for example, a `Point` is the `position` of a `Building`). There are a number of convenience geometry-valued properties that are defined in the GML core schemas. These geometry properties are used to include geometry objects that are defined in the Geometry schemas. Many new types and elements have been added to the GML 3.0 Geometry schemas. These are covered in *Volume II: GML. A Technical Reference Guide* and in the *GML Version 3.00 OpenGIS® Implementation Specification* (http://www.opengis.org/docs/02-023r4.pdf).

Reference

http://www.opengis.org/docs/02-023r4.pdf (October 15, 2003).

Additional reference

http://schemas.opengis.net/gml/3.0.1/base/feature.xsd (October 15, 2003).

Chapter 4

GML core and application schemas

GML provides a framework for describing geographic objects and does not define specific geographic objects, such as a Road or River. To define specific geographic objects, a database administrator or data modeller needs to create GML application schemas, which are essentially 'vocabularies' that define the structure of GML data. As discussed in Chapter 3, GML application schemas are based on the GML core schemas. In order to create and maintain GML application schemas, it is necessary to be familiar with one or more of the GML core schemas.

4.1 About the GML core schemas

There are only three core schemas in GML 2: Feature, Geometry and XLinks. By comparison, 25 additional core schemas have been added to GML 3.0. Many of these new schemas can be grouped into different object classes, some of which comprise multiple schemas. Descriptions of all of the GML core schemas are provided in Chapter 10 and Appendix A.

The GML core schemas provide the framework to create additional GML application schemas to describe application domains that make use of geographic information. Different application domains may use different core schemas, and it is rare that any application schema will require all of the core schemas. Database administrators and data modellers usually need to be familiar with only a subset of the listed core schemas. An application schema can be thought of as an XML language for a particular domain, such as ITS, transportation or forestry.

4.2 About GML application schemas

Once you decide to create a set of GML application schemas for your domain, you need to make decisions about the content of these schemas. Given the variety of available geographic information, even within a particular domain, this process can be quite challenging. Table 1 shows a list of feature types that might be created for six different domains. When you examine these feature types, several issues become apparent, including the following:

- The same features may be used in different domains. For example, Highway is a feature type in both tourism and forestry.

Geography Mark-up Language (GML). R. Lake, D. S. Burggraf, M. Trninić, L. Rae © 2004 Galdos Systems Inc.
Published by John Wiley & Sons, Ltd   ISBNs: 0-470-87153-9 (HB); 0-470-87154-7 (PB)

Table 1   Sample feature types for different domains

Domain	Feature Types
Transportation	Road, Street, Highway, Service Station, Bridge, Intersection, Route, Tunnel, Event, Airport, Aqueduct, Cableway, Lock, ParkingSite, Pipeline, Railroad, Trail, OnRamp, OffRamp, Highway, RunawayRamp, Causeway, Cut, Embankment, Barrier, Gate, SnowShed
Tourism	Road, Street, Highway, PointOfInterest, Restaurant, Museum, ArtGallery, ServiceStation, Hospital, Hotel, Motel, Bridge, NationalPark, RegionalPark, Chateau, Castle, Viewpoint, Fort, Square, Plaza, Farm
Agriculture	Road, Street, TruckingRoad, Farm, FarmHouse, Field, ProcessingPlant, FeedLot, GrainElevator, RailLine, MachineryDepot, SeedCompany, FertilizerCompany, Abattoir, Hospital, ServiceStation
Electrical utility	Road, Street, NuclearPowerStation, Reactor, Cooling Tower, SubStation, SwitchingYard, DistributionLine, PowerLine, UtilityCorridor, Transformer, ElectricalTower, UtilityPole, Hospital, PowerConsumer, CoalFiredStation, GasTurbineStation, HydroelectricStation, Dam, Reservoir, AccessRoad
Air transportation	ControlTower, FlightZone, Airport, Runway, Taxiway, AirCorridor, NavigationBeacon, Radar, ApproachLights, Road, Street, Highway, Tower, CommunicationTower, MunicipalAirport, InternationalAirport
Forestry	Road, Street, Highway, ForestAccessRoad, CutBlock, Clearcut, SawMill, PulpMill, Bridge

- A feature may be classified as a different feature in different domains. For example, a Road can be a TruckingRoad or a ForestAccessRoad. These different classifications may apply for only part of the extent of the feature. For example, a TruckingRoad might turn into a ForestAccessRoad.

- Even when you establish a common vocabulary within a domain, different people may disagree about the classification, that is, there will always be differences of opinion as to the correctness of the vocabulary. For example, 'this should be a Road not a Highway'.

- You can arrange the feature types into a classification hierarchy. For example, an ElectricalTower and a CommunicationTower are both towers. And, a ForestAccessRoad and a TruckingRoad are both kinds of roads.

- The same feature name – such as Road – may have a different meaning, depending on the domain.

- Different feature names – such as Road and Street – may have the same meaning in different schemas.

- The same physical object can have different names in different natural languages, such as Rue and Chemin instead of Street and Path.

On the basis of this list of issues, it is clear that the process of creating GML application schemas is inexact. This is not a fault of the process nor of GML. It simply reflects the nature of reality and of our relationship to it.

The content of a GML application schema is application specific with different schemas developed for different domains. Each GML application schema defines a list of types of geographic objects. To create an application schema, you use eXtensible Mark-up Language (XML) schema, import the required GML core schema(s) and follow a few simple rules, as shown in Figure 1. These rules are covered in *Volume II: GML. A Technical Reference Guide* and the *GML Version 3.00 OpenGIS® Implementation Specification* (http://www.opengis.org/docs/02-023r4.pdf).

4.2.1 GML application schemas and namespaces

GML application schemas always have an associated namespace, a concept that was introduced in Chapter 3. In XML, namespaces are determined by the W3C Namespaces specification (http://www.w3.org/TR/xml-names11/). Namespaces in GML are also based on this specification. A namespace is essentially an Internet address that is represented by a prefix, which is placed before a GML element or attribute to make it globally unique. For example, the gml prefix represents the http://www.opengis.net/gml namespace. Note that only GML core schemas can use the GML namespace.

The domain part of the namespace 'address' is owned by a particular organization. For example, the Open GIS Consortium (OGC) owns the opengis.net domain. Every domain indicates the authority responsible for the associated GML

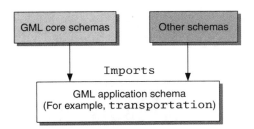

Figure 1 GML application schemas build on GML core schemas.

application schema. Different organizations – such as Great Britain's Ordnance Survey or the US Geological Survey (USGS) – can create their own schemas and associate them with a namespace for which they are responsible.

For example, by using a namespace prefix, such as `osgb:Road` (for GB Ordnance Survey), the `Road` feature is completely unambiguous. Note that since each feature is defined in a schema with a declared namespace, you do not have to worry about naming conflicts. Note that the prefix letters (for example, `osgb`) are completely arbitrary and are not the namespace (for example, http://www.ordnancesurvey.co.uk/xml/namespaces/osgb).

4.2.2   GML application schema networks

GML application schemas can be shared on the Internet and can import or include one another. For example, a Utility application schema can import and use elements and types from Road Network and Land Parcel application schemas to define new Utility-specific features. Figure 2 shows a potential network of GML application schemas.

In addition to importing the Land Parcel and Road Network Schemas, the Electrical Utility Schema imports the GML core schemas. These schemas may also import other schemas. Note that whenever you create a new GML application schema, it does not have to import all of the GML core schemas, but only those that are required for your specific application.

GML is designed specifically to support the development of interconnected and distributed schemas of geographic objects. The OGC is responsible for the GML core schemas. In addition to developing and maintaining the schemas, the OGC also provides them online 24 hours a day, seven days a week. The schemas are located at http://schemas.opengis.net/gml/. As new GML application schemas are created, a similar level of service will be required for different domains. Eventually, there will be rich networks of GML application schemas covering many application domains.

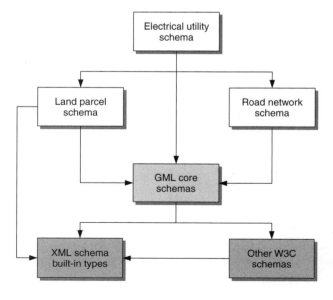

Figure 2 A potential network of GML application schemas.

4.3 Chapter summary

With GML 2, only three core GML schemas are used to create GML application schemas. With GML 3.0, there are 28 core schemas, including the Feature and Geometry schemas. Since most GML application domains require only a few of the core schemas, database administrators and data modellers do not necessarily need to be familiar with all of the core GML schemas to create GML application schemas.

To create GML application schemas, you need to follow a few rules and import the core GML schemas. These rules are detailed in Chapter 11 and in the *GML Version 3.00 OpenGIS® Implementation Specification*. From a management perspective, you need to know the following about GML application schemas:

1. Application schemas are human and machine-readable.
2. Application schemas provide 'vocabularies' that identify the key objects of interest in a given application domain.
3. Before you create a new application with GML, you need to create application schemas or use those already in existence.
4. Once you have created the appropriate GML application schemas, you need to establish a standardized system for managing these application schemas.
5. As more GML application schemas are created, different schema networks will become available over the Internet.
6. Application schemas provide the framework for building geospatial web services that are discussed in the following chapters.

Whenever you create a new GML application schema, it must include a target namespace that belongs to your organization. This ensures that you create element and type names that are unique and do not conflict with those from other application schemas. The following chapters examine how these schemas can be deployed in the real world.

References

http://www.opengis.org/docs/02-023r4.pdf (October 15, 2003).

http://www.w3.org/TR/xml-names11/ (October 15, 2003).

Chapter 5

Technical issues for deploying GML

While GML is a very exciting technology, it is also very new, and those interested in GML deployment need to develop a clear plan of action. For example, you need to select the best approach for developing GML application schemas and you need to be aware of the current limitations of GML. This chapter explores many of the technical issues that you need to consider before you attempt to deploy a project in GML. Examples of different deployment solutions are also provided.

5.1 GML deployment issues

Before you decide to use GML for your domain, you need to be aware of the various technical issues that can arise when you deploy GML in the real world. By carefully considering the following issues, you will have a clearer idea of how to approach GML deployment for your domain.

5.1.1 How do you develop and deploy GML application schemas?

Since GML is still a relatively new technology, it is unlikely that you will find GML application schemas on the Internet that suit your specific requirements. Therefore, you probably need to develop new GML application schemas for your domain. It is not too difficult, however, to develop GML application schemas, and if your domain already has existing models, schema development can be partially or completely automated.

Before you create the schemas, you need to determine where they will reside on your system and how software developers, database administrators and data modellers can access them. This is not necessarily an issue if you are deploying GML from a single geospatial web service. Most applications, however, require several geospatial web services. You should also consider deploying the GML application schemas on a web site that is visible to your data users or through a schema registry. Chapter 6 provides more information about geospatial web services and schema registries.

5.1.2 What about data compression?

GML is a text-based language and, without data compression, the data volumes can be substantial. Many GML-based applications will require gigabytes of

Geography Mark-up Language (GML).   R. Lake, D. S. Burggraf, M. Trninić, L. Rae © 2004 Galdos Systems Inc.
Published by John Wiley & Sons, Ltd   ISBNs: 0-470-87153-9 (HB); 0-470-87154-7 (PB)

disk storage. If you use data compression, you can expect significant compression ratios ranging from 5:1 to 10:1. Data compression, however, is not always the best solution, given that extensive memory and CPU resources are still required to store large volumes of data on the client application.

Coder–decoders (Codecs) offer a very promising solution to this problem. Although Codecs express the GML data in a binary format for transport that is similar to standard data-compression techniques, they also provide standard XML Application Programming Interfaces (APIs) to navigate the data. This provides an effective data-compression ratio of approximately 4:1 to 10:1 and significantly reduces the client application's processing and memory resource requirements. ExpWay's BinXML is one example of a product that uses this approach.

5.1.3 How will your applications process GML data?

It is important to decide how your applications will process GML data. At present, there are few native GML applications. Currently, you can use data-conversion mechanisms, such as Safe Software's Feature Manipulation Engine (FME), to convert your data. This is probably a temporary solution, since most GIS applications will have built-in GML support and some commercial products will also provide native GML processing.

5.1.4 What is the overall function of GML in your system?

It is equally important to consider the overall function of GML in your system. GML is often used as a common distribution format in which the GML data is converted and stored in another format, such as Oracle 9i Spatial or ESRI ArcSDE, before it is redistributed again as GML. As a data distributor, you might decide to provide data in the standard GML format and then let your customers convert the data to the proprietary format of their choice. Many organizations are exploring this option, including Great Britain's Ordnance Survey and Geomatics Canada.

Some vendors already provide data-loader software that can take GML data and download it into Oracle or other databases. One such example is Snowflake Software's Go Loader. Similar loader products for other data stores are expected to become natively available in 2004. If you plan to redistribute the data after processing, you might want to consider using a native GML data-storage method to minimize data-conversion costs.

If your systems have a purely relational-database architecture, you might not achieve desired performance levels with GML data, given the nested nature of the data. To properly deploy GML applications, developers need to explore the appropriate options for handling this data, such as moving to an object-relational-database architecture or native XML database. This issue is common to all users working with XML data and is not unique to GML.

GML is expected to play an increasingly important role in geospatial web services, with GML being used for data transport and as part of web service operations. For example, you can configure an OGC Web Feature Service (WFS) to replicate GML data to another OGC WFS. This is called transactional replication, and it is a useful tool for synchronizing data between different web feature services.

Another role for GML is in data integration. If you build or maintain spatial data warehouses, you are probably familiar with the considerable problems faced by data integrators, and the great difficulty of synchronizing the warehouse with the various data sources. By converting your data to GML prior to integration, you can greatly reduce the cost and complexity of data integration.

5.1.5   Planning for the future

Since GML is a new and pervasive technology, it is important to develop a long-term view of how it will be used. For example, you might need to consider the following questions:

- If you are currently deploying GML as an interoperability solution, how will that look in two or three years?
- Will interconversion tools still be required, or will you move towards a more native GML environment?
- Will your system become more distributed or more centralized?

Like any technology, GML is not the solution to every problem. This is especially true today, given that GML is still a relatively new technology and has a considerable impact on data-processing resources. Before you begin a project with GML, it is critical to match your deployment objectives to what is currently feasible with the technology. For example, you can

- have interactive browser-based applications that deal with many hundreds of features,
- deliver large quantities of features in batch-mode format and import this data into existing database products,
- use GML-based WFS transactions to provide change-only updates of geographic databases over the web.

You cannot reasonably expect to use GML to create interactive browser-based applications with many thousands of features without Codec or compression technology. Nor is it currently feasible to use GML to distribute millions of features over the Internet using XML text. Note, however, that available processing and network resources are expanding quickly. Dedicated and public networks with high bandwidth (for example, greater than 2 Gbits/s) are becoming increasingly widespread, as are Wi-Fi networks with speeds up to 54 Mbits/s. The current issues respecting GML performance are likely to be overcome in the near future.

GML is a new technology that is expected to evolve quite substantially in the next five years, and therefore a pilot project phase should precede most real-world deployments. This pilot project phase can provide you with the opportunity to assess and resolve software problems and to determine if it is reasonable to deploy a full-scale system in GML.

5.2   GML deployment examples

In the near future, the following approaches are expected to be used to deploy GML data:

- Data acquisition and integration for a spatial data warehouses.
- Geographic data distribution in bulk transfer mode.
- Geographic data distribution in interactive mode via an OGC Web Feature Service.
- Transactional replication of geographic data via a network of geospatial web services.

In the following pages, examples are provided to illustrate each of these approaches. These examples address some of the different technical issues discussed earlier in this chapter.

5.2.1 Spatial data warehouse example

Figure 1 shows how GML can be used as part of a data warehouse solution where all of the data is still centralized and manually integrated. In this example, a 'GML Connector' component – such as a commercial WFS add-on to a proprietary GIS or spatial database – is used to convert data from the various proprietary GIS sources.

The data is then transferred in GML format to the data-integration system, which provides tools for data integration, including feature-level editing, schema

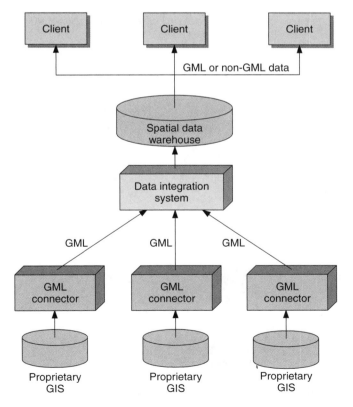

Figure 1 GML-facilitated data integration for a data warehouse.

mapping and data presentation for quality assurance checks. As the figure illustrates, it is more efficient to integrate data in GML than to work with the various legacy formats individually.

Once the data from each region is integrated in GML, the data can then be loaded into the warehouse using a native GML database, such as Galdos Cartalinea, or a conventional database product, such as Oracle 9i Spatial. The spatial data warehouse then distributes the data in GML or non-GML format to the various clients.

5.2.2   Bulk transfer data-distribution example

In Figure 2, geographic feature data is distributed in bulk transfer mode. GML data is stored in a spatial data warehouse and then transferred to a data-distribution system that distributes the data to different GML gateways. These gateways convert the data and send it to various data consumers via FTP, HTTP, CD-ROM or DVD.

Bulk transfer – which is still the most practical method for distributing very large amounts of GML data – is currently used, for example, for all initial data downloads by Great Britain's Ordnance Survey. To put this in perspective, consider the following example. Suppose a customer wants all of the GB Ordnance Survey data for the United Kingdom. Since this data comprises more than 400 million features, multiple DVDs are required to transfer the data (http://www.ordnancesurvey.co.uk/media/news/2001/nov/osmastermaplaunch.html). In contrast, a City of Vancouver dataset with 75,000 features includes

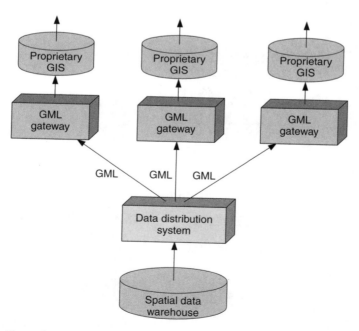

Figure 2 Bulk distribution of data in GML format.

building footprints, road segments, rivers, bridges and coastline, and can easily fit on a 128-MB memory stick without compression.

Over the next two years, it is likely that most GML data will be distributed in initial bulk loads. Subsequent updates or additions to the data will be distributed with change-only or transactional replication methods, which are discussed further below.

Note that GML is particularly suited to the bulk distribution of data, in that it naturally preserves all of the relationships between features. This is not the case, for example, with a relational query in which all relationships are lost in the construction of a 'flat' query result.

5.2.3 OGC Web Feature Service (WFS) distribution example

Figure 3 shows an OGC WFS distributing GML data. In this case, users provide requests for data from web clients, proprietary GIS products or native GML applications, and the response is provided in GML. Examples of such WFSs include Galdos Cartalinea, Ionic Software's Red Spider and CubeWerx Software's CubeStore WFS. Chapter 6 provides a more detailed discussion on OGC WFSs.

Future GML systems will likely involve a combination of data load and interactive data requests. With this approach, customers can request initial downloads using FTP, HTTP, CD-ROM or DVD and then use interactive operations to retrieve all subsequent changes. The best way to accomplish this is through transactional replication using WFS data requests.

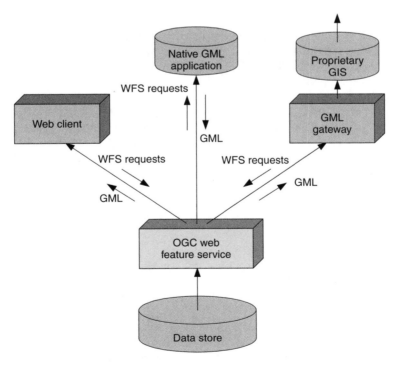

Figure 3 Interactive distribution of geographic data in GML format.

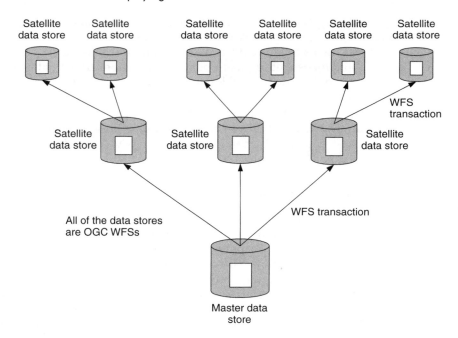

Figure 4 Transactional replication of updated GML data.

5.2.4   Transactional replication example

Figure 4 shows a generic example of transactional replication between a master geospatial data store (OGC WFS) and various satellite data stores (also assumed to be OGC transactional WFSs). Whenever a change is made on the master data store, it is propagated to the other satellites. The white squares represent the updated data.

The satellites do not necessarily contain all of the data in the master data store, but only data for a particular field or region of interest. For example, one satellite might have data for a particular city or for utilities in a province. These satellites can also have their own satellites for which they are the master data store. When data updates are made to the master data store, they are also propagated (as WFS transactions) to all the satellites that require these changes. Each satellite data store may also contain data not provided by replication from a master data store.

5.3   Chapter summary

GML is simply another technology, and like any other technology, you need to understand the relevant technical and business issues and establish a plan before launching a complex project. The most important issues to consider are as follows:

- You need to develop application schemas and determine how to deploy them. Once you have these schemas, you need to store them somewhere. For most applications, it is recommended that you use a web site or a schema registry.

- Given that many GML applications require extensive data storage, you might need to consider compressing the data. In addition to traditional data-compression methods – which do not address the memory and CPU requirements for the consuming application – Codecs should also be considered.

- You need to determine how your applications will process GML data. Currently, there are not many native GML applications for processing GML data. You can use conventional data-conversion mechanisms or GML data gateways. Within the next one to two years, many applications will have built-in GML support.

- You need to establish the overall function of GML in your system. For example, you can simply use GML to distribute the data and then convert it to another format. Data-loader software is already available for converting GML data. To store the GML data, you can use relational databases, object-relational databases or native XML databases. If your system has a purely relational-database architecture, GML applications (as with other XML applications) may impose performance penalties. GML will play an important role in the deployment of geospatial web services. It is also possible to use GML to reduce data-integration costs by converting all data to GML before integrating the data.

- You need to develop a long-term plan for using GML. You need to consider how GML will be deployed in the future and whether it fits your system's requirements. Consider the current limitations of GML. Although you can easily use GML to develop applications that process many features, it is not currently feasible to have applications that process extremely large quantities of features. Nor is it feasible to distribute millions of features at one time over the Internet. To properly address these long-term issues, it is recommended that you include a pilot project phase in your GML deployment plan.

The following GML deployment solutions are likely to be successful in the near future: data integration for a spatial data warehouse, data distribution in bulk transfer mode, data distribution via an OGC WFS and transactional replication from one WFS to another. With a spatial data warehouse, geographic data is obtained from various proprietary GIS sources, converted to GML, integrated in a data-integration system and stored in a spatial data warehouse for distribution to various clients.

In bulk transfer, GML data from a spatial data warehouse is sent in bulk to a data-distribution system. The data is distributed to various GML gateways, which convert and send the data to the appropriate proprietary GIS systems. As an alternative to the data-distribution system, an OGC WFS can be used to distribute geospatial data. With this approach, the WFS distributes data from a data store to various kinds of clients. Each client can also send requests back to the WFS using GML.

Transactional replication is another solution for handling subsequent updates to data. With transactional replication, data updates are automatically propagated from a master WFS to all designated-satellite-web-feature services that are part of the network. With this approach, it is easy to ensure that various WFSs in a network are kept in sync with each other.

All of these solutions are being considered for GML deployment in various domains. Chapter 7 discusses GML deployment in different domains, such as local government, utilities, natural resources, disaster response and location-based services.

Reference

http://www.ordnancesurvey.co.uk/media/news/2001/nov/osmastermaplaunch.html (October 21, 2003).

Additional references

http://www.expway.com/ (October 21, 2003).
http://www.safe.com/ (October 21, 2003).
http://otn.oracle.com/products/spatial/htdocs/data_sheet_9i/9iR2_spatial_ds.html (October 21, 2003).
http://www.esri.com/software/arcgis/arcinfo/arcsde/index.html (October 21, 2003).

http://www.ordnancesurvey.co.uk/ (October 21, 2003).
http://www.geocan.nrcan.gc.ca/ (October 21, 2003).
http://www.snowflakesoft.co.uk/ (October 21, 2003).
http://www.galdosinc.com/ (October 21, 2003).
http://www.ionicsoft.com/ (October 21, 2003).
http://www.cubewerx.com/ (October 21, 2003).

Chapter 6

GML and geospatial web services

As discussed in Chapter 1, a web service is an application that accepts and processes requests from other applications across a network, such as the Internet. A web service can also be described as an exchange of messages between services and their consumer clients. The client sends a request message to the service, and the service sends a response message back to the client. The communication is between software applications, that is, the client and the service.

Geospatial web services are web services that deal with geographic information. These web services can provide access to geographic information stored in a database, perform simple and complex geographic computations, and return messages that contain geographic information. Figure 1 provides a very top-level illustration of the interaction between a client and a geospatial web service.

In this simple example, a client sends a request to a geospatial web service that returns a response message to the client. The geospatial web service retrieves data for the response message from a geographic database. Note that a data-access web service is not a database but a set of interfaces that provide access to data stored in a database. When a client sends data updates to a geospatial web service, the data is updated in the database. This means that a web service hides how the data is actually stored. The underlying store could be ESRI ArcSDE, Oracle, another relational or object-relational DataBase Management System (DBMS), a collection of flat files or a native XML DBMS, such as X-Hive.

In the example shown in Figure 1, a request message is submitted via a PDA. Note, however, that many types of clients can be used, including cellular phones and portable or desktop computers. Figure 2 shows the interaction between multiple clients and services.

In addition to geospatial web services, this figure also shows a service and a type registry. The service registry stores service descriptions and the type registry stores GML application schemas. All of these different services are part of a geospatial data network. There are many different kinds of geospatial web services, including those that are currently being developed by the Open GIS Consortium (OGC). These OGC web services include the OGC Web Feature Service (WFS), the OGC Web Coverage Service (WCS), the OGC Web Map Service (WMS) and the OGC Catalog Service.

Geography Mark-up Language (GML).   R. Lake, D. S. Burggraf, M. Trninić, L. Rae © 2004 Galdos Systems Inc.
Published by John Wiley & Sons, Ltd   ISBNs: 0-470-87153-9 (HB); 0-470-87154-7 (PB)

Figure 1 High-level concept of geospatial web services.

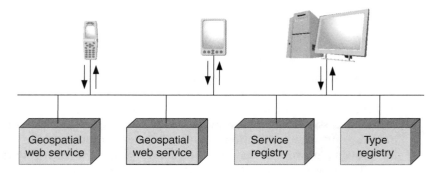

Figure 2 Interaction between multiple clients and services.

As more geospatial web services are created, many will become part of a spatial infrastructure called the Geo-Web, which is a global network of interconnected geographic information sources and processing services. To successfully integrate different services on the Geo-Web or as part of another kind of spatial infrastructure, there is a need for a common architecture. The basic elements of this common architecture are outlined in the *OpenGIS Reference Model* (http://www.opengis.org/docs/03-040.pdf).

As this common service architecture develops, a rich variety of services will become available. Many of these services will deal with specific commercial and business operations in various sectors, such as urban planning, forestry and disaster management. These services will also perform various functions, such as locating, displaying and processing data. In addition, they will also provide access to other services and data. The geospatial aspects of these services will be supported by GML and by the other OGC and W3C standards discussed in this book.

6.1 GML and HTML

To understand the role of GML in relation to geospatial web services, consider a comparison between GML and HTML, as shown in Table 1. With geospatial web services, the request and response messages can contain GML data, including features and other geographic objects. In addition, GML data can be stored in the geographic databases, which provide the data for the services. The services can also access schema registries that contain GML application schemas, which define the structure and content of the GML data. Schema registries are discussed in more detail in Section 6.3.

Table 1   HTML and GML

HTML	GML
Is a language for expressing the appearance of pages on the World Wide Web.	Is a language for describing the world in terms of its geographic features.
Played a key role in the creation of the World Wide Web.	Will play a key role in the creation of the Geo-Web.
Is simply a language and requires supporting HTML-enabled technology, such as web browsers and web servers, to display content on the World Wide Web.	Is simply a language and requires supporting GML-enabled technology, such as geospatial web services, to provide and update geographic information over the Geo-Web.

6.2   Current OGC geospatial web services

As of January 2004, the OGC is involved in developing the following kinds of geospatial web services: WFS, WCS, WMS and Catalog Service. GML is used to implement all of these services.

6.2.1   Web Feature Service (WFS)

The OGC WFS – one of the keys to GML deployment – became an adopted specification in September 2002. The OGC interface provides standardized access to a feature store and enables users to create, update or retrieve GML feature data, locally or across the Internet. When a client sends a request for information to an OGC WFS, the service sends a response message that provides geographic feature data in GML. The WFS can thus be viewed as a GML data server.

The WFS responds to OGC Common Query Language (CQL) expressions that request data. Two possible examples of CQL expressions are: 'Find all towns inside this state' and 'Find all roads with three lanes that intersect the provincial border'. GML is used to express parts of the request – such as GML feature names, GML geometries and GML properties – in combination with the OGC CQL filter grammar. The response from the WFS is encoded in GML.

A client can determine the types of features supported by a given WFS through the capabilities interface. This can enable a WFS client to determine, for example, that a given WFS can serve geographic information on roads, power lines, substations, transformers, dams and power stations. The client can also request a detailed description of any group of these feature types. This description is provided by the WFS as a GML application schema. The WFS specification is available at http://www.opengis.org/docs/02-058.pdf.

6.2.2   Web Coverage Service (WCS)

The OGC is working on the definition of a Web Coverage Service (WCS). In OGC terminology, a coverage describes a property's distribution over a portion of the earth's surface. Coverages can also describe a property's distribution in time.

Some common examples of coverages include

- Digital Elevation Models (DEMs). Figure 3 shows an example of a DEM, which is a kind of coverage that is used to represent terrain relief.
- Remotely sensed imagery, such as satellite images and aerial photographs
- Soil type distribution or soil surveys
- Rock core samples
- Distribution of road pavement type along a road.

A WCS is an OGC web service that can provide or update coverage data on request. As with the OGC WFS, GML is used to encode parts of the request message and the returned coverage data. According to Topic 6 of the *OpenGIS® Abstract Specification* (http://www.opengis.org/docs/00-106.pdf) and GML, a coverage is a special kind of feature; consequently, a WCS can also be considered as a special kind of WFS. For a more technical discussion regarding the WCS specification, please see http://member.opengis.org/tc/archive/arch02/02-024.pdf.

6.2.3   Web Map Service (WMS)

A WMS is an OGC web service that provides maps on request. A map is a graphical visualization of geographic data, and this data can be represented in or described by GML. The OGC is currently working on a revision to the OGC WMS that takes features returned from a WFS, styles these features into a graphical format and returns the resulting map to the requesting client. The WMS specification is available online at http://www.opengis.org/docs/01-068r2.pdf.

6.2.4   Catalog Service

A Catalog Service is an OGC web service that supports the storage, retrieval and management of metadata, which is data that describes other data or services.

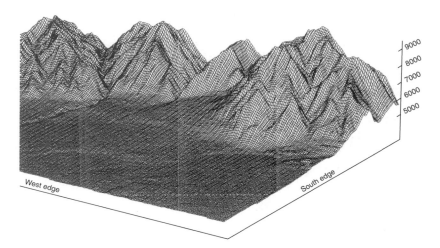

Figure 3   Digital Elevation Model example.

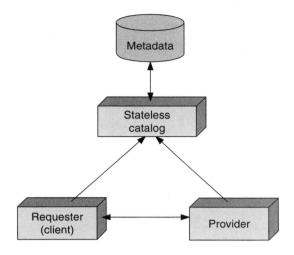

Figure 4 Catalog Service as matchmaker.

Application schemas are one kind of metadata that can be supported by a Catalog. Other kinds of metadata supported by the Catalog include service offers, associated data descriptions and interface definitions. A service offer is a description of a web service that includes an association to a specific address and the message encoding, plus a description of the request and response messages. The Web Services Description Language (WSDL) is a key part of the mechanism used to describe service offers.

> *Note:* Note that the OGC Catalog is also referred to as a Registry or Registry/Repository, and the supporting OGC web service is often called a Web Registry Service (WRS) or Catalog Service for the web.

The OGC Catalog is a key component in a common service architecture that manages shared resources and facilitates the discovery of resources within an open and distributed system. The Catalog basically plays the role of a matchmaker between a client and a service provider, as shown in Figure 4. A service provider publishes service-related metadata to a Catalog (registry), and then a client attempts to find resources of interest by querying the Catalog (registry). The Catalog Service specification is available online at http://www.opengis.org/docs/02-087r3.pdf.

6.3   Schema registries

As discussed in Chapter 4, GML application schemas will be increasingly deployed on the Internet as the standard framework for sharing geographic data. As more schemas become available, it will be increasingly necessary to provide more sophisticated mechanisms for managing schemas. Schema registries – a

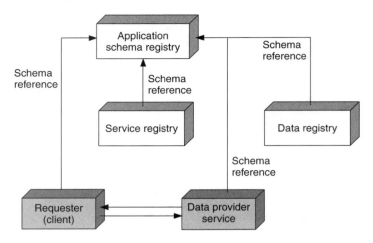

Figure 5 GML application schema registry and other services.

deployment of the OGC Catalog – will fill this need by providing:

- online access to the GML application schemas for the appropriate domain,
- administrative functions for the schema administrator, such as authoring and versioning,
- classification and other means for helping a user locate a schema.

Schema registries will be the source for all of the GML application schemas that are supported by various geospatial web services. As illustrated by the different schema references in Figure 5, clients, data providers and other services use the schemas that are stored on the GML application schema registry. These schemas provide the clients and providers with the feature types for request and response messages. Note that the schema registry and service registry are both metadata registries that can be part of a single Catalog.

6.4 GML and web-service description

As discussed earlier in this chapter, a web service exchanges messages with a client. The client sends a request message, such as 'send me particular kinds of data'. Then the server responds with a response message or a fault, if it is not possible to return any data. In order to send and receive these messages, there is a need for technologies that describe and carry these messages. The WSDL, HyperText Transfer Protocol (HTTP) and Simple Object Access Protocol (SOAP) can be used for this purpose.

6.4.1 WSDL and web services

The W3C WSDL is emerging as the standard way to describe web service interfaces. WSDL provides an XML grammar to describe the request and response messages that are sent to and from a service. In this context, the request is referred to as an input message and the response as an output message. The following example shows the general structure of this XML grammar:

```
Interface
    Operation-1
        Input Message
        Output Message

    Operation-2
        Input Message
        Fault Message

    Operation-3
        Output Message
```

In the above example, an interface comprises three different operations, which can be described in terms of input, output and fault messages. Note that all three types of message are not always part of an operation. For example, Operation-1 has an input message and an output message, while Operation-3 only has an output message.

GetRoute is an example of an operation that can be described using WSDL. This operation includes an input message that requests information about a specific route and an output message that returns information about the route, as shown below:

```
Operation-1: GetRoute
    Input Message (StartPoint,EndPoint);
    Output Message (Route);
```

In WSDL, XML Schema elements and types are used to describe the arguments and parts in each message. Given that GML is based on XML Schema, you can use GML application schema elements and types to describe the geographic parts of WSDL messages. For example, you could create a GML application schema that defines geographic features for the forest industry. A WSDL description of a service could then use these features – such as ForestStand, ForestCover and CutBlock – to provide forestry data in an output message to a client. The following example shows how this can be encoded in WSDL:

```
Operation-1: GetStandInformation
    Input Message (StandID);
    Output Message (ForestSpecies,BH);
```

Figure 6 shows the role that GML application schemas play in providing elements and types for WSDL messages. Chapter 19 provides a more technical discussion of WSDL in relation to GML. Additional information is also available at http://www.w3.org/TR/wsdl.

6.4.2 WSDL, SOAP and HTTP

In addition to using GML application schema elements and types to describe the content of input and output messages, as discussed above, WSDL also provides an implementation description that describes how the WSDL message is carried. The implementation description identifies the protocol that is used to receive and send

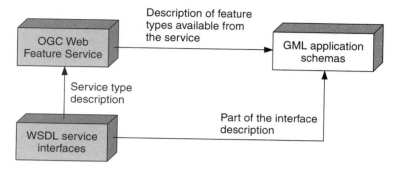

Figure 6 Role of GML application schemas in WSDL service description.

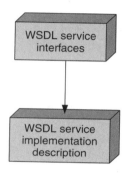

Figure 7 WSDL implementation description.

the messages. This is shown in Figure 7. HTTP – a familiar protocol for anyone who uses the Internet – is one of the protocols supported by WSDL.

Many web services use SOAP to send and receive messages. SOAP messages are sent over another communications protocol, such as HTTP. SOAP provides an envelope for transporting an XML payload between communicating processes, such as a client and a server. Since it is XML-based, SOAP is an ideal technology for carrying messages that contain GML data. For a more technical discussion about SOAP, please see http://www.w3.org/TR/SOAP/.

Figure 8 shows the relationship between the different protocols. WSDL provides the web service description for a message, and SOAP provides the framework for carrying a message. GML application schemas provide the elements and types for encoding GML data within the SOAP framework, and HTTP is the transport protocol that is used to send the message.

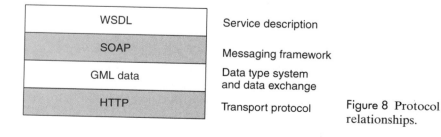

Figure 8 Protocol relationships.

6.5 Chapter summary

Geospatial web services process request and response messages that contain geographic information. These web services can interact with various clients – including cellular phones, PDAs and desktop computers – and other services. Many geospatial web services are being developed and will become part of a common spatial infrastructure called the Geo-Web. GML and other OGC standards will be used to support the geospatial aspects of these web services.

The relationship between GML and geospatial web services can be compared to that between HTML and web servers. With HTML, web servers and other supporting technologies are used to distribute HTML-encoded information over the World Wide Web. With GML, geospatial web services and supporting technologies are used to distribute GML-encoded information over the Geo-Web. HTML, however, is used primarily to format the visual layout of web pages, while GML is actually used to encode geographic information.

The OGC is currently developing a number of different geospatial web services, including the OGC WFS, WCS, WMS and Catalog Service. All of these services provide access to different kinds of geographic data or metadata. The WFS provides access to geographic feature data, the WCS to coverage data, the WMS to maps and the Catalog Service to metadata, such as GML application schemas. The request and response messages between clients and the different services can also be encoded in GML. GML application schemas can be stored in schema registries, which are a deployment of the OGC Catalog Service. The schema registry provides a mechanism for managing application schemas.

For clients to interact with a web service, they need to know the structure of the request and response messages that the service supports. WSDL provides an XML grammar for describing these messages in terms of input and output, and GML can be used to describe the geospatial parts of these messages. This is independent of the implementation. WSDL also provides an implementation description that describes how the message is carried.

References

http://www.opengis.org/docs/03-040.pdf (October, 2002).

http://www.opengis.org/docs/02-058.pdf (October 15, 2003).

http://www.opengis.org/docs/00-106.pdf (October 15, 2003).

http://member.opengis.org/tc/archive/arch02/02-024.pdf (October 15, 2003).

http://www.opengis.org/docs/01-068r2.pdf (October 15, 2003).

http://www.opengis.org/docs/02-087r3.pdf (October 15, 2003).

http://www.w3.org/TR/wsdl (October 15, 2003).

http://www.w3.org/TR/SOAP/ (October 15, 2003).

Chapter 7

Real-world deployment examples

GML can be applied in most domains of interest that use geographic information, such as

- local, regional and national government services
- utility network planning and monitoring
- resource development and environmental protection
- disaster management
- Location-Based Services (LBS) and Intelligent Transportation Systems (ITS)
- geospatial intelligence.

Whenever different domains decide to deploy GML, they need to determine how they are going to model their GML data and then create GML application schemas based on these models. As with any technology, different individuals have specific responsibilities when it comes to deploying and using GML-enabled systems. These roles and their associated organizations and responsibilities can be broken down into four categories, as shown in Table 1.

7.1 GML in local government

In the context of this book, local government includes the governments of municipalities and small regions, such as counties in North America, communes in France or prefectures in Japan. Although local governments have different responsibilities, they are typically concerned with the following general functions:

- Land-use planning
- Road planning
- Land parcel and other 'tenure' administration
- Local utilities planning and maintenance (for example, water, gas, electricity and telephone).

One of the more common objectives for the local government is to provide services to the local population and staff. These services often include

Geography Mark-up Language (GML).   R. Lake, D. S. Burggraf, M. Trninić, L. Rae © 2004 Galdos Systems Inc.
Published by John Wiley & Sons, Ltd   ISBNs: 0-470-87153-9 (HB); 0-470-87154-7 (PB)

Table 1 Breakdown of organizations and their responsibilities

Role	Organization Type	Responsibility
System and Business Analyst, Data Modeller, Database Administrator (DBA)	Data Holders, Data Providers, Web Service Providers	Create abstract models of business processes and objects. These models are used by DBAs to create GML application schemas. Create, modify and adapt GML application schemas, possibly based on abstract models that have been developed by systems and business analysts.
Developer	Commercial GIS Vendors, Application Software Vendors, System Integrators and Service Providers	Develop GML-aware technology components, such as a WFS. Build and maintain spatial infrastructures using commercial GIS components.
Technical Administrator	Data Providers, Data Holders	Maintain application schemas, service offers, CRS and other GML features and object definitions in online registries that are part of a spatial infrastructure, such as the Geo-Web.
General End-User	Employees, Field Personnel, Administrators, Planners, Agents and Consumers	Access geospatial web services through client devices, such as cellular phones, to accomplish specific functions. Most of these users are unaware of GML.

the following:

- Dispatching support personnel for water, sewage or other utility problems
- Dispatching fire or police services
- Registering land titles
- Analysing and displaying demographics to be used by municipal services
- Scheduling bus and transit services
- Managing traffic
- Managing land-use planning and decision making
- Establishing and managing citizen information services
- Managing local government assets.

GML plays a direct role in delivering these services over the Internet through geospatial web services, some of which were discussed in Chapter 6. Data modellers – who are employed by the local government – create GML application

Table 2   Sample feature types for a municipal government

Domain	Features Types
Municipal Government	LandParcel, Road, Street, WaterMain, WaterMainAccessPoint, SewerMain, SewerMainAccessPoint, PublicPark, Bridge, FireHydrant, FireStation, PoliceStation, Fire, PoliceEvent, PoliceCruiser, FireTruck, Subdivision, BicyclePath, BusStation, BusStop, CityHall, RecreationCenter

schemas that define geographic feature types. Table 2 lists possible features that a data modeller might create for a municipal government system.

GML application schemas, and GML data based on these schemas, can be stored in geospatial web services that are maintained by the local government. With standardized and centralized GML data, the local government can meet its objectives more easily. The following examples discuss possible applications of GML to different local government systems.

EXAMPLE 1   Bus-Scheduling Service

In this scenario, users connect to a bus-scheduling service through their landline or cellular telephones. They enter a bus stop and bus route number and are provided with the next time that a bus is expected at that stop. Five minutes before the bus is scheduled to arrive, the customer receives a telephone call that announces the bus's arrival.

In this service, the street network, bus stops and bus routes are encoded in GML and stored via a GML Web Feature Service (WFS) for transit features. The WFS receives the customer requests and generates the time of arrival for the caller. This service also queues reminder calls for the customer. GML is used to describe the messages sent to the web service, and these descriptions can be found through an OGC-compliant service offer registry. Bus-scheduling service developers can access the registry to obtain the appropriate schemas needed to develop the telephone-client end of the service.

EXAMPLE 2   Asset-Management Service

With this service, government employees use PDAs or cellular telephones to access and report information about the location, extent and other characteristics of the government's fixed and mobile assets. Note that the information captured for asset management is part of an integrated information base that can be accessed by many other applications, such as security, facility planning and service-call applications. GML is used to capture this information. External applications can access this GML data and calculate associated asset

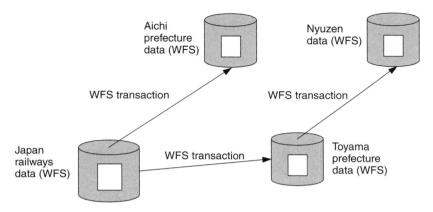

Figure 1 Integrated network of GML databases for Japan Railways.

valuations. WFS transactions – for example, insert or update feature – that employ GML can be used to transmit information about government assets captured in the field to a central or local-office WFS. Some of these WFS transactions are discussed in Chapter 19. Note that since these transactions can easily be transmitted over the wired or wireless Internet, they can be provided by manual (for example, PDA) or automated devices.

EXAMPLE 3   Integrating with Other Governments and the Private Sector

Local governments do not exist in isolation. Many other organizations – such as electrical, gas, telecommunications and security companies – are interested in the data that is owned and maintained by local governments. Regional and national governments are also interested in this data. Conversely, local governments are interested in data that is owned by other governments and private-sector companies.

In addition to local government, GML can also be used to model and store data for various companies and levels of government. This leads to a network of interconnected GML databases (WFSs), as shown in Figure 1. In this figure, data that belongs to Japan Railways, a national government organization, is replicated from one GML database to another, thus synchronizing this data between the various parties at all times. This synchronization is made possible because GML and the web feature services that store the GML data are extensible and standardized. Note that any update to the national database is automatically propagated to the local government databases affected by the update.

7.2 GML and utilities

GML is also very useful for utility companies, including companies that provide gas, oil, water, data and telephone services. All of these organizations would benefit greatly from the emergence of the GML-driven Geo-Web. While many of these organizations have built proprietary geographic systems, they are receiving a lot of economic pressure to move to systems that are based on open standards.

There are many advantages to establishing open-standards-based systems. For example, these systems reduce the cost of data development and maintenance.

Table 3   Sample feature types for an electrical utility company

Domain	Features Types
Electrical Utilities	`DistributionCorridor, TransmissionLine, TransmissionTower, SwitchingStation, Substation, PowerPlant, Transformer, Switch`

They also enable companies to participate in a rapidly changing world where the boundaries between different companies are becoming more complex and are constantly evolving.

Utility companies can use GML application schemas to define the geographic objects that describe their industry. Table 3 shows a list of geographic features that a data modeller might create for an electrical utility schema. Once these features are encoded in GML, they can be stored in GML databases and made securely accessible to utility planners, asset managers, external contractors and power distributors.

The diversity of the utility industry demands a highly flexible information structure. As a result, standards such as GML are essential for describing geographic features. The following examples discuss possible scenarios in the utilities sector that can benefit from the use of GML.

EXAMPLE 1   Gas Company Sells Electricity

Consider an example of a gas company that wants to use its existing sales, billing and service infrastructure to distribute non-gas products, such as water or electricity. To make this cost effective, the gas company needs to integrate various information systems. In this example, we assume that the gas company is not acquiring the electrical company's assets; it is simply acting as a seller of the electricity.

To provide an effective level of service, the gas company needs online access to the information about the electrical distribution system. Because of the different formats and models for the geographic information, it can be difficult to integrate the information. With the universal adoption of GML, the gas company can read and display an electrical company's geographic features as easily as their own features. Given that it does not need any new software, the gas company avoids spending millions of dollars on custom software development and integration. GML and the Geo-Web provide a standard data encoding and architecture that enables utility companies to reorganize their affairs in a manner that makes the most sense to their business.

EXAMPLE 2   Critical Infrastructure Protection

Modern industrial societies are very vulnerable to the disruption of their utility infrastructure. A modern city can only go for a few days without its water, gas and electrical systems. As a result, natural and manmade disasters can be extremely costly and have significant social impact.

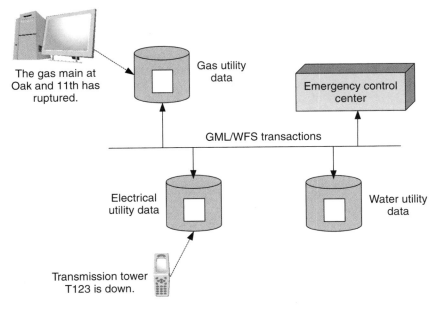

Figure 2 Wide-area update of various utility databases.

What happens when disasters occur in areas that do not have a standards-based information infrastructure? For example, a city's utility companies have different data structures and systems that do not communicate effectively with one another; administrators cannot have a clear picture of what is happening during a critical situation. As a result, they cannot make timely and effective decisions.

Even if each utility has purchased geographic information software from different vendors, they can share a common standard by using GML. This allows information to be exchanged, processed and displayed regardless of which utility company the information belongs to. Civilian and defence authorities can access all of the relevant information and rapidly put together a complete and coherent picture of what is happening.

GML standardization has other benefits. Through the associated WFS specification, GML enables the remote update of geographic features over the Internet. Updates are immediately reflected in the database available to the emergency-management authorities. Figure 2 shows an example of a network of three utility company databases. In this figure, a message about a gas main rupture is sent to the gas utility database, and another message about a transmission tower is sent to the electrical utility database.

7.3   GML and natural-resource management

Industries that develop and manage natural resources – such as petroleum and natural gas, forest products, minerals and water – have been active users of geographic information for many years. Nonetheless, natural-resource industries suffer from the same fragmentation of geographic information that occurs in other domains. These industries can also use GML to integrate their geographic data. Table 4 lists the different feature types that data modellers might create in different natural-resource sectors.

Table 4   Sample feature types for natural-resource management

Sector	Feature Types
Forest Sector	Road, Lake, River, TerrainSurface, Bridge, ForestStand, CutBlock, SelectiveCutArea, ForestTenure, SampleSite, ForestAccessRoad, Bridge, ProcessingFacility
Mining and Mineral Exploration Sector	Road, Lake, River, TerrainSurface, Bridge, MineralDeposit, SampleSite, Lithology, OreDeposit, ProcessingFacility
Oil Industry	Road, Lake, River, TerrainSurface, Bridge, DrillHole, OilWell, SeismicTrack, WellHead, OffshorePlatform, SupportVessel, Field, RockFormation, Rig

Note that certain geographic features, such as Road and Lake, are shared by all of the different sectors. These features – which are also known as 'background' features – present an important problem that can be resolved with standardized GML deployment. By establishing a Geo-Web with integrated natural-resource GML data, it is possible to avoid the current situation, in which multiple organizations map the same features in different formats. As a result, each sector can significantly cut data-development and maintenance costs. Consider the following examples regarding the application of GML to forest-management and environmental-protection systems.

EXAMPLE 1   Forest Management – Private and Public-Sector Partnerships

In many parts of the world, small-scale corporations, individuals and local organizations are responsible for acquiring forest information and extracting forest resources. There is a growing need to provide these different organizations with distributed services to assist in forest-resource planning and management. Companies and governments are re-orienting themselves to provide these services, which can include

- harvest planning
- harvest or cut authorization
- silviculture planning
- execution and monitoring
- tenure administration.

To provide these services, all of the organizations involved need to adopt standards over wide geographic areas. By using GML-based vocabularies, it is possible to use simple

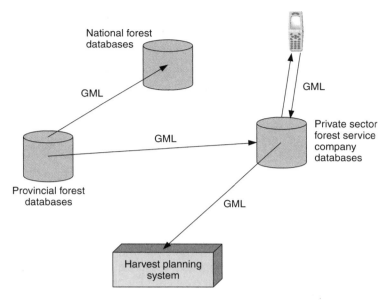

Figure 3 Integrated forestry database network.

Internet-based software to access sector-specific information from various distributed data providers. An integrated forest information database network can then be continuously developed and used as the basis for web-based information services in both the private and public sector. Figure 3 shows an example of an integrated forestry database network. Field personnel can send requests via WFS transactions to a particular database, which is connected to other forestry databases and systems. All updates to the data are replicated throughout the network.

EXAMPLE 2   Environmental Protection – Monitoring Water Quality

Monitoring water quality is a complex business. It is not simply a matter of measuring and reporting point-source data, since the real quality and safety of the water supply may be compromised before such measures are effective. To properly monitor water quality, point-source measurements must be combined with other measurements. These other measurements include the monitoring and evaluation of existing and potential pollution sources, plus all the environmental changes that might affect water quality.

Consider the development of a water-quality service for a local government. This service provides an integrated picture of the region's water supply, which can be used by government authorities and professionals in the water-management industry. Table 5 lists some sample feature types for a water-quality service.

Some of these feature types are geographic features, such as rivers and streams, while others relate to the water-management sector. Note that the measurement data and geographic features may come from many sources including

- local, regional and national government departments
- farms and agricultural processing plants

Table 5   Sample feature types for a water-quality service

Service	Feature Types
Water-Quality Service	`River, Stream, Lake, Pond, Dam, Weir, Bridge, Road, SewagePlant, SewageField, SepticTank, WaterReservoir, WaterIntakePoint, SewageDischargePoint, ChemicalOutFalls, MaterialOutFalls, SurfaceTerrain, Salinity, ChlorophyllConcentration, DissolvedOxygen`

- many kinds of chemical and manufacturing concerns
- facilities for the treatment of waste water, hazardous materials, garbage and sewage.

GML 3.0 provides the standardization and extensibility that is required to integrate this complex range of data. Value objects, observations and units of measure – which are discussed in Chapter 16 – support measured data. The OGC is also developing GML application languages for sensors and measurements.

7.4 GML and disaster management

Disasters have a number of common characteristics. Although some disasters only directly impact a small area, the majority involve many individuals and agencies over a wide geographic area. This diversity presents significant problems for the authorities who must respond to disaster events.

Although specific kinds of geographic features are associated with disasters, disaster-management applications require access to geographic features from many different domains. As a result, standardized geographic data is critical to responding effectively to disasters. Table 6 shows some of the existing standardized 'background' feature types that data modellers might use to design flood and earthquake managements systems in the same urban area.

GML can play a significant role in the development of disaster-management systems. First, GML can be used to develop a common standard for all geographic features, facilitating access to the diverse geographic features impacted by the event. GML can also be used to define geographic features for specific kinds of disasters, such as floods, earthquakes, terrorist incidents and aircraft crashes. GML application schemas that define these feature types can be created, published on

Table 6   Sample feature types for disaster response

Database	Feature Types
Existing Regional Databases	`Road, Bridge, River, Lake, Dam, Building, MunicipalBoundary, FloodBoundary, ImpactArea, SandBagDepot, DykeBreach`

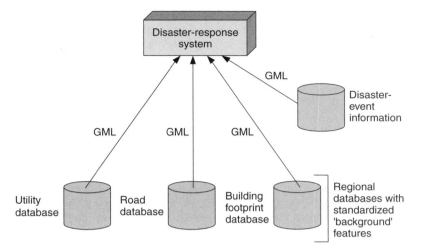

Figure 4 Integrated database network for a disaster-response system.

the Internet in registries and used as the basis for a wide variety of disaster-management web services. Figure 4 shows a possible infrastructure for a disaster-response system.

In this figure, a disaster-response system collects existing GML feature data from municipal databases for utilities, roads and building footprints. The system also collects disaster-event information from another database that contains feature data for a specific type of disaster event. Note that the same municipal databases would also be used for everyday municipal administration.

Disasters are also events that evolve over time. For example,

- flood boundaries change over time
- roads or bridges are destroyed, disconnected or blocked
- forested regions are burned or buried beneath tons of rock or snow
- aquatic and marine resources are killed by spreading oils slicks or toxic plumes.

To properly model these events, the corresponding geographic features need to have properties that account for changes over time. GML 3.0 provides dynamic features that can capture both the time-variant and time-invariant properties of a geographic feature. Chapter 14 provides more information about dynamic features. The following examples show how GML can be used in two different disaster-response systems.

EXAMPLE 1   Earthquake-Response System

In this example, GML is used in an earthquake-response system at the regional government level. This system provides municipal, regional and national authorities with information

Table 7   Sample feature types for an earthquake-response system

System	Feature Types
Earthquake Response	Road, River, DebrisChute, Earthquake, Aftershock, EmergencyCenter, Epicenter, Building, Dam, Bridge, ImpactArea

that authorities can use to plan various types of earthquake-response measures after an earthquake occurs.

In this example, it is assumed that the municipalities and local governments involved have an established GML-enabled spatial infrastructure. Additional GML application schemas are created to describe the earthquake event and its aftermath. Table 7 lists some of the sample feature types that data modellers might create for an earthquake-response system.

The earthquake-response system maintains data on earthquake events and distributes critical information – such as the earthquake impact area – to key municipal, regional and national information systems. The responding agencies have field personnel that provide input to create or update features impacted by the earthquake. These changes are communicated in GML, using WFS messages, to the earthquake-response system.

Although the response system can still have a proprietary back-end storage system, such as a GIS, the geographic data is sent in GML format. This standardized format further enables wide-area information updates. In addition to data updates, this response system can also access portrayal and mapping services that extract information from the network of GML databases. The services then use the extracted information to compose visual presentations that can be displayed on various end-user devices, such as cellular phones and desktop computers. Note that GML 3.0 provides schema components that support the visual presentation of geographic data. These components – which build on W3C SVG and Synchronized Multimedia Integration Language (SMIL) – are described in detail in Chapter 18.

EXAMPLE 2   Storm-Surge Planning System

Various parts of the world are threatened by hurricanes or typhoons. Storm surges – large waves that can flood coastal and inland regions to depths in excess of 3 m – are often the most dangerous result of these events. A GML-enabled disaster-response system can be used to respond to storm-surge disasters.

In this example, the storm-surge planning system assesses the geographic areas that can be impacted by a potential storm surge. The system is also designed to assist in the development of preventative and protective measures to reduce damage and loss of life.

As with other disaster-response systems, the data modellers need to ensure that the storm-surge planning system has access to standardized 'background' geographic features. The data modellers also need to provide access to a DEM, which is encoded in GML as Coverage data (Coverages are discussed in Chapter 17). The system also requires access to a database with storm-surge features. Table 8 lists sample feature types that data modellers might provide for the storm-surge planning system.

To conduct a storm-surge analysis with this system, planners can perform critical tasks, such as specifying the water depth over a given area or retrieving a flood-level time history

Table 8   Sample feature types for a storm-surge planning system

System	Feature Types
Existing Regional Systems	`Road, River, SewageTreatmentPlant, SewageOutfall, PotableWaterIntake, Reservoir, Lake, Coastline, Building`
Storm-Surge Planning System	`FloodBarrier, Dykes, PumpingStation, SandBagDepot, HazardousMaterial, FloodedArea, RoadBlockage, DownedPowerLine, StormSurge, ImpactArea`

from a storm-surge simulation program. In each case, the system can then apply the flooded area 'map' and determine the specific features that are likely to be impacted. Because the planners can also access data from the impacted municipalities and regions, they can import any desired set of features to perform this analysis.

Once the storm-surge analysis is complete, planners can attach specific properties to GML features in the municipal and local government databases. These properties indicate the potential impact of a storm-surge event on different features. The planners can also add new features – which represent preventative emergency measures – to these databases.

7.5   GML and location-based services (LBS)

In the context of this book, the term *LBS* refers to services that provide information related to an individual's location. These services are of interest to various end-users, including sales personnel, customer-support staff, tourists and the general consumer. All of these individuals can access LBS to retrieve information – such as traffic, weather or driving directions – about the area in which they are located. GML can play a significant role in deploying LBS by providing standardized geographic feature types such as those listed in Table 9.

GML application schemas for LBS are already under development and include the OGC's Open Location Services (OpenLS) initiative (http://member.opengis. org/tc/archive/arch03/03-006r1.pdf) and G-XML for Location Services from DPC in Japan (http://gisclh.dpc.or.jp/gxml/contents-e/index.htm). GML 3.0 provides various components that offer particular support for LBS. These components include temporal elements, dynamic features, user-defined coordinate reference systems, units of measure dictionaries and a variety of new geometry and topology types. GML's extensibility is also of importance to specialized LBS applications.

Table 9   Sample GML feature types for location-based services

Service	Feature Types
Location-Based Services	`Person, Service, PointOfInterest, Hotel, Building, Road, Route`

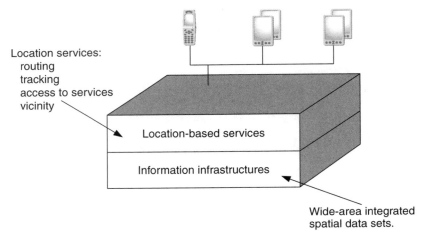

Location services:
   routing
   tracking
   access to services
   vicinity

Location-based services

Information infrastructures

Wide-area integrated
spatial data sets.

Figure 5 Long-term vision for LBS and integrated geospatial data.

As GML becomes more widespread, LBS developers can reuse existing government, utility, natural-resource and disaster-management databases of standardized GML features. This kind of standardization and data sharing is essential for providing accurate and timely LBS. Figure 5 illustrates the long-term vision of LBS. As shown in this figure, various mobile devices access the services to retrieve location-specific routing, tracking, service access and vicinity information. The services are implemented on top of an existing information infrastructure that comprises a globally integrated network of geospatial data, including a standardized set of GML-defined features. Information from location-based services is supplied as GML data to the end-user device or middle-tier processing services.

In addition to having limited bandwidth for the visual display of information, the typical LBS user often has limited attention available for any kind of information processing. In the future, it is likely that the presentation of LBS information will involve a wide variety of devices – including vehicle navigation systems, PDAs, cellular telephones and desktop computers – and presentation modes. GML's separation of presentation and content is ideally suited to LBS applications because GML data can be easily 'styled' for presentation in various forms, including graphics or as voice instructions.

Although there are still issues regarding the bandwidth for XML transmission and processing in portable devices, these performance concerns are gradually disappearing due to the emergence of the following:

- Higher bandwidth wireless networks
- Increased horsepower for portable computing devices
- Efficient codecs and compression schemes (for example, KDDI Xeus and ExpWay's BinXML).

7.6 Chapter summary

Regardless of the application domain, different individuals have specific roles in deploying and using GML-enabled systems. These roles can be broken down into four categories: data modellers, developers, technical administrators and end-users. Data modellers are responsible for establishing models of geographic data and creating GML application schemas based on these models. Developers are responsible for developing GML-aware components and using these components to build and maintain spatial infrastructures, such as the Geo-Web. The technical administrator is responsible for maintaining GML metadata and data that is stored in online registries and repositories. The general end-user accesses the GML data to perform a specific task. Most of these end-users do not deal directly with the GML encodings, but view the data as text or as a graphic, such as a map.

GML can be deployed in many different domains, including local government, utilities, natural-resource management, disaster-response and location-based services. Whenever a new GML system is deployed, feature type definitions need to be created or imported from existing GML application schemas. These schemas – and the instances created from elements and types defined in the schemas – can be stored in registries and various geospatial web services.

In all of these domains, a common architecture is required to ensure that the data is standardized and consistent. Given that most geographic data is not field specific, many domains need to access data and metadata from external sources. By establishing a standardized Geo-Web of geospatial web services and registries for various domains, it will be possible to establish a truly global network of geographic data that can be shared by all domains. GML and a common service architecture will make this Geo-Web a reality.

References

http://member.opengis.org/tc/archive/arch03/03-006r1.pdf (October 20, 2003).

http://gisclh.dpc.or.jp/gxml/contents-e/index.htm (October 20, 2003).

Additional references

http://www.opengis.org/docs/02-023r4.pdf (October 15, 2003).

http://member.opengis.org/tc/archive/arch03/03-007r1.pdf (October 20, 2003).

PART II

GML: A TECHNICAL REFERENCE GUIDE

Chapter 8

Basic concepts of GML

GML is an XML-based language that is used to encode geographic information. It has been developed through the efforts of many individuals within the OpenGIS Consortium (OGC). As an XML-based language, GML-encoded information can be easily transported over the Internet.

GML is well suited for encoding the geographic information that is processed by geospatial web services. As more geospatial web services emerge, the Geo-Web – a globally integrated web of geographic information – will become available. GML also provides the mechanisms for linking information in the Geo-Web.

GML is not the only XML-based language that has been developed for describing geographic objects. The Database Promotion Center (DPC) in Japan has also developed a mark-up language called G-XML. Following a series of discussions, the DPC and the OGC have worked together to converge these two languages. As a result, GML 3.0 has added new sections based on G-XML, and the DPC-based G-XML Version 3.0 is now based on GML as a GML application schema for location-based services.

8.1 GML and the XML technology family

The eXtensible Mark-up Language (XML) comprises different languages that allow you to specify the structure of XML documents, including the following:

XML 1.0 and Document Type Definitions (DTD). GML 1.0 was based on XML 1.0, and DTDs were used to describe the contents of instance documents in GML 1.0, which used different data definition languages through the use of profiles. The GML 1.1 and 1.2 profiles were based on the XML DTD, while profile 1.3 was based on RDF and RDFS.

XML Schema. XML Schema has been used to create GML schemas in all the versions of GML since version 2.0, which contains three core GML schemas. Many new core GML schemas were added in GML 3.0.

XML Linking Language (XLink) and XML Pointer Language (XPointer). XLink and XPointer are used in GML to create links that express associations between geographic features and other GML elements. While most of the current uses of XLL (Xlink) and XPointer in GML use

---

Geography Mark-up Language (GML).   R. Lake, D. S. Burggraf, M. Trninić, L. Rae © 2004 Galdos Systems Inc.
Published by John Wiley & Sons, Ltd   ISBNs: 0-470-87153-9 (HB); 0-470-87154-7 (PB)

simple links and simple pointers (for example, bare name Xpointers), there are no restrictions on their usage in GML. GML application schema developers are free to use all of the features and capabilities of both XPointer and XLL.

Resource Description Framework (RDF) and RDF Schema (RDFS). GML 1.0 included RDF in profile 1.3 but RDF has not been used in any of the newer versions of GML. However, GML continues to borrow ideas from RDF. Recently, there has been significant interest in using RDF and XML Topic Maps (XTM) to provide additional semantics to GML. An XTM topic, for example, can correspond to a GML feature element name (such as Road), and XTM might then provide a classification hierarchy for GML.

eXtensible Stylesheet Language Transformations (XSLT). You can use XSLT to perform many useful tasks with GML data, such as mapping schemas or generating graphical maps. XSLT's rule-based and declarative programming style make it a popular technology for styling GML data to SVG. XSLT engines, which are widely available, are being distributed in modern web browsers, PDAs and even cellular phones. Commercial tools exist for creating XSLT stylesheets for styling GML to SVG.

Scalable Vector Graphic (SVG). SVG provides a means of generating graphical representations of GML geographic features. For example, SVG can be used to create maps based on GML data. SVG is particularly well suited to this purpose because of the following reasons:

- It is an XML grammar.
- It provides a user-interaction model that can be readily scripted in any modern web browser.
- Plug-ins such as Adobe, Corel and Bitflash are available for most operating systems, browser platforms, cellular telephones and PDAs. A mobile profile of SVG has been defined, with basic and tiny subsets, and is currently being implemented.
- SVG provides advanced graphics features such as 'text on path', alpha blending, raster operations, polygonal stroking and filling, image and symbol support and vector-based animation.

Web Services Description Language (WSDL). WSDL is useful for describing geospatial web services, which often have GML-encoded components. WSDL is the web equivalent of the Common Object Request Broker Architecture (CORBA) or Distributed Component Object Model (DCOM) Interface Description Language (IDL) and is used to describe the interfaces to a web service.

WSDL provides two kinds of descriptions: abstract and implementation. The abstract component describes the interface as a set of operations with each operation consisting of input and output messages. Since the description of the message elements and types makes use of XML Schema, geospatial web services can make use of GML application schema components. The implementation description binds the abstract interfaces to specific data transports – for example, HTTP POST and SOAP/HTTP – and identifies the service endpoint addresses. Note that a WSDL description can be further enhanced through the use of extensibility elements.

Simple Object Access Protocol (SOAP). SOAP can be used for sending GML-based geographic requests, especially in relation to geospatial web services. SOAP is ideal for this purpose as it is intended for transmitting XML payloads. In addition, SOAP can send attachments, which can be used, for example, to carry the binary file component of a coverage using the GML file encoding.

8.2   GML instances and schemas

Like XML, GML is not simply a collection of elements and attributes used to encode geographic documents. It is also a language that provides mechanisms for structuring and defining complex models of geographic information that can be applied in a wide range of geospatial applications.

To encode geographic information in GML, there are two different kinds of encoding files: schemas and instances. GML schemas are metadata files, written in XML Schema, that define the structure and content of GML instances. GML instances are the individual occurrences of data that conform to one or more schemas.

For example, a data modeller in the transportation industry might create a GML application schema with a `Bridge` element whose content model defines how to encode any bridge. If data collectors for a transportation application want to create a GML encoding for a particular `Bridge` feature, such as the Golden Gate Bridge in San Francisco or Hijiribashi in Japan, they can create a data instance of that `Bridge`.

Before data modellers can create the `Bridge` feature in an application schema, they need to have an existing framework for creating schemas. In GML, there are two kinds of schema files:

- **GML Application Schemas**. These schemas are not provided by the core GML schemas and are typically created by database administrators for various domains. A GML application schema can be viewed as an application-specific XML language for a vertical domain like ITS, transportation or natural resources. The transportation schema discussed above is an example of a possible GML application schema.

- **GML Core Schemas**. These schemas, which are the core part of GML, provide the necessary framework for creating GML application schemas for different domains of interest.

Figure 1 shows the relationship between GML core schemas, application schemas and instances. As shown in the figure, the GML core schemas provide the framework for domain-specific GML application schemas, which, in turn, provide the framework for encoding GML data instances.

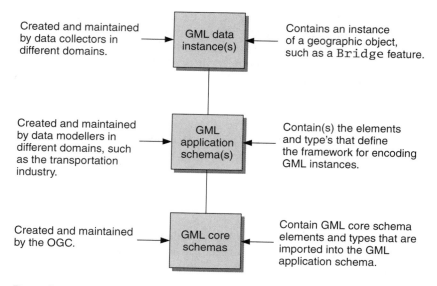

Figure 1   GML schemas and instances.

8.3   Dictionaries

Dictionaries are normative instance documents that contain a collection of dictionary entries about a particular area of interest. In GML 3.0, dictionaries are primarily used for defining coordinate reference systems, units of measure, value types and temporal reference systems. For example, a coordinate reference system dictionary comprises a collection of coordinate reference system definition entries.

Although access to dictionaries can be restricted to a particular individual or organization, it is also possible to make them available throughout a particular field of interest or to make them universally accessible over the Internet. The organization that owns the domain where the dictionary is located is regarded as the authority for the dictionary. This authority may be further subdivided through the complete path Uniform Resource Identifier (URI) that refers to the dictionary document. To reference a dictionary, use an anyURI attribute to reference the entry by its unique identifier. The rules for referencing and defining different kinds of dictionaries are covered in Chapters 15 and 16.

Online dictionaries are likely to play an increasing role in the development of the Geo-Web. Dictionaries of units of measure, location keywords and coordinate reference systems will be deployed using the OGC Catalog Service (also known as the Web Registry Service (WRS)). For an example of a dictionary for coordinate reference systems, go to http://crs.opengis.org/crsportal.

8.4   Introducing features and geometries

In GML 2, only three core GML schemas were used to create GML application schemas. These three schemas are as follows:

- **Feature**. In GML, the concept of a feature is based on the feature model from the ISO/OGC view of geographic information. This is covered in the *OpenGIS® Abstract Specification: Topic 5 – Features*, which is available online at http://www.opengis.org/docs/99-105r2.pdf. Features can be concrete physical objects, such as roads or rivers, or they can be abstract objects, such as political boundaries or health districts. The feature schema does not define specific features, such as roads or rivers, but provides basic types and elements for creating feature types in GML application schemas.

- **Geometry**. This schema is used to create geometry types that describe the geometric characteristics of features. For example, a `Point` is a simple geometry type. All of the geometries in GML 2.0 are simple geometries in that they are composed of linear elements only. The GML 3.0 geometry schemas, on the other hand, offer a much richer collection of geometric objects.

- **XLinks**. This schema provides attributes for referencing external data, such as dictionaries. This schema was added in GML 2.0 because the W3C had not published a normative schema for XLinks (also known as the eXtensible Linking Language (XLL)).

With GML 3.0, there are 28 core schemas, all of which are discussed in this guide. Chapter 10 provides an overview of all of the core schemas in GML 3.0 and Chapter 11 covers the rules for creating application schemas. The *Geography Mark-up Language (GML) 2.0* specification is available online at http://www.opengis.org/docs/01-029.pdf.

8.5   What are the differences between GML 2.0 and GML 3.0?

GML releases 2.1 through 2.1.2 were essentially to correct minor errors in the original schemas. GML 3.0, on the other hand, represents a significant step forward in terms of content and greatly extends the applicability of GML.

GML 2.0 and GML 3.0 share the same model respecting features and their properties (including the use of remote and in-line properties). They also express feature relationships in the same manner. GML 2.0 geometries are a subset of GML 3.0 geometries. GML 3.0 also provides support for many new kinds of objects, including

- temporality
- dynamic features
- coordinate reference systems
- units of measure
- coverages
- many new geometries, including three-dimensional and non-linear geometries
- topology
- default styling
- a general mechanism for metadata.

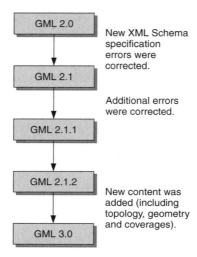

Figure 2 Changes from GML 2.0 to 3.0.

Figure 2 shows the evolution from GML 2.0 to 3.0. GML 3.0 is almost entirely backwards compatible with GML 2.0, and therefore it is possible to regard GML 3.0 as essentially the same as GML 2.0 but with a much richer set of component building blocks.

8.6   XML and GML – GML is XML

GML instance documents are well-formed XML documents, that is, they follow the rules for well-formed XML documents covered in *Extensible Markup Language (XML) 1.0 (Second Edition)* (http://www.w3.org/TR/REC-xml). In addition, GML instance documents must be XML Schema valid, which means they must comply with a GML application schema. The *Worked Examples CD* contains many examples of GML instance documents and their corresponding GML application schemas.

GML instance documents also include XML namespace declarations. The following example shows a namespace declaration in a GML instance document that imports a `Transportation.xsd` schema from an example namespace at http://www.ukusa.org/.

```
<?xml version="1.0" encoding="UTF-8"?>
<app:RoadInfrastructure
    ⌐xmlns:app="http://www.ukusa.org/app"
    ⌐xmlns:gml="http://www.opengis.net/gml"
    ⌐xmlns:xsi="http://www.w3.org/2001/XMLSchema-instance"
    ⌐xsi:schemaLocation="http://www.ukusa.org/app
    ⌐Transportation.xsd">
```

The document also imports the namespaces represented by the `gml` and `xsi` prefixes. For more information about namespaces, please refer to http://www.w3.org/TR/1999/REC-xml-names-19990114/or http://www.w3.org/TR/xml-names11/. Namespaces – which are used to create unambiguous names for GML objects and properties – are a required feature of any GML application schema. GML application schemas must declare a target namespace. This is discussed in Chapter 11.

Note that GML uses elements to describe objects and their properties. There are a few cases in GML in which XML attributes are used as element modifiers, for example, `xlink:href`, `uom`, `frame` and `srsName`. These attributes are discussed throughout this book but are detailed in Chapters 9 and 11. Note that in most circumstances, GML application schema developers do not need to use attributes in their schema definitions.

8.7   XML Schema and GML

While GML 1.0 was based on DTDs and RDF, since GML 2.0, GML has been based firmly on XML Schema. To understand GML, you need to understand XML Schema, especially in terms of how it is used to write GML instances and GML application schemas. Note that GML has additional rules for writing GML application schemas, and therefore not every valid XML Schema is a valid GML application schema. These rules are discussed in Chapter 11.

There are many available sources that provide a comprehensive description of XML Schema, including *XML Schemas* by Chelsea Valentine, Lucinda Dykes and Ed Tittel (Valentine *et al.*, 2002). For online sources of XML Schema, see http://www.xml.com/.

This section contains a brief discussion of the key aspects of XML Schema that apply to GML. To understand many of the concepts that form the GML model, you need to be familiar with the following terms, some of which have specific meanings in XML Schema.

Type in GML. The term 'type' can cause some confusion in GML, given that it has more than one meaning. GML features, geometries and other GML objects are encoded as XML elements. The name of each element – for example, `Road`, `River` and `Point` – is also referred to as the semantic type of the GML object. In XML Schema, each element name has an associated content model (also called the structural type), which is indicated in XML Schema by the `type` attribute. In other words, the terms 'feature type' or 'feature type name' refer to the semantic type or element name (for example, `Road`) that is used in GML instance documents. This is not to be confused with the structural type or XML Schema content model, as defined by the `type` attribute in the element declaration – for example `RoadType`. Note that structural types are only in schemas and do not appear in GML instances. In the sections of this book that discuss GML application schemas, the term 'element' is often used instead of 'type' to refer to the semantic type of a feature (or other GML object), and 'content model' is often used to refer to the structural type.

Data type (general). In a programming language, a data type is a set of data values that have shared predefined characteristics. Examples of data types include integer, character, string and pointer. Usually, a limited number of data types come built into a language. The language usually specifies the range of values for a given data type, how the computer processes these values and how they are stored. Most modern programming languages – for example, Java, C# and C++ – provide mechanisms for programmers to create their own data types. Variables in a programming language are said to be of a particular data type if the possible values of that variable are constrained to those of the data type.

Lexical type. A set of values for a string variable. In XML Schema, all simple data types are strings and hence have lexical representations.

XML Schema type. The type of an XML element or attribute. For an attribute, this is always a simple lexical type. For an element, this may be a lexical type or a structural type consisting of further elements and/or attributes. The XML Schema type is also called a content model. It is declared in XML Schema using either `simpleType` or `complexType` elements.

Complex type. Written in XML Schema as `complexType`, this declares a structural type that can have attributes and child elements. Only elements, and not attributes, can have complex types. In addition, complex types can be derived from one another using either extension or restriction. A complex type that uses derivation is said to have complex content. A complex type that has only text content and attributes is said to have simple content.

Simple type. Written in XML Schema as `simpleType`, this declares a structural type that only has text content. A `simpleType` cannot contain attributes or elements. Attributes must always have content models of `simpleType`. Elements may have content models of either `simpleType` or `complexType`.

Content model. A description of an element type's content and structure. The name of an element's associated content model is always denoted by the `type` attribute in an element declaration unless the content model is anonymous.

Element. Elements are used to encode both GML objects and their properties. Properties are always the child elements of GML objects and they cannot be other GML objects. Note that element names for objects and properties are visible in GML instances.

Substitution groups. In XML Schema, a substitution group is a collection of XML elements that are substitutable for a specific named element in the group. This element is referred to as the 'head' element of the substitution group. Note that membership in a substitution group is not necessary for an object to be one of the GML object classes because membership is determined entirely by the complex type that the object's content model derives from.

Substitution groups are used in GML to create template patterns especially for collection, bag or set structures. For example, if you wanted to say 'any feature can go here' when you define a property in a GML application schema, you would use a substitution group, such as `_Feature`.

Note that in XML Schema, the fact that A is a subtype of B (in the XML Schema sense) does not imply that an instance of A can be substituted for by an instance of B. Note further that substitution groups apply to elements and not to content models. In GML, the head element of a substitution group is usually declared abstract – meaning that it cannot appear in an instance document – and is denoted by a preceding underscore, for example, _Feature. The meaning of this head element is 'any object of this type', that is, _Feature can be read as 'any feature'.

XML Schema inheritance. The type of a GML object is determined by the ultimate inheritance of the object's content model from one of the core GML object content models. This inheritance is XML Schema inheritance. For example, GML-aware software determines if a given element is a feature by determining if the content model of that feature ultimately derives from AbstractFeatureType.

Use of any URI. One of the built-in simple types of XML Schema is anyURI. This is a string type that complies with the lexical restrictions of a URI. The following attributes in GML are based on anyURI: xlink:href, srsName, uom and frame. These attributes are used to reference remote data. Note that srsName, uom and frame reference remote dictionary entries and can be attached to objects or properties. On the other hand, an xlink:href attached to a GML property indicates that the value of the property is that which is pointed to by the href attribute.

8.8   Chapter summary

As an XML-based mark-up language for geographic information, GML provides a format that can be easily transported over the Internet, thus making GML a key component in the emergence of the Geo-Web. In GML, there are data instances and metadata files that describe the structure and content of GML data instances. The metadata files are called schemas and these are further divided into GML core schemas, which are provided by GML 3.0, and GML application schemas, which are created by database administrators. The GML core schemas provide the framework for creating GML application schemas.

Dictionaries are instance documents that contain dictionary entries for coordinate reference systems, units of measure, values and temporal reference systems. These dictionaries are referenced from other GML instance documents using an anyURI attribute. Dictionaries can be deployed through OGC Catalog Services, which are also known as registries or repositories.

While GML 2.0 has the feature, geometry and Xlinks schemas, GML 3.0 contains 28 core schemas. The model is the same respecting features and their properties; however, GML 3.0 also supports many new kinds of objects, including dynamic features, coverages and coordinate reference systems.

GML instance documents are well-formed XML, include XML namespace declarations and must be XML Schema valid with respect to a GML application schema. Because XML Schema is used to create GML application schemas, you need to be familiar with XML Schema to understand GML. Some important concepts – many of which are from XML Schema – that you need to understand include: content model, semantic type, complex type, simple type, substitution groups and XML Schema inheritance. Please ensure that you are familiar with these terms before continuing to read this book.

References

http://crs.opengis.org/crsportal (October 20, 2003).

http://www.opengis.org/docs/99-105r2.pdf (October 20, 2003).

http://www.opengis.org/docs/01-029.pdf (October 20, 2003).

http://www.w3.org/TR/REC-xml (September 20, 2003).

http://www.w3.org/TR/1999/REC-xml-names-19990114/ (October 20, 2003).

http://www.w3.org/TR/xml-names11/ (October 20, 2003).

VALENTINE, C., DYKES, L., and TITTEL, E. (2002) *XML Schemas.* Sybex Inc., Alamed, CA.

http://www.xml.com/ (October 17, 2003).

http://www.finetuning.com/ (October 17, 2003).

Additional references

http://www.w3.org/TR/REC-xml (September 20, 2003).

http://www.w3.org/TR/xmlschema-0/ (September 20, 2003).

http://www.w3.org./TR/xptr/ (September 20, 2003).

http://www.w3.org/XML/Linking (September 20, 2003).

http://www.w3.org/TR/rdf-schema/ (September 20, 2003).

http://www.w3.org/TR/REC-rdf-syntax/ (September 20, 2003).

http://www.w3.org/TR/xslt (September 20, 2003).

http://www.w3.org/TR/SVG11/ (September 20, 2003).

http://www.w3.org/TR/wsdl (September 20, 2003).

http://www.w3.org/TR/SOAP/ (September 20, 2003).

Chapter 9

Introducing the GML model and GML features

To understand GML, it is necessary first to understand the object-property model. The GML object-property model is partially based on the subject-property model of the Resource Description Framework (RDF) (http://www.w3.org/TR/REC-rdf-syntax/). The RDF subject-property model can be used as a reference point for understanding the GML model.

In the object-property model, objects are declared as XML elements, for example, <app:Road>. The element children of these object elements are the properties of the object, for example, <app:numLanes> for the number of lanes in the Road. Properties are assigned to a GML object in the content model of the GML object in a GML application schema. This is covered in Chapter 11.

Some of the most important kinds of objects in GML are features, geometries, topologies, coverages, observations and Coordinate Reference System (CRS) definitions. The kinds of objects in GML are denoted by the set of abstract XML schema elements and types that are contained in the core GML schemas. These abstract schema elements and types are covered in Chapter 10.

This chapter discusses the object-property model and provides examples of GML instances to illustrate how this model is encoded. Guidelines for encoding features, properties, feature collections and feature relationships are also provided. Additional concepts relating to features – such as geometry properties, coverages and units of measure – are also included.

9.1 About GML objects and properties

In GML 2, features and geometries were the only GML objects (http://www.opengis.org/docs/01-029.pdf). As a result, in earlier discussions, the GML model was called the feature-property model instead of the object-property model. In GML 3, there are many different kinds of objects, and, therefore the GML model is now called the object-property model. This model is encoded in GML by declaring an object as an XML element and then assigning properties to that object.

Every GML object is described by properties and the values of these properties. Figure 1 shows that properties contain values that describe the object. The

Geography Mark-up Language (GML). R. Lake, D. S. Burggraf, M. Trninić, L. Rae © 2004 Galdos Systems Inc.
Published by John Wiley & Sons, Ltd   ISBNs: 0-470-87153-9 (HB); 0-470-87154-7 (PB)

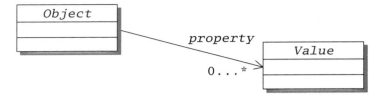

Figure 1 Simple object-property model.

object itself does not contain these values, but has a property that contains a value. Properties typically have only one value. Arrays are an exception to this rule.

In a GML instance, objects and properties can be encoded as shown in the following:

```
<app:ObjectName gml:id="..">
    <app:propertyName1>value1</app:propertyName1>
    <app:propertyName2>value2</app:propertyName2>
    <app:propertyName3>value3</app:propertyName3>
</app:ObjectName>
```

In this example, an object called *ObjectName* contains three properties, each of which has a value. The object and the properties are all XML elements, and the property elements are contained within the object element. These property elements are considered child elements of the object. In GML, as in XML, a child element can only have one parent.

Note that the objects are named using the UpperCamelCase convention, that is, the first letter is upper case and properties are named using lowerCamelCase. This reflects the naming conventions for properties and objects used throughout GML. UpperCamelCase implies that each word in a concatenated string of words starts with an upper case letter. In lowerCamelCase, the first letter of the first word is in lower case, and all subsequent words start with a letter in upper case.

The values of the properties in the above example are not expressed as attributes, but as values inside the opening and closing property tags. Alternative ways of encoding property values are discussed later in this chapter.

Note: Note that app and gml are prefixes that represent the namespaces of the object and properties. A namespace is a concept from XML that is used to assign unique identities to sets of elements and attributes. In this chapter, the app prefix refers to a generic application namespace. The gml prefix refers to the GML namespace, http://www.opengis.net/gml.

In GML, the value of a property can also be another object, as shown in Figure 2. In GML, an object cannot directly contain another object, but must have a property whose value is the object. The following example shows how to encode an object with a property whose value is an object called ObjectName2.

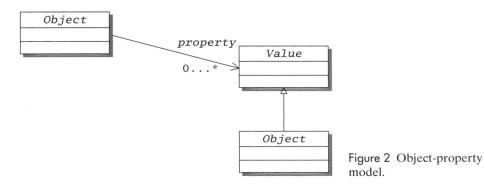

Figure 2 Object-property model.

```
<app:ObjectName1 gml:id="..">
   <app:property1>...</app:property1>
   <app:property2>
      <app:ObjectName2 gml:id="..">...</app:ObjectName2>
   </app:property2>
   <app:property3>...</app:property3>
</app:ObjectName1>
```

In GML, objects that are values of properties can also have their own properties, which can then have objects that also have properties. This nesting of objects and properties can continue indefinitely. Before discussing other aspects of the object-property model, let's examine how this model relates to features, which were at the core of GML 2 and remain a fundamental part of GML 3.0.

9.2 Features and properties

In GML, features can be concrete physical objects, such as roads or rivers, or they can be abstract objects, such as political boundaries or health districts. The GML feature concept is based on the Feature model from the Topic 5 of the *OpenGIS®* *Abstract Specification* (http://www.opengis.org/docs/99-105r2.pdf).

In contrast to legacy GIS approaches, a feature is not defined primarily as a geometric object, but as a meaningful object that might have some properties that are geometric and other properties that are not. For example, a GolfCourse feature can have a property that describes its physical extent and another property that describes its name.

Note that specific features like bridges are not defined in GML. These feature types are defined in application schemas created by data modellers. Application schemas are covered in Chapter 11. A feature is described by its properties, which express characteristics of the feature. The following is a simple GML example of a Bridge feature with properties:

```
<app:Bridge gml:id="..">
   <app:span>400</app:span>
   <app:height>50</app:height>
   <app:mobility>DrawBridge</app:mobility>
</app:Bridge>
```

In this example, `app:Bridge` refers to a `Bridge` feature in the namespace that has the `app` prefix (in these examples, `app` is the prefix for the example namespace http://www.ukusa.org/). The elements `span`, `height` and `mobility` are all properties of the `Bridge` feature. The specific `Bridge` instance is identified by the ID attribute, `gml:id`. The `Bridge` feature is an object whose content model is derived from the abstract base GML feature type. Inheritance and abstract base GML types are discussed in Chapter 10.

Note: In GML 3, every object instance inherits the `gml:id` attribute to indicate a unique identifier for each object. In previous versions of GML, the `fid` attribute was used to indicate the unique identifiers for GML features and `gid` was used for GML geometries. Although GML 3 is backwards compatible, it is recommended that you use the `gml:id` attribute when you create instances for features and geometries in GML 3. The `fid` attribute is deprecated in GML 3. It is also recommended that all GML objects be assigned ids.

Note that in GML, you can identify the properties of a feature simply by looking at the instance document. You do not have to refer to the associated XML schema. In other words, GML instances are human readable, which is an inherent advantage of the GML object-property pattern.

The above `Bridge` example can be read as equivalent to the following expressions:

```
span(Bridge)=400   or Bridge. span=400
height(Bridge)=50 or Bridge.height=50
mobility(Bridge)="DrawBridge" or
    ⌐Bridge.type="DrawBridge"
```

9.2.1   Remote properties

In GML, properties can have in-line values or remote values that are not in-line children of the property element. The following example shows an instance with in-line properties:

```
<app:Bridge gml:id="..">
   <app:span>400</app:span>
   <app:height>50</app:height>
   <app:mobility>DrawBridge</app:mobility>
   <gml:centerOf>
      <gml:Point gml:id="P1"   srsName="..">
         <gml:pos>...</gml:pos>
      </gml:Point>
   </gml:centerOf>
</app:Bridge>
```

In addition to the `span`, `height`, and `mobility` properties, the `Bridge` feature also has a `centerOf` property, which is a geometry property whose value is a geometry object called `Point`. This `Point` in turn has a `pos` property, which contains the coordinate values of the `Point`. These coordinates specify the position of the `Point` in the CRS, which is denoted by the `srsName` attribute. This optional attribute is covered in detail in Chapter 15. GML provides many geometry elements and types, which are discussed in Chapter 12.

The following example shows an encoding of a remote-property value:

```
<app:Bridge gml:id="B1">
    <app:span>400</app:span>
    <app:height>50</app:height>
    <app:mobility>DrawBridge</app:mobility>
    <gml:centerOf>
        <gml:Point gml:id="P1" srsName="..">
            <gml:pos>100.1 200.3</gml:pos>
        </gml:Point>
    </gml:centerOf>
</app:Bridge>
<app:Tower gml:id="T1">
    <app:height>90</app:height>
    <gml:centerOf xlink:href="#P1"/>
</app:Tower>
```

The `Tower` feature references a point on the `Bridge` feature by attaching an `xlink:href` attribute to the `centerOf` property. The `xlink:href` attribute, from the XLink schema, contains the remote-property value, P1, which is the unique identifier for the `Point` geometry that is the value for the `centerOf` property of the `Bridge` feature. In this example, both the `Bridge` and `Tower` features are located in the same instance document.

In GML, remote-property values can reference data instances that are located within the same instance document, as shown in the above example, or in another document that is stored in the same machine or somewhere else on the Internet. In the above example, the '#' in front of P1 indicates that the reference points to an instance that is located within a document. Because a file name is not included with the '#', this indicates that the reference is within the same document. This syntax is from XPointer, where it is called a bare name XPointer. For more information about XPointer, please see http://www.w3.org/XML/Linking. Although it is similar in appearance to an HTML reference, note that `xlink:href` references are not HTML hyperlinks.

The following example shows a remote-property value that references an instance located in another document in the same path:

```
<app:Tower gml:id="T1">
    <app:height>90</app:height>
    <gml:centerOf
        ⌐xlink:href="TransportationBridges.xml#P1"/>
</app:Tower>
```

In this example, the `xlink:href` attribute points to a P1 instance that is located in another document called `TransportationBridges.xml`. The following example shows a remote-property value that references an instance in a document that is located elsewhere on the Internet:

```
<app:Tower gml:id="T1">
    <app:height>90</app:height>
    <gml:centerOf
```

```
    ⌐xlink:href="http://www.ukusa.org/
    ⌐TransportationBridges.xml#P1"/>
</app:Tower>
```

Note that GML allows xlink attributes to appear only on GML properties. A property with an xlink:href is considered a remote property, regardless of where the xlink:href points; for example, it could be within the same file. The xlink:href on a property is interpreted as pointing to the value of the property, and the value is interpreted as though it were inserted in-line.

9.2.2   Parents and children

Note that features in GML cannot be direct children of one another, because property elements are the direct children of features. For example, the following encoding is invalid in GML, because both House and Kitchen are GML features.

```
<app:House gml:id="..">
    <app:Kitchen gml:id="..">...</app:Kitchen>
</app:House>
```

The following example shows how to properly encode Room features as the values of properties of the House feature:

```
<app:House gml:id="..">
   <app:room>
       <app:Kitchen gml:id="..">...</app:Kitchen>
   </app:room>
   <app:room>
       <app:Den gml:id="..">...</app:Den>
   </app:room>
</app:House>
```

In this valid GML example, the room property represents the target role of the Kitchen and Den with respect to the House.

9.2.3   Role of attributes in GML

In GML instances, properties of features and other objects are expressed by elements. For example, the following is not valid GML:

```
<app:Bridge gml:id=".." span="400" height="40"/>
```

Although it is not necessarily invalid to use the attributes as shown above, GML processing software will not interpret these attributes as properties of the GML feature. The correct way to express the above in GML is

```
<app:Bridge gml:id="..">
   <app:span>400</app:span>
   <app:height>40</app:height>
   ...
</app:Bridge>
```

Within GML only a few attributes are typically used, such as gml:id, srsName and xlink:href. All of these attributes are used to reference remote data. Note that attributes have no meaning in GML application schemas.

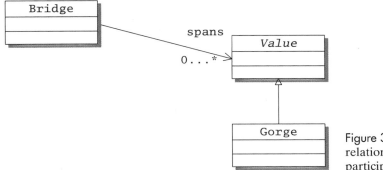

Figure 3 A simple relationship with two participants.

9.3 How are feature relationships expressed?

Since the value of a GML property can be another object, it is possible to interpret a property that connects two objects as representing a relationship between the objects. Features can be objects that relate to other objects. For example, a Bridge spans a Gorge. The spans property provides the relationship between the Bridge and Gorge features. This is a simple relationship that involves two participants.

In simple relationships, GML can express the roles of either participant in the relationship. For example, you can also say River is spannedBy Bridge. GML also supports complex relationships between multiple participants. GML 3.0 also includes a topology model that provides a formal mathematical way of expressing certain kinds of spatial relationships. This topology model is covered in Chapter 13.

Simple relationships are expressed with GML properties where the property name designates the role of the target participant with respect to the source in the relationship. Consider the 'Bridge spans Gorge' example, as shown in Figure 3. The spans property is used to indicate the relationship between the Bridge and the Gorge, where Bridge is the source and Gorge is the target.

The following example shows how to encode this simple relationship in GML. Note that a Gorge instance is the value of the spans property and that the value is encapsulated by the property. In other words, it is an in-line property.

```
<app:Bridge gml:id="B1">
   <app:span>400</app:span>
   <app:height>50</app:height>
   <app:mobility>DrawBridge</app:mobility>
   <gml:centerOf>
      <gml:Point gml:id="P1" srsName="..">
         <gml:coordinates>...</gml:coordinates>
      </gml:Point>
   </gml:centerOf>
   <app:spans>
      <app:Gorge gml:id="G1">
         <app:width>250</app:width>
      </app:Gorge>
   </app:spans>
</app:Bridge>
```

```
<app:Tower gml:id="T1">
   <app:height>90</app:height>
   <gml:centerOf xlink:href="#P1"/>
</app:Tower>
```

The name of the property expresses the relationship name or target role name in the relationship. This is typical of most properties that define relationships in GML. Note also that the `spans` property has the `app` namespace prefix, that is, it is not defined in the core GML 3 schemas, but in an application schema. Remote properties can also be used to encode relationships between features, as shown in the following example:

```
<app:Bridge gml:id="B1">
   <app:span>400</app:span>
   <app:height>50</app:height>
   <app:mobility>DrawBridge</app:mobility>
   <gml:centerOf>
      <gml:Point gml:id="P1" srsName="..">
         <gml:pos>...</gml:pos>
      </gml:Point>
   </gml:centerOf>
   <app:spans xlink:href="#G1"/>
</app:Bridge>
<app:Tower gml:id="T1">
   <app:height>90</app:height>
   <gml:centerOf xlink:href="#P1"/>
</app:Tower>
<app:Gorge gml:id="G1">
   <app:width>250</app:width>
   <app:spannedBy xlink:href="#B1"/>
</app:Gorge>
```

Note that there is a separate `Gorge` instance that is independent of the `Bridge` instance. While the `Bridge` instance has a feature relationship where the `Bridge` `spans` the `Gorge`, the `Gorge` is `spannedBy` the `Bridge` in the `Gorge` instance. The `Gorge` is the source participant, and the `Bridge` is the target. Note that in GML, such relationships can be partial; that is, it is not necessary to provide both roles in the relationship.

9.4 Features and feature collections

When geospatial data is encoded in GML, a data instance usually contains, a collection of objects, all of which can have different kinds of properties. In GML, feature collections are the most commonly used object collections. While feature collections contain a number of GML features, they are also GML features themselves. Consider the following example of a feature collection called `State` that contains two `County` features, both of which are values of the `featureMember` property:

```
<app:State gml:id="G1">
   <gml:description>Counties of the state of California
      ↳</gml:description>
```

```
<gml:name>State of California</gml:name>
<gml:boundedBy> .. </gml:boundedBy>
<gml:featureMember>
   <app:County gml:id="c1">
      <app:birthRate>...</app:birthRate>
      <app:population>...</app:population>
   </app:County>
</gml:featureMember>
<gml:featureMember>
   <app:County gml:id="c2">
      <app:birthRate>...</app:birthRate>
      <app:population>...</app:population>
   </app:County>
</gml:featureMember>
...
</app:State>
```

A feature collection can have properties that apply to the feature collection as a whole, that is the feature collection behaves as a feature. This is consistent with the definitions of the *OpenGIS® Abstract Specification* (http://www.opengis.org/docs/ 99-105r2.pdf). Note that `featureMember` is a property of the feature collection and it expresses the membership relationship between the enclosed or referenced feature and the feature collection.

Just as features in GML cannot be direct children of one another, features cannot be direct children of feature collections. The `featureMember` property is used to describe the relationship between the `State` feature collection and the `County` features.

Note that in GML, properties usually contain only one value. GML however provides the `featureMembers` property, which is used to contain arrays of more than one value, as shown in the following example. The `featureMembers` property contains two `County` features, as opposed to the previous example, which had two separate `featureMember` properties, each containing one `County` feature.

```
<app:State gml:id="G1">
   <gml:description>Counties of the state of California
      ↵</gml:description>
   <gml:name>State of California</gml:name>
   <gml:boundedBy>...</gml:boundedBy>
   <gml:featureMembers>
      <app:County gml:id="c1">
         <app:birthRate>...</app:birthRate>
         <app:population>...</app:population>
      </app:County>
      <app:County gml:id="c2">
         <app:birthRate>...</app:birthRate>
         <app:population>...</app:population>
      </app:County>
   </gml:featureMembers>
   ...
</app:State>
```

In the above examples, all of the features in the `State` feature collection are in-line properties. The following example shows a feature collection with remote features:

```
<app:State gml:id="G1">
    <gml:description>Counties of the state of California
       ⌣</gml:description>
    <gml:name>State of California</gml:name>
    <gml:boundedBy>...</gml:boundedBy>
    <gml:featureMember xlink:href="#c1"/>
    <gml:featureMember xlink:href="#c2"/>
    <gml:featureMember xlink:href="#cn"/>
</app:State>
```

By using `xlink` attributes, a feature collection can mix both in-line and remote feature members. Note that a feature collection can also contain a number of feature collections, through the `featureMember` property.

Feature collections also have a mandatory `boundedBy` property that contains either an `Envelope` or `Null` object. Note that in GML 2, the `boundedBy` property contained a `Box` instead of an `Envelope`. `Box` is deprecated in GML 3.0. If none of the features in a feature collection have spatial properties, then the value of the feature collection's `boundedBy` property is `Null`. If the feature collection or any of its feature members have geometry properties, then the value of `boundedBy` should be an `Envelope`.

If a feature collection has temporal properties, the `boundedBy` property can contain an `EnvelopeWithTemporalPeriod` object instead of `Envelope`. This is covered in Chapter 14. Note that an individual feature can also have an optional `boundedBy` property. For more information about the `boundedBy` property, please read Section 9.5.3.

9.5   More about features

Specific features are not defined in the GML core schemas, but in GML application schemas, usually created by data modellers or database administrators. Feature types must derive ultimately from `AbstractFeatureType`. The `feature.xsd` schema provides components for defining the following kinds of elements in application schemas:

- Features
- Feature collections
- Feature members
- Feature properties.

Chapter 11 covers the rules for creating features, collections and properties in GML application schemas.

9.5.1   How do I describe the geometric characteristics of a feature?

A feature can have multiple geometric descriptions that can be of varying dimensions. For example, a feature can have linear, point and area descriptions at the same time. A feature's geometry description is determined by its geometry-valued properties, that is, properties having a geometry value. The name of the geometry-valued property describes the role of the geometry in relation to the feature. Note that it is not accurate to say that a feature has geometry; for example, there are no point features, linear features or area features. Instead, features have geometry-valued properties that express the geometric characteristics of a feature. GML provides the following built-in convenience geometry-valued properties, which are defined in `feature.xsd` and `geometryAggregates.xsd`:

- `centerOf`
- `position`
- `extentOf`
- `edgeOf`
- `centerLineOf`
- `multiCenterOf`
- `multiPosition`
- `multiCenterLineOf`
- `multiEdgeOf`
- `multiExtentOf`
- `multiCoverage`
- `location`.

Note that `location` is a special case of a geometry-valued property, because its value can also be an address or a narrative description of location. These geometry-valued properties – which are also discussed in the *GML Version 3.00 OpenGIS® Implementation Specification* (http://www.opengis.org/docs/02-023r4.pdf) – are used to relate a feature to a geometry object, as illustrated in the following example:

```
<app:Bridge gml:id="..">
...
    <gml:centerOf>
       <gml:Point gml:id="P1" srsName="..">
          <gml:pos>...</gml:pos>
       </gml:Point>
    </gml:centerOf>
    <gml:centerLineOf>
       <gml:LineString gml:id="L1" srsName="..">
          <gml:coordinates>...</gml:coordinates>
       </gml:LineString>
    </gml:centerLineOf>
</app:Bridge>
```

The `Bridge` feature has `centerOf` and `centerLineOf` properties, whose values are geometry objects from the GML geometry schemas. These geometry objects, `Point` and `LineString`, have `pos` and `coordinates` properties that contain the coordinate values of the `Point` and `LineString`. In GML 3, many new geometry objects have been added to the geometry schemas. Chapter 12 discusses these objects in more detail.

9.5.2   What about digital elevation models (DEMs) and other surfaces?

GML provides a special subtype of feature called a coverage, which is a function that describes the distribution of some property over a geographic or temporal region. An example of a coverage is the distribution of the surface elevation in a Digital Elevation Model (DEM). Note that GML coverages can include both text and binary value models. The interplay between features and coverages is both subtle and complex. Coverages, and their relation to features, are covered in detail in Chapter 10.

9.5.3   What is the `boundedBy` property of a feature or feature collection?

The `boundedBy` property is intended to provide information about the approximate extent of a feature. The value of the `boundedBy` property can be `Null` or an `Envelope`. The `Envelope` can contain `coordinate` or `pos` properties with coordinate values that indicate the spatial extent of the envelope of the feature or feature collection. Note that in GML 3.1, the `Envelope` can also contain a `posList` property. In GML 2, the value of the `boundedBy` property was a `Box` object, which is deprecated in GML 3. The `boundedBy` property can be encoded in GML 3 as follows:

```
<gml:boundedBy>
    <gml:Envelope srsName="..">
        <gml:pos>0 0</gml:pos>
        <gml:pos>200 200</gml:pos>
    </gml:Envelope>
</gml:boundedBy>
```

A `Null` object is normally used if the feature or feature collection has no geometric properties. Note that `Null` provides a concrete set of possible explanations as to why the value is not present, including `inapplicable`, `missing`, `template` `unknown` and `withheld`. If a feature has geometric properties and a `boundedBy` property is supplied, its value should be an `Envelope` object, however, the data creator can still decide to provide a `Null` value with one of the above-mentioned explanation values (for example, `inapplicable`).

There are two kinds of `Envelope`: a spatial envelope, `Envelope`, and a spatiotemporal envelope, `EnvelopeWithTimePeriod`. A spatial `Envelope` is a geometry object that is defined implicitly by two of its opposite corner points whose coordinates are specified relative to the CRS that is referenced in the `srsName` attribute attached to the `Envelope`. The two corner points must lie in the same CRS. Note that the `Envelope` can have any dimension. `EnvelopeWithTimePeriod` is typically used to bound a time-dependent feature – for example, a feature that derives from `DynamicFeatureType`. Temporal envelopes and dynamic features are covered in detail in Chapter 14.

Note that, unlike feature collections, the `boundedBy` property is optional for features. There are cases, such as with a Web Feature Service (WFS), where it is a best practice to provide the property for all features and use `Null` if no `Envelope` can be supplied.

Note further that a `boundedBy` property value is intended to be a minimum bounding *n*-dimensional rectangle for the feature in question. If the feature has only one geometry-valued property whose value is a `Point`, the value of the `boundedBy` should be an `Envelope` with two equal points.

9.5.4   Does every feature have a location?

`AbstractFeatureType` has an optional `location` property, and the value of this property is one of the following:

- Any geometry (for example, `Point`, `LineString` and `Polygon`)
- A location keyword (for example, 'Holland')
- A location string (for example, '`corner of 5`[th] `Avenue and Main Street`').

Note that although the `location` property is optional, all user-defined features inherit the property unless it is removed by restriction. This means that all GML features potentially have a `location`. Database applications that are GML aware – such as a WFS – can be expected to accept an update, create or insert operation for this optional `location`, even if it is not declared in the application schema. Specific restrictions on this requirement may be put in place by a WFS and other data-access services. WFSs and GML are discussed in more detail in Chapter 19. Note that it is likely that the `location` property will be removed from GML features in GML 3.1.

9.5.5   How do I specify units of measure for feature properties?

Often GML properties do not specify the units of measure, as in the following `Bridge` example that was discussed earlier in the chapter:

```
<app:Bridge gml:id="B1">
  <app:span>400</app:span>
  <app:height>50</app:height>
  <app:mobility>DrawBridge</app:mobility>
  ...
</app:Bridge>
```

The `span` and `height` properties provide values, but it is not clear what the units of measure are. To specify properties with units, you need to use the uom attribute, which references a unit of measure dictionary definition, which can be defined with elements from `valueObjects.xsd` and its associated schemas. Note that units of measure were not supported in GML 2. The following example shows two properties that include uom attributes:

```
<app:Bridge gml:id="B1">
  <app:span
    ⌣uom="http://www.ukusa.org/measures.xml#metre">400
    ⌣</app:span>
  <app:height uom="http://www.ukusa.org/measures.xml#ft">
    ⌣50</app:height">
  <app:mobility>DrawBridge</app:mobility>
  ...
</app:Bridge>
```

In this toy example, both the `span` and `height` properties point to a units of measure dictionary that provides specific measures for the values. This Uniform Resource Indicator (URI) is provided to illustrate the reference, and should not be used in real world instances. When you use the `uom` attribute, you typically need to reference existing units of measure dictionaries or ones that are created by data collectors in your domain of interest. Chapter 16 covers units of measure dictionaries and the associated schemas. Note that the dictionary definitions can be given in-line in the GML data instance or via a web service (for example, the OGC Catalog Service) or simply as an external dictionary file.

9.6 Chapter summary

GML is based on the object-property model, in which objects are described by properties and property values. In GML 2, this model was called the feature-property model. In GML 3, there are many different kinds of objects, including geometries, coverages, time objects, observations, default styles, coordinate reference systems and topologies.

Geographic features are still at the core of GML 3, and to understand the role of properties in the object-property model, it is useful to look at examples of simple features with properties. A `Bridge` feature, for example, can have simple `span`, `height` and `mobility` properties that all contain values. Property values can be contained inside of property tags or they can be referenced remotely with the `xlink:href` attribute.

The rules for using properties in GML instances are as follows:

1. The property elements are *always* children of the object element. An object cannot have another object as a child element, but can have a property whose value is another object.

2. The first letter of a property name is usually lower case (for example, `centerOf`), while the first letter of an object name is usually upper case (for example, `Bridge`). This is simply a naming convention and should not be relied upon by GML software.

3. The values of the object's properties must be one of the following:
   - Simple data types, such as integers or strings
   - Another GML object, such as a feature or a geometry
   - An object that is located elsewhere, in the same document or in another document, and pointed to by the `xlink:href` attribute on the property.

4. The value of the `xlink:href` attribute points to the value of the property. A property with an `xlink:href` attribute is the equivalent of a UML association. Note that `xlink:href` attributes only have meaning in GML when attached to GML properties. In XLink terms, a GML object cannot be a link element.

Properties can also be interpreted as representing relationships between different objects, particularly with features. For example, a `Bridge` feature `spans` a `Gorge`, and the `Gorge` can be `spannedBy` the `Bridge` feature. These simple relationships can be expressed as in-line or remote properties. Although

GML does allow for the expression of more complex relationships between multiple features, this is not covered in this book.

A feature collection is essentially a collection, or group, of GML features that is itself a feature. For example, a `City` feature collection can contain multiple features instances, such as `Road`, `Bridge` and `Building`, or a `State` feature collection can contain multiple `County` feature instances. The `featureMember` property is used to denote features within a feature collection. Each `feature-Member` property can only contain one feature. The `featureMembers` property can be used to contain an array of features.

Note that feature collections can also include other feature collections. When feature collections are defined in GML application schemas, their content models need to derive from the abstract feature collection type from the GML 3 feature schema. These abstract types are covered in Chapter 4, along with other rules for creating GML application schemas.

The GML 3.0 core schemas do not define specific features, such as `Bridge` and `County`. These features are defined in application-domain-specific schemas. The `feature.xsd` provides a number of convenience geometry-valued properties that express the relationship between a feature and a geometry object. For example, the `centerOf` property provides the relationship between a `Bridge` feature and a `Point` geometry.

Features can have an optional `boundedBy` property that provides information about the approximate spatial extent of a feature. This property – which is mandatory for feature collections – can have an `Envelope` or `Null` object. `Envelope` can have two `pos` properties or a `coordinates` property. Note that in GML 3.1, `Envelope` will also have a `posList` property. If the feature or feature collection has temporal properties, it is possible for the `boundedBy` property to contain an `EnvelopeWithTimePeriod` instead of an `Envelope`.

Features and feature collections also have an optional `location` property whose value can be any geometry object, a location keyword or a location string. Note that all user-defined features can have a `location` property, unless the property is restricted in the new feature's content model. The `location` property is also covered in Chapter 12.

GML 3 supports coverages, which are a subtype of feature for describing the distribution of properties over a geographic region. It is also possible to reference unit of measure dictionary entries from feature properties. The `uom` attribute is used to reference these dictionary entries. Coverages are discussed in Chapter 17, and unit of measure dictionaries and the `uom` attribute are covered in Chapter 16.

References

http://www.w3.org/TR/REC-rdf-syntax/ (September 25, 2003).

http://www.opengis.org/docs/01-029.pdf (February 20, 2001).

http://www.opengis.org/docs/99-105r2.pdf (October 17, 2003).

http://www.w3.org/XML/Linking (September 20, 2003).

http://www.opengis.org/docs/02-023r4.pdf (October 15, 2003).

Additional reference

http://www.w3.org/TR/rdf-schema/ (September 25, 2003).

Chapter 10

GML core schemas overview

In GML 3.0, there are 28 core schemas that can be used to create GML application schemas for various domains of interest. It is helpful to group these GML core schemas into different object classes, some of which comprise multiple schemas. Table 1 lists the different object classes in GML and their associated schemas.

The GML core schemas provide the framework for creating additional GML application schemas for different domains that make use of geographic information. Most data modellers or database administrators do not need to be familiar with all of the above-listed schemas to create GML application schemas for their specific domain of interest. If you need to import all of the schemas, you can import `gml.xsd`, which is a 'wrapper' schema that includes all of the GML core schemas. Chapter 11 covers the different issues about importing GML core schemas into application schemas. Note that all of the GML schemas are available at http://schemas.opengis.net/gml and that the examples in this book are based on the core schemas at http://schemas.opengis.net/gml/3.0.1/.

10.1 About the GML base schema

The `gmlBase.xsd` schema imports `xlinks.xsd` (from the W3C namespace) and includes `basicTypes.xsd`. The `xlinks.xsd` schema provides all of the elements, types and attributes that are required for referencing remote properties. Various `simpleContent` types – including `booleanOrNull`, `stringList`, `MeasureType` and `CoordinatesType` – are provided by `basicTypes.xsd`. These types are used throughout the different GML core schemas. These types can also be used in the creation of GML application schemas.

With the exception of `basicTypes.xsd`, all of the GML core schemas depend on `gmlBase.xsd`, which provides the _GML element and the `AbstractGMLType`. All GML objects derive ultimately from `AbstractGMLType`. The `gmlBase.xsd` schema defines components that establish the GML model and syntax. In particular, the schema provides the following:

- The root type from which GML objects are derived (`AbstractGMLType`)
- Pattern and base components for GML properties

Geography Mark-up Language (GML). R. Lake, D. S. Burggraf, M. Trninić, L. Rae © 2004 Galdos Systems Inc.
Published by John Wiley & Sons, Ltd ISBNs: 0-470-87153-9 (HB); 0-470-87154-7 (PB)

Table 1 GML 3.0 object classes

Object Classes	Purpose	Associated Schemas

GML base	Provides the root of the GML class hierarchy and other required constructs for creating GML objects.	`gmlBase.xsd`
		`xlinks.xsd`
		`basicTypes.xsd`

Feature	Provides constructs for defining features and feature properties in GML.	`feature.xsd`

Geometry	Consists of five separate schemas that define the geometry objects for GML. It is based on ISO/TC 211 19107.	`geometryBasic0d1d.xsd`
		`geometryBasic2d.xsd`
		`geometryPrimitives.xsd`
		`geometryAggregates.xsd`
		`geometryComplexes.xsd`

Topology	Provides constructs for defining topology objects. It is based on ISO/TC 211 19107.	`topology.xsd`

Reference systems	Consists of a set of schemas that provide elements and types for defining spatial reference systems.	`referenceSystems.xsd`
		`coordinateReferenceSystem.xsd`
		`coordinateSystems.xsd`
		`datums.xsd`
		`operations.xsd`
		`dataQuality.xsd`

Temporal	Provides constructs for defining objects that deal with time and for modelling features that change over time. Also provides elements for encoding temporal reference systems.	`temporal.xsd` `dynamicFeature.xsd`
Coverages	Defines elements and types for supporting coverages. It is based on ISO/TC 211 19123.	`coverage.xsd` `grids.xsd`
Default style	Provides the `defaultStyle` property that can be attached to a feature collection or a feature and provides constructs for modelling style descriptors that can be applied to a GML data set via the `defaultStyle` property.	`defaultStyle.xsd`
Direction	Provides types that can be used to express direction in various ways.	`direction.xsd`

Units, measures, values and observations	Provide the foundation for encoding unit of measure dictionaries, for encoding measured quantities, for creating user-defined quantities and values, and for defining observations.	`units.xsd`
		`measures.xsd`
		`value.xsd`
		`observation.xsd`

- Pattern and base components for object collections and arrays
- Components for associating metadata with GML objects
- Components for supporting general definitions and dictionary entries.

These components are all discussed in detail in Section 7.2 of the *GML Version 3.00 OpenGIS® Implementation Specification* (http://www.opengis.org/docs/02-023r4.pdf). This book only provides an overview of some of these components.

10.1.1   Components for defining objects in GML

The gmlBase.xsd defines the abstract type, AbstractGMLType, from which all GML objects derive – for example, all geographic features, coverages, observations, topologies, geometries and Coordinate Reference Systems (CRS). The _Object element is the root XML object, which can be considered an object of 'AnyType'. This element is used to create the root abstract type, Abstract-GMLType, from which new concrete GML object types are derived, indirectly and directly. The following schema fragment shows the content model for Abstract-GMLType and the element declaration for the _GML abstract element, which represents any GML object that has identity.

```
<element name="_GML" type="gml:AbstractGMLType"
   ⌐abstract="true" substitutionGroup="gml:_Object"/>

<complexType name="AbstractGMLType" abstract="true">
   <sequence>
      <element ref="gml:metaDataProperty" minOccurs="0"
         ⌐maxOccurs="unbounded"/>
      <element ref="gml:description" minOccurs="0"/>
      <element ref="gml:name" minOccurs="0"
         ⌐maxOccurs="unbounded"/>
   </sequence>
   <attribute ref="gml:id" use="optional"/>
</complexType>
```

The above fragment shows the pairing of the _GML element and Abstract-GMLType. This pairing illustrates a basic pattern used in GML schemas. Note that this is only one of the patterns that can be used, and that other patterns are often recommended for creating GML application schemas. These patterns are discussed in *Developing and Managing GML Application Schemas: Best Practices* (http://www.geoconnections.org/developersCorner/devCorner_devNetwork/components/GML_bpv1.3_E.pdf).

In the above pattern, each GML object type is represented by a global element declaration, which contains an element name and type. The type is defined in a separate content model. The name of the element is a semantic name and the substitution group is an abstract element, such as _Object.

Note that AbstractGMLType has the following three optional properties: metaDataProperty, description and name. All of these properties have a minimum occurrence of '0', which means that each property is optional. The metaDataProperty and name properties have unbounded as the value of the maxOccurs attribute. This means that the property can occur an unlimited number of times. The `description` property, on the other hand, has a maximum occurrence of '1', and therefore this property can occur only once within its parent element.

AbstractGMLType has an optional `gml:id` attribute that allows objects to have unique identifiers. Given that all GML objects derive from Abstract-GMLType, the above schema fragment implies that in general GML objects have these three optional properties. Although this is usually the case, it depends on how different objects – including user-defined and GML core schema objects – are derived from AbstractGMLType.

Note: Database and web service application implementers should note that, unless they specify otherwise, GML data consumers can expect that any optional property can be given a value, which means that the values of these optional properties can be updated or modified.

10.1.2   Components for defining GML properties

In GML, properties are instantiated as child elements of objects. As discussed in Chapter 9, property values can be expressed in two ways, as in-line values and by reference with an `xlink:href` attribute. The gmlBase.xsd schema provides the AssociationAttributeGroup element, which allows for remote references based on the xlinks.xsd schema.

The AssociationType is a type that illustrates a basic pattern for defining property elements, as shown in the following schema fragment:

```
<element name="_association" type="gml:AssociationType"
   ⌐abstract="true" substitutionGroup="gml:_property"/>

<complexType name="AssociationType">
   <sequence>
      <element ref="gml:_Object" minOccurs="0"/>
   </sequence>
   <attributeGroup ref="gml:AssociationAttributeGroup"/>
</complexType>
```

The abstract element name is _association with a type of Asso-ciationType whose content model includes both the _Object and the AssociationAttributeGroup, which means that an element of this type may have a content element or `xlink` attributes, or both. Note that the above example simply provides a convenient pattern for constructing property elements. Unlike GML objects, properties are not required to derive from an abstract type. Nor do they have to be substitutable for _property. On the contrary, substitution for _property should be avoided. Note that _property is likely to be removed from GML 3.1. Chapter 11 includes guidelines for creating user-defined GML properties.

> *Note:* It is not recommended that GML schema parsers or other type-detection software rely on type inheritance or substitution groups for property detection. Properties are always children of GML objects and should be detected as such. GML properties are not typed any more than associations in UML are typed. GML objects, on the other hand, are detected and can be classified by their position in the GML type (or class) hierarchy.

10.1.3   Components for defining object collections

The concrete `BagType` and `ArrayType` are provided for creating generic object collections and arrays. The `gmlBase.xsd` schema also provides the `member` element, which is used as an object collection member property in GML instances. The `member` element is of `AssociationType` type, that is, `Association-Type` is the name of the content model of the `member` element. Note that the content model for the `featureMember` property discussed in Chapter 9 follows the `AssociationType` pattern, except that it can only contain elements from the `_Feature` substitution group, instead of `_Object`. Feature collections and the `featureMember` property are further discussed in Section 10.2.

10.1.4   Components for associating metadata with GML objects

As discussed above, the base GML object (`_GML`) has the built-in metadata properties, such as `name` and `description`, that are automatically inherited by every GML object. If you want to include additional metadata in your application schemas, you need to use the abstract `_MetaData` element and `metaDataProperty`.

In GML instances, different GML objects – such as features, geometries or topologies – can have `metaDataProperty` elements that contain or refer to metadata. As defined in the `gmlBase.xsd` schema, this property has an optional `about` attribute, which carries a URI that points to the content to which the metadata is applied. The property's value can be a bare name XPointer that refers to a single feature identified by the `gml:id` attribute, or it can be an XPointer expression that refers to multiple features. A typical use of the `about` attribute, for example, is within a feature collection, where it is used to assign different metadata property values to particular members of the collection.

To define metadata properties and objects, you need to create or use an existing metadata schema. The rules for doing this are covered in Chapter 11. The `gmlBase.xsd` also provides a concrete metadata element, `GenericMeta-Data`, which acts as a convenience container for metadata. This generic element is used in cases in which it is not possible to substitute a metadata object for `_MetaData`. Note that this element is likely to be deprecated in GML 3.1. Note further that ISO is developing an XML Schema (ISO TC/211 – 19139) for metadata that will be compliant with GML types.

10.2   Substitution and derivation by restriction and extension

The core GML schemas provide a number of patterns in which substitution head elements are used. In most cases, the pattern has the following form:

```
<SomeObject>
  <someProperty>
    <SomeOtherObject>...</SomeOtherObject>
  </someProperty>
</SomeObject>
```

The following schema fragment shows the element declarations and type definitions for *someProperty* and *SomeOtherObject*:

```
<element name="someProperty" type="xxx:SomePropertyType">

<complexType name="SomePropertyType">
  <complexContent>
    <sequence>
      <element ref="xxx:_SomeAbstractObject">
    </sequence>
  </complexContent>
</complexType>

<element name="SomeOtherObject"
  ⌐type="xxx:SomeOtherObjectType"
  ⌐substitutionGroup="xxx:_SomeAbstractObject">

<complexType name="SomeOtherObjectType">
  ...
</complexType>
```

In the above example, *_SomeAbstractObject* is the head of a substitution group of elements. Only objects, such as *SomeOtherObject*, that are made substitutable for *_SomeAbstractObject* can appear as values of the property *someProperty*.

Consider the following example of a State feature collection with a state-Member property whose value is a County:

```
<app:State>
  <app:stateMember>
    <app:County>...</app:County>
  </app:stateMember>
</app:State>
```

The substitution mechanism can be used in a number of different ways in the GML application schema definitions for State, stateMember and County. Two possible variants of the substitution mechanism are discussed below.

10.2.1   Variant I

The following schema fragment shows the simplest and most permissive pattern for defining State, stateMember and County:

```
<element name="State" type="gml:FeatureCollectionType"
  ⌐substitutionGroup="gml:_FeatureCollection"/>

<element name="stateMember" type="gml:FeaturePropertyType"
  ⌐substitutionGroup="gml:featureMember"/>

<element name="County" substitutionGroup="gml:_Feature">
  <complexType>
```

```
          <complexContent>
              <extension base="gml:AbstractFeatureType"/>
          </complexContent>
      </complexType>
  </element>
```

With this fragment, a new content model is not required for the `State` feature collection, because the substitution of `stateMember` for `feature-Member` allows the `State` feature collection to have a `stateMember` property. The `featureMember` property is one of the elements in the feature collection content model (`AbstractFeatureCollectionType`, from which the concrete `FeatureCollectionType` derives). The substitution for `feature-Member` enables GML-aware software to detect that `stateMember` is a feature membership association, that is, it associates a feature collection with its feature members. Note also that the `stateMember` property can contain a `County` object because `County` is substitutable for `_Feature`, and `featureMember` can contain any object from the `_Feature` substitution group.

10.2.2   Variant II

There are circumstances in which Variant I is not restrictive enough, in that any feature in the `_Feature` substitution group can be enclosed within a `stateMember` property, and, consequently, this property has essentially the same content as a `featureMember` property. The following schema fragment shows a more restrictive version of the `stateMember` property:

```
<element name="State" type="gml:FeatureCollectionType"
  ⌐substitutionGroup="gml:_FeatureCollection"/>

<element name="stateMember" type="app:StatePropertyType"
  ⌐substitutionGroup="gml:featureMember"/>

<complexType name="StatePropertyType">
   <complexContent>
      <restriction base="gml:FeaturePropertyType">
         <sequence>
            <element ref="app:_StateFeature"/>
         </sequence>
      </restriction>
   </complexContent>
</complexType>

<element name="_StateFeature"
  ⌐type="app:AbstractStateFeatureType" abstract="true"
  ⌐substitutionGroup="gml:_Feature"/>

<complexType name="AbstractStateFeatureType">
   <complexContent>
      <extension base="gml:AbstractFeatureType"/>
   </complexContent>
</complexType>

<element name="County" type="app:CountyType"
  ⌐substitutionGroup="app:_StateFeature"/>
```

```
<complexType name="CountyType">
   <complexContent>
      <extension base="app:AbstractStateFeatureType"/>
   </complexContent>
</complexType>
```

In the above fragment, the content model of the `StatePropertyType` derives by restriction from `FeaturePropertyType` and replaces the abstract `_Feature` object with an abstract `_StateFeature` object. Note that if derivation by extension is used instead of restriction, the `StatePropertyType` contains both the `_StateFeature` and `_Feature` objects.

In GML instances that are based on the above schema, the `stateMember` property can only enclose objects from the `_StateFeature` substitution group, such as `County`. Note that the `CountyType` content model still derives (indirectly) from `AbstractFeatureType`, but now `County` is in the substitution group headed by `_StateFeature`.

In the above schema fragment (as well as in Variant I), the `State` feature collection can still have a `featureMember` property. The following schema fragment shows a `State` feature collection with a new content model that overrides the `featureMember` property, replacing it with the `stateMember` property.

```
<element name="State" type="app:StateType"
   ⌐substitutionGroup="gml:_FeatureCollection"/>

<complexType name="StateType">
   <complexContent>
      <restriction base="gml:FeatureCollectionType">
         <sequence>
            <element ref="gml:metaDataProperty"
               ⌐minOccurs="0" maxOccurs="unbounded"/>
            <element ref="gml:description"
               ⌐minOccurs="0"/>
            <element ref="gml:name" minOccurs="0"
               ⌐maxOccurs="unbounded"/>
            <element ref="gml:boundedBy"/>
            <element ref="gml:location" minOccurs="0"/>
            <element ref="app:stateMember"/>
         </sequence>
      </restriction>
   </complexContent>
</complexType>
```

Note that the `StateType` content model derives by restriction from `FeatureCollectionType`, replacing the `featureMember` property with `stateMember`. Because the content model derives by restriction, it is necessary to restate any properties that you wish to include from the original content model, for example, `metaDataProperty`, `boundedBy` and `location`. The overall result is that a `State` feature collection can now only contain `County` features as members.

10.3   Abstract elements and types in GML 3.0

In addition to the root abstract elements and types discussed above, GML also provides a number of abstract types and elements that are used to create different kinds of objects in GML. Abstract types are complex types whose `abstract` attribute is set to `true`. Typically, abstract type names all begin with the word 'Abstract', such as `AbstractFeatureType`, `AbstractGeometryType` and `AbstractTopologyType`. Abstract types are used in GML to denote kinds of GML object content models. Elements of abstract type cannot be instantiated. In GML schemas, concrete types derive from the content models of these abstract types. Elements of concrete type can be instantiated. Table 2 lists some of the abstract GML elements and their associated abstract types.

Most of these abstract elements are substitutable for the _GML element from `gmlBase.xsd`; that is, the substitution group in the element declaration is _GML. The element declaration also includes the associated abstract type. The following schema fragment shows the element declaration for _Feature in the `feature.xsd` schema.

```
<element name="_Feature" type="gml:AbstractFeatureType"
    ⌐abstract="true" substitutionGroup="gml:_GML"/>
```

Note that _GML is not the substitution group for all of the abstract elements listed in Table 2. For example, coverages are also substitutable for the _Feature element. Figure 1 shows the hierarchy of the GML object classes and the abstract elements for each class.

The observation class does not have an abstract element, but the elements declared in `observation.xsd` are substitutable for _Feature. As the figure shows, most of the object classes are substitutable for _GML, which is substitutable for _Object.

Note: In processing GML application schemas, GML-aware software determines the root type – as shown in Table 2 – by tracing the inheritance (XML Schema) hierarchy. This is referred to as type detection and is a basic function in being able to support GML data. A simple schema parser should be able to read a GML application schema and determine the kind of GML object for every object declared in the schema.

Table 2   Abstract elements and types

Abstract Element Name	Meaning	Associated Abstract Type
gml:_Feature	Any GML feature	gml:AbstractFeatureType
gml:_Geometry	Any GML geometry	gml:AbstractGeometryType
gml:_Value	Any GML value	gml:AbstractValueType
gml:_Topology	Any GML topology	gml:AbstractTopologyType
gml:_CRS	Any GML CRS	gml:AbstractCRSType
gml:_TimeObject	Any GML temporal object	gml:AbstractTimeType
gml:_Coverage	Any GML coverage	gml:AbstractCoverageType
gml:_Style	Any GML style	gml:AbstractStyleType

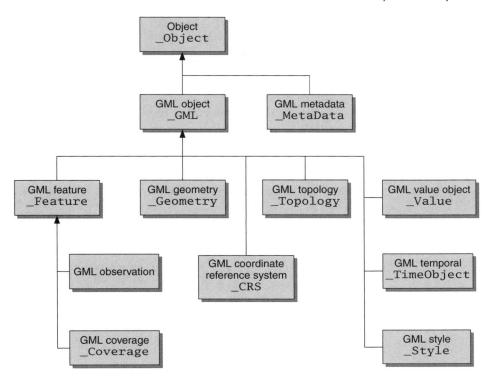

Figure 1 Hierarchy of GML object classes.

10.4 Chapter summary

GML 3.0 has 28 core schemas that provide the framework for creating GML application schemas. In this guide, these schemas are grouped into different object classes. The schemas that you use depend on your application domain. The `gml.xsd` schema is provided for application developers that want to import all of the GML 3.0 core schemas.

The base schemas – `gmlBase.xsd`, `basicTypes.xsd` and `xlinks.xsd` – provide the basic types and components for defining GML objects, properties, object collections, metadata and dictionary entries. All GML objects must derive, indirectly or directly, from `AbstractGMLType`. Properties can derive from the abstract `AssociationType`, but they are not restricted to the same pattern as GML objects. However, to define a property with a remote value, the property should either derive from `AssociationType` or contain the `AssociationAttributeGroup` attribute. Components are also provided for defining object collections and arrays.

In addition to the `name` and `description` properties, `AbstractGMLType` also provides a generic metadata property that can contain or reference metadata. Any application domain can create its own metadata, which is typically defined in a separate metadata application schema.

In GML, the substitution mechanism is often used to allow an element's content model to have various kinds of elements as content. For example, if

SomeObject has *someProperty*, and the content model of *someProperty* has a *_SomeAbstractObject* element, then, in a GML instance, *someProperty* can contain any object that is substitutable for *_SomeAbstractObject*. In other words, *_SomeAbstractObject* must be at the head of the substitution group for the object.

If a content model derives by extension from another content model, the new content model also inherits all of the elements in the original content model in addition to any new elements. This can be problematic if you want the new element to have a more specialized function. Derivation by restriction can be used to restrict out any elements from the original content model that are not needed for the new content model. If restriction is used, however, it is necessary to restate all of the elements that you still want the new content model to inherit from the original.

GML 3.0 also provides abstract elements and types for defining the different kinds of GML objects. For example, the content models for all GML features must derive from `AbstractFeatureType`, and `_Feature` is the global element at the head of substitution groups in feature element declarations. All of the GML abstract types derive, directly or indirectly, from `AbstractGMLType`. GML software determines the kinds of GML objects in a GML application schema by tracing the XML Schema inheritance hierarchy back to the base types in the GML core schemas, for example, `AbstractFeatureType`.

References

http://schemas.opengis.net/gml/3.0.1/ (October 15, 2003).

http://www.opengis.org/docs/02-023r4.pdf (October 15, 2003).

http://www.geoconnections.org/developersCorner/ devCorner_devNetwork/components/GML_bpv1.3_ E.pdf (October 22, 2003).

Additional references

VALENTINE, C., DYKES, L., and TITTEL, E. (2002) *XML Schemas*. Sybex Inc., Alameda, CA.

http://www.iso.org (October 15, 2003).

Chapter 11

Developing and managing GML application schemas

Chapter 9 introduced the GML object-property model and provided fragments of document instances to illustrate how to encode features and properties in GML. In GML 3.0, there are many different kinds of instance documents, including feature collections, dictionaries and topological complexes. All of these documents need to be supported by one or more corresponding application schemas. That is, the application schemas define the different elements that are instantiated in GML instance documents. For example, to represent a `Bridge` feature, you need to define a `Bridge` feature type in an application schema.

To create application schemas, schema authors need to observe the following rules:

- The application schema must declare a target namespace.
- The application schema must import the appropriate core schemas from GML 3.0.
- All GML objects must derive directly or indirectly from the corresponding abstract type, for example, `AbstractFeatureType`.
- All GML object collections must derive from the corresponding abstract collection type, for example, `AbstractFeatureCollectionType`.
- Properties can be declared as global elements or as local elements within an object's content model. The type of a property may derive from or follow the pattern of `AssociationType`.
- Objects defined in an application schema must conform to the rules respecting the base types from which these objects derive, as defined in the GML 3.0 specification (http://www.opengis.org/docs/02-023r4.pdf).

Note that, in addition to features, GML application schemas define many different kinds of GML objects, including geometries, topologies and Coordinate Reference Systems (CRSs). In this chapter, most of the examples pertain to features and feature collections. This chapter also covers the schema dependencies and additional rules about importing multiple GML core schemas. Metadata schemas and schema registries are also discussed.

Geography Mark-up Language (GML).   R. Lake, D. S. Burggraf, M. Trninić, L. Rae © 2004 Galdos Systems Inc.
Published by John Wiley & Sons, Ltd   ISBNs: 0-470-87153-9 (HB); 0-470-87154-7 (PB)

11.1   How do I create an application schema that defines features?

This is the most common task that must be performed by an organization that wishes to work with GML data. In the future, application schema development may simply involve picking schema components off the Internet. Until this option becomes available, however, you need to know how to construct GML application schemas yourself. The task is not a difficult one, as you will see shortly.

To illustrate the rules for creating GML application schemas, this section contains examples for creating feature types, feature collections and properties of features. For additional information about defining schema types for other GML object classes, refer to the appropriate chapter. For example, Chapter 12 covers the creation of geometry types, and Chapter 15 covers the creation of CRS types. The rules for creating application schemas are also covered in Clause 8 of the *GML Version 3.00 OpenGIS® Implementation Specification* (http://www.opengis.org/docs/02-023r4.pdf).

There are many different applications that you can use to create GML application schemas, including Altova's XMLSpy, which provides a graphical user interface for modelling XML Schemas. Appendix D contains a tutorial for using XMLSpy to create the transportation schema covered in this section.

11.1.1   Declaring a target namespace

Every application schema must declare a target namespace, which is the namespace in which the application schema types are located. A target namespace is declared in the application schema using the `targetNamespace` attribute of the `schema` element from XML Schema, as shown below:

```
<schema targetNamespace="http://www.ukusa.org/app"
    ⌐xmlns:app="http://www.ukusa.org/app"
    ⌐xmlns:xsd="http://www.w3.org/2001/XMLSchema"
    ⌐xmlns="http://www.w3.org/2001/XMLSchema"
    ⌐xmlns:gml="http://www.opengis.net/gml"
    ⌐elementFormDefault="qualified">
```

The above example contains a URI (Uniform Resource Identifier) for a generic organization. The target namespace identifier is a URI that is controlled by the application schema owner's organization. Note that the target namespace cannot be the GML namespace (http://www.opengis.net/gml). Namespace prefixes are also declared for the target namespace and for other namespaces that are referenced from the schema. For example, the namespace prefix for the above-mentioned target namespace is `app`.

> *Note:* A declared feature element in GML does not need to reside in the same namespace as the associated content model. By using different namespaces, you can re-use existing type definitions for different application schemas.

11.1.2   Importing the feature schema

For application schemas that only need to define feature (and geometry) types, you need to do one of the following: import `feature.xsd` directly, or import

one of the GML schema documents that includes `feature.xsd`, as discussed
in Section 11.3. The following schema fragment shows how to import the `fea-
ture.xsd` directly:

```
<import namespace="http://www.opengis.net/gml"
    ⌣schemaLocation="http://schemas.opengis.net/gml/3.0.1/
    ⌣base/feature.xsd"/>
```

In this example, the path to `feature.xsd` – as indicated by the `schemaLo-
cation` attribute – is a URI reference to a remote repository on the OGC web
site. The path can also refer to a local copy of the document. The schema location
might also be referenced using a Uniform Resource Name (URN) or Persistent
URL (PURL). URNs are discussed in Chapter 15. See http://www.faqs.org/rfcs/
rfc2141.html for more information on the use of URNs, and see http://www.purl.
org/for PURLs.

Note that the `<import>` element specifies that the components described
in `feature.xsd` are associated with the GML namespace, http://www.opengis.
net/gml. To ensure XML Schema validity, this namespace identifier must match
the target namespace specified in the schema that is being imported.

11.1.3   Defining features

To define a new feature in a GML application schema, you declare the feature as a
global element, and the content model of the global element must derive, directly
or indirectly, from `AbstractFeatureType`, as shown in the following:

```
<element name="Bridge" type="app:BridgeType"
    ⌣substitutionGroup="gml:_Feature"/>

<complexType name="BridgeType">
    <complexContent>
        <extension base="gml:AbstractFeatureType">
            <sequence>
                <element name=".." type=".."/>
                <element name=".." type=".."/>
            </sequence>
        </extension>
    </complexContent>
</complexType>
```

As shown in the above example, to create the `Bridge` feature, you can do
the following:

1. Create a complex type called `BridgeType` that derives from `Abstract-
   FeatureType`.
2. Create the `Bridge` element from the `BridgeType` content model.
3. Specify that the `Bridge` element is substitutable for `_Feature`. The
   substitution is not required in order to declare that `Bridge` is a feature,
   but it is required if you wish to include the `Bridge` in an element that
   has `_Feature` as a value. For example, the content model for the GML-
   defined `featureMember` property has the `_Feature` element as its only

value, and, therefore, it can only contain elements from the _Feature substitution group.

Note that by having the `BridgeType` derive from `AbstractFeature-Type`, this specifies that the `Bridge` is a feature, and, therefore, all GML-aware software can detect that it is a feature. As discussed in Chapter 10, there are two ways to derive types: by extension and by restriction. Note that, in most cases, derivation is by extension, and if you are new to GML, it is advised that you use derivation by extension.

As mentioned above, it is also possible for the feature to derive indirectly from the `AbstractFeatureType`. For example, you might define an `AbstractBridgeType` for bridges in your application schema and then create a `SuspensionBridge` feature that derives from the `AbstractBridgeType`. When the `AbstractBridgeType` is defined, it must derive from `Abstract-FeatureType` (or another abstract type that derives, directly or indirectly, from `AbstractFeatureType`).

Note: The term *type* can be confusing in GML, particularly with regard to GML application schemas. Consider the element declaration in the above `Bridge` example. The `name` attribute indicates the semantic GML type, `Bridge`, and the `type` attribute indicates the `Bridge` feature's content model, `BridgeType`. Note that throughout this book, the semantic GML type is often called an element. The term 'feature type' refers to the semantic type, not the structural type (that is, the type definition).

In GML instances, the name of a feature element, such as `Bridge`, is provided by the name in the element declaration. Note that GML 3.0 currently supports a limited number of attributes, including `gml:id`, `xlink:href`, `uom`, `srsName`, and `frame`. The children of a feature are always properties that describe the feature, and these properties can only be encoded as child elements. A feature cannot have other GML objects as immediate children. In other words, features must follow the object-property model discussed in Chapter 9.

Note further that the above schema fragment shows a `Bridge` with a named type (`BridgeType`). According to *Developing and Managing GML Application Schemas: Best Practices*, use named types only if you expect to derive from that type (http://www.geoconnections.org/developersCorner/devCorner_devNetwork/components/GML_bpv1.3_E.pdf). For example, if you want to use `BridgeType` as a base type for other bridge type definitions, then it should be a named type. Otherwise, anonymous types should be used to 'hide' the type definition, as shown in the following schema fragment:

```
<element name="Bridge" substitutionGroup="gml:_Feature">

    <complexType>
       <complexContent>
          <extension base="gml:AbstractFeatureType">
             <sequence>
                <element name=".." type=".."/>
```

```
                    <element name=".." type=".."/>
                </sequence>
            </extension>
        </complexContent>
    </complexType>
</element>
```

For illustrative purposes, most of the schema examples in this book have named types, but in many cases, this is not necessary.

11.1.4 Defining feature collections

Feature collections must be declared as global elements whose content models, derive directly or indirectly from AbstractFeatureCollectionType, as shown below:

```
<element name="City" type="app:CityType"
    ⌐substitutionGroup="gml:_FeatureCollection"/>

<complexType name="CityType">
    <complexContent>
        <extension base="gml:AbstractFeatureCollectionType">
            <sequence>
                <element name=".." type=".."/>
                <element name=".." type=".."/>
            </sequence>
        </extension>
    </complexContent>
</complexType>
```

AbstractFeatureCollectionType is an abstract type that provides a template for the feature collection consisting of name and description properties, an id attribute, a mandatory boundedBy property and an unbounded number of featureMember properties that contain or point to the members of the collection. Issues related to the restriction of featureMember are covered in Chapter 10.

If you want to create a new feature collection without user-added content, it is not necessary to derive from AbstractFeatureCollectionType. Instead, you can set the type attribute to FeatureCollectionType, as shown in the following element declaration for a State feature collection:

```
<name="State" type="FeatureCollectionType"
    ⌐substitutionGroup="gml:_FeatureCollection"/>
```

In this case, the State feature collection cannot have any user-added content and can only contain the properties specified in the FeatureCollectionType content model. If you want the feature collection to contain new content, then the content model should derive, by extension or restriction, from AbstractFeatureCollectionType.

Note that the feature collections created in this section are also features, because they derive indirectly from AbstractFeatureType. To create a feature 'collection' that is not itself a feature, you should just follow the ArrayAssociationType pattern.

11.1.5   Defining a feature's properties

To define a feature's properties, include the property elements inside of the feature's content model. For example, simple and complex property elements can be declared in the BridgeType content model, as shown below:

```
<element name="Bridge" type="app:BridgeType"
  ⌐substitutionGroup="gml:_Feature"/>

<complexType name="BridgeType">
   <complexContent>
      <extension base="gml:AbstractFeatureType">
         <sequence>
            <element name="span" type="integer"/>
            <element name="height" type="integer"/>
            <element ref="gml:centerLineOf"/>
            <element ref="app:mobility"/>
         </sequence>
      </extension>
   </complexContent>
<complexType>
```

The span, height and mobility properties are simple properties, while centerLineOf is complex. The ref attribute is used for global properties to indicate that the property's element declaration and type definition are located elsewhere. In the above schema fragment, the centerLineOf property is defined in feature.xsd and the mobility property is defined in an application schema that contains the BridgeType definition. The element name and type definition for the mobility property can appear as follows:

```
<element name="mobility" type="app:MobilityType"/>

<simpleType name="MobilityType">
   <restriction base="string">
      <enumeration value="Fixed"/>
      <enumeration value="DrawBridge"/>
      <enumeration value="HorizontalSwingBridge"/>
      <enumeration value="UniversalSwingBridge"/>
      <enumeration value="TelescopingBridge"/>
      <enumeration value="LiftBridge"/>
   </restriction>
</simpleType>
```

The value of this user-defined mobility property is an enumeration; that is, it is a restricted string whose value can be one of the items in an enumerated list. As shown in the above schema fragment, the enumerated values for the mobility property are Fixed, DrawBridge, HorizontalSwingBridge, UniversalSwingBridge, TelescopingBridge and LiftBridge. Simple type use is covered in detail in Section 2.3 of the *XML Schema Primer* at http://www.w3.org/TR/xmlschema-0/.

GML properties can be simple or complex. Simple properties can only have text content and cannot have attributes or enclose other elements. Features can have different kinds of complex properties whose values can be expressed remotely or in-line. If this discussion seems too complex at first reading, please refer to the XMLSpy tutorial in Appendix D for a more visual explanation of these concepts.

11.1.6 Remote and in-line properties

In a GML data instance, if you want to encode a remote property, the `xlink:href` attribute must be used, as shown in the following example of the `app:Bridge` feature:

```
<app:Bridge gml:id="B1">
   <gml:centerLineOf xlink:href="#L1"/>
</app:Bridge>
```

In GML, the `centerLineOf` property has the following element declaration and complex type definition:

```
<element name="centerLineOf" type="gml:CurvePropertyType">

<complexType name="CurvePropertyType">
   <sequence>
      <element ref="gml:_Curve" minOccurs="0"
         ⌐maxOccurs="1"/>
   </sequence>
   <attributeGroup ref="gml:AssociationAttributeGroup"/>
</complexType>
```

As shown in the above complex type definition for `CurvePropertyType`, the property can contain an element that is substitutable for `_Curve`, such as `LineString`. The `maxOccurs` attribute specifies that the property can contain one instance of that element, while `minOccurs="0"` specifies that it is possible for the property to not contain the element. Note that in GML, the default value of `maxOccurs` is 1, and, therefore, the `maxOccurs` attribute is typically included only if the value is not 1.

The `CurvePropertyType` complex type definition also includes the `AssociationAttributeGroup` attribute, which contains association attributes, such as `xlink:href`. This allows for the geometry value to be referenced remotely. The properties defined in GML follow the same pattern as the one shown for `CurvePropertyType`, in that they can contain an element that is another object, and they also have association attributes. By following this pattern, it is possible for properties to have either remote or in-line values.

In application schemas, it is possible to restrict a property to only have a remote value, as shown in the following definition for a user-defined property called `app:remoteCenterLineOf`:

```
<element name="remoteCenterLineOf"
   ⌐type="app:RemoteCurvePropertyType">

<complexType name="RemoteCurvePropertyType">
   <attributeGroup ref="gml:AssociationAttributeGroup"/>
</complexType>
```

You can also restrict a property to only have in-line content, as shown in the following example of a user-defined property called app:inlineCenterLineOf:

```
<element name="inlineCenterLineOf"
    ⌐type="app:InlineCurvePropertyType">

<complexType name="InlineCurvePropertyType">
    <sequence>
        <element ref="gml:_Curve" minOccurs="1"
            ⌐maxOccurs="1"/>
    </sequence>
</complexType>
```

Note that the attribute group has been removed, and the value of minOccurs is 1, which forces the property to always contain the object.

11.1.6.1   Constraining remote and in-line properties

Note that in order for properties to reference remote values in GML data instances, the property type must support xlink attributes from the AssociationAttributeGroup. This is accomplished by deriving from AssociationType or including AssociationAttributeGroup. For example, in the following definition for PointPropertyType, the in-line content is a Point object.

```
<complexType name="PointPropertyType">
    <annotation>
        <documentation>A property that has a point as its
            ⌐value domain  can either be an appropriate
            ⌐geometry element encapsulated in an element of
            ⌐this type or an XLink reference to a remote
            ⌐geometry element (where remote includes geometry
            ⌐elements located elsewhere in the same
            ⌐document). Either the reference or the contained
            ⌐element must be given, but neither both nor
            ⌐none.</documentation>
    </annotation>
    <sequence>
        <element ref="gml:Point" minOccurs="0"/>
    </sequence>
    <attributeGroup ref="gml:AssociationAttributeGroup">
        <annotation>
            <documentation>This attribute group includes the
                ⌐XLink attributes (see xlinks.xsd). XLink is
                ⌐used in GML to reference remote resources
                ⌐(including those elsewhere in the same
                ⌐document). A simple link element can be
                ⌐constructed by including a specific set of
                ⌐XLink attributes. The XML Linking Language
                ⌐(XLink) is currently a Proposed
                ⌐Recommendation of the World Wide Web
                ⌐Consortium. XLink allows elements to be
                ⌐inserted into XML documents so cas to
                ⌐create sophisticated links between resources;
                ⌐such links can be used to reference remote
```

```
              ↳properties. A simple link element can be used
              ↳to implement pointer functionality, and this
              ↳functionality has been built into various
              ↳GML 3 elements by including the
              ↳gml:AssociationAttributeGroup.
              ↳</documentation>
        </annotation>
     </attributeGroup>
  </complexType>
```

You cannot use XML Schema to prohibit property types from having both `xlink` attributes and in-line content. It is possible, however, to record a constraint on usage of the property type in an `<annotation>` element as part of the property element declaration, as shown in the above complex type definition for `PointPropertyType`.

Note: If a property is an array-valued property (for example, `feature-Members`), the content of the property can be an unbounded list of the same object. According to GML best practices, array-valued properties should only have in-line content. For example, `featureMembers` cannot reference features remotely.

Another solution is to use a formal language, such as Schematron, which was used in some of the GML schemas. Schematron is a language for constraining arbitrary patterns in XML documents. For more information about Schematron, refer to http://www.ascc.net/xml/resource/schematron/Schematron2000.html.

The following schema fragment shows the abstract `hrefOrContent` rule that is defined in `gmlBase.xsd`. Any schema that imports or includes `gml-Base.xsd` can use the following abstract Schematron rule.

```
<annotation>
   <appinfo
      ↳source="urn:opengis:specification:gml:
      ↳schema-xsd:gmlBase:v3.00">
      <sch:schema>
         <sch:title>Schematron validation</sch:title>
         <sch:ns prefix="gml" uri=
            ↳"http://www.opengis.net/gml"/>
         <sch:ns prefix="xlink" uri=
            ↳"http://www.w3.org/1999/xlink"/>
         <sch:pattern name="Check either href or content
            ↳not both">
            <sch:rule abstract="true" id="hrefOrContent">
               <sch:report test="@xlink:href and
                  ↳(*|text())">Property element may not
                  ↳carry both a reference to an object
                  ↳and contain an object.</sch:report>
               <sch:assert test="@xlink:href |
                  ↳(*|text())">Property element must
                  ↳either carry a reference to an  object
                  ↳or contain an object.</sch:assert>
```

```
          </sch:rule>
        </sch:pattern>
      </sch:schema>
    </appinfo>
    <documentation>GML base schema for GML 3.0
      ⌣Components to support the GML encoding model.
      ⌣The abstract Schematron rules can be used by
      ⌣any schema that includes
      ⌣gmlBase.</documentation>
</annotation>
```

The following schema fragment shows how an element declaration may include the above `hrefOrContent` rule to limit the property to act in either by-value or by-reference mode, but not both.

```
<element name="baseCurve" type="gml:CurvePropertyType">
  <annotation>
    <appinfo>
      <sch:schema>
        <sch:pattern>
          <sch:rule context="gml:baseCurve">
            <sch:extends rule="hrefOrContent"/>
          </sch:rule>
        </sch:pattern>
      </sch:schema>
    </appinfo>
    <documentation>This property element either
      ⌣references a curve via the XLink-attributes or
      ⌣contains the curve element.</documentation>
  </annotation>
</element>
```

11.1.6.2   Feature relationships and associations

Properties of GML features can be used to express relationships between features. As discussed in Chapter 9, you can have a `Bridge` feature that `spans` a `Gorge`. The following example shows an element declaration and complex type for a user-defined `spans` property.

```
<element name="Bridge" type="app:BridgeType"
  ⌣substitutionGroup="gml:_Feature"/>

<complexType name="BridgeType">
  <complexContent>
    <extension base="gml:AbstractFeatureType">
      <sequence>
        <element name="span" type="integer"/>
        <element name="height" type="integer"/>
        <element ref="gml:centerLineOf"/>
        <element ref="app:mobility"/>
        <element name="spans" type="GorgePropertyType"
          ⌣minOccurs="0"/>
      </sequence>
    </extension>
  </complexContent>
</complexType>
```

The content model for the spans property can be defined as follows:

```
<complexType name="GorgePropertyType">
   <sequence>
      <element ref="app:Gorge" minOccurs="0"/>
   </sequence>
   <attributeGroup ref="gml:AssociationAttributeGroup"/>
</complexType>
```

In the above example, the spans property expresses the relationship of the target feature to the source feature; Bridge is the source feature and Gorge is the target. When you model a relationship between features, you can also define a property of the target feature that expresses its relationship to the source feature. For example, the Gorge feature can be spannedBy the Bridge, as shown below:

```
<element name="Gorge" type="app:GorgeType"
   ⌣substitutionGroup="gml:_Feature"/>

<complexType name="GorgeType">
   <complexContent>
      <extension base="gml:AbstractFeatureType">
         <sequence>
            <element name="width" type="integer"/>
            <element name="depth" type="integer"/>
            <element ref="gml:centerLineOf"/>
            <element name="spannedBy"
               ⌣type="app:BridgePropertyType"
               ⌣minOccurs="0"/>
         </sequence>
      </extension>
   </complexContent>
</complexType>
```

BridgePropertyType is used in the above example to indicate the target of the property. In many cases, the property type name should reflect the name of the property (for example, SpannedByPropertyType).

The complex type for the spannedBy property can be defined as follows:

```
<complexType name="BridgePropertyType">
   <sequence>
      <element ref="Bridge" minOccurs="0"/>
   </sequence>
   <attributeGroup ref="gml:AssociationAttributeGroup"/>
</complexType>
```

Note that Appendix D provides guidelines for using XMLSpy to model feature relationships in GML application schemas.

11.1.7   Global and local properties

When you define simple and complex properties in GML, these new properties can be declared as global elements or as local elements within feature content models. For example, the app:spans property in the previous section was defined as a global property. The following fragment shows how app:spans can be encoded as a local property:

```
<element name="Bridge" type="app:BridgeType"
    ⌐substitutionGroup="gml:_Feature"/>

<complexType name="BridgeType">
    <complexContent>
        <extension base="gml:AbstractFeatureType">
            <sequence>
                . . .
                <element name="spans"
                    ⌐type="app:GorgePropertyType"/>
            </sequence>
        </extension>
    </complexContent>
<complexType>
```

Unless you anticipate re-using a property, in most cases it is best to encode it as a local element. All GML objects on the other hand must be encoded as global elements.

11.1.7.1   Deriving from `AssociationType` or following the `AssociationType` pattern

As discussed in Chapter 10, the definition for `AssociationType` provides a pattern that can be used to define properties. The following schema fragment shows the complex type for `AssociationType`:

```
<complexType name="AssociationType">
    <sequence>
        <element ref="gml:_Object" minOccurs="0"/>
    </sequence>
    <attributeGroup ref="gml:AssociationAttributeGroup"/>
</complexType>
```

In this chapter, all of the examples of user-defined property definitions follow this pattern, as in the `GorgePropertyType` definition from the previous section. As with `AssociationType`, the `GorgePropertyType` definition contains an object and the `AssociationAttributeGroup`, which allows the property to have association attributes, such as `xlink:href`.

> *Note:* The `AssociationAttributeGroup` contains the `remoteSch-ema` attribute in addition to the `xlink` attributes. This is intended to contain an XPointer expression that points to a schema fragment that describes the content model for the remote property value. For more information about XPointer, see http://www.w3.org./TR/xptr/.

Another way to define a property is to derive from `AssociationType`, as shown below:

```
<element name="spans" type="app:GorgePropertyType"/>

<complexType names="GorgePropertyType">
   <complexContent>
      <restriction base="gml:AssociationType">
         <sequence>
            <element name="Gorge" type="app:GorgeType"
               ⌐minOccurs="0"/>
         </sequence>
      </restriction>
   </complexContent>
</complexType>
```

11.1.8   About attributes

You can also specify that a property or feature should include other GML-defined attributes, such as `srsName`, `uom` or `frame`. For example, the properties for the `Bridge` feature can also have a `uom` attribute that provides a reference to a units definition in a unit of measure dictionary. The following fragment shows a definition for a `height` property with a `uom` attribute:

```
<complexType name="HeightType">
   <simpleContent>
      <extension base="integer">
         <attribute name="uom" type="anyURI"
            ⌐use="optional"/>
      </extension>
   </simpleContent>
</complexType>
```

Note that in the earlier examples of the `Bridge` feature, `height` was a simple property. If you want a property to include attributes, it cannot be a simple type. For example, `HeightType` is a complex type with simple content. The `height` property can either be a global or local property, and its element declaration must be a complex type, as in the following type definition, in which `height` is a global property:

```
<complexType name="BridgeType">
   <complexContent>
      <extension base="gml:AbstractFeatureType">
         <sequence>
            <element ref="app:height"/>
            ...
         </sequence>
      </extension>
   </complexContent>
<complexType>
```

Features and other objects can also have attributes. The following schema fragment shows an example of a type definition for a topology object with a `weight` attribute. Topology objects are covered in detail in Chapter 13.

```
<element name="weightedEdge"
  ⌐type="gml:WeightedEdgePropertyType">
<complexType name="WeightedEdgePropertyType">
   <sequence>
      <element ref="gml:Edge" minOccurs="0"/>
   </sequence>
   <attribute name="weight" type="integer" default="1"/>
   <attributeGroup ref="gml:AssociationAttributeGroup"/>
</complexType>
```

When you define new objects or properties with attributes, if they are of complex type, the attributes must be included as shown in the above example.

Note: You cannot use XML attributes to express properties of GML application objects.

11.2   How do I create `metaData` schemas?

If you want to assign metadata to objects in your application schemas, you need to create a new metadata schema or use an existing one. A metadata schema can be user-defined or based on some enterprise, national or international standard. A metadata schema defines a root metadata package element and a set of properties, which are the metadata properties of any object whose `metaDataProperty` contains or references an instance of the metadata package. Figure 1 shows an example of a `BridgeDistrict` feature collection with a `metaDataProperty` whose value is a `BridgeMetaData` package that is defined in a `Bridge` metadata schema.

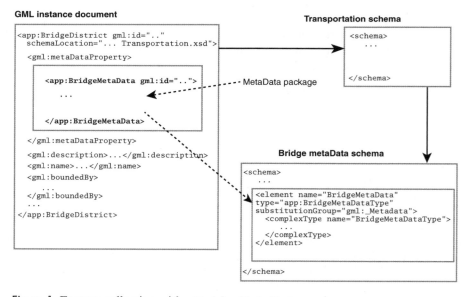

Figure 1 Feature collection with a `BridgeMetaData` package.

Note that in the above `BridgeDistrict` instance, the value of the `schemaLocation` attribute is the `Transportation.xsd` schema. This schema contains the definitions for the other user-defined objects and properties used in the GML instance. The metadata schema that contains the `BridgeMeta-Data` definition is included or imported by the `Transportation.xsd`. This is only one of the ways in which a metadata schema can be imported into a GML instance document. In most cases, however, the metadata definitions should either be part of, or imported or included by, an application schema that contains definitions for the other user-defined objects and properties used by the instance document.

11.2.1 Defining a `metadata` package

The package element must be substitutable for `_MetaData`. The content model of the package element must be a complex type that derives by extension from `AbstractMetaDataType`. The following example shows an element declaration for the `BridgeMetaData` package from Figure 1:

```
<element name="BridgeMetaData"
    ⌐type="app:BridgeMetaDataType"
    ⌐substitutionGroup="gml:_MetaData"/>
```

11.2.2 Defining properties for a `metadata` package

The properties in metadata packages are defined in the same way as other GML properties, as discussed in Section 11.1.5. Properties can either be references to global properties or they can be defined in the metadata package's content model, such as `app:BridgeMetaDataType`. For the `Bridge` metadata package, the content model can be encoded as follows:

```
<complexType name="BridgeMetaDataType">
    <complexContent>
        <extension base="gml:AbstractMetaDataType">
            <sequence>
                <element name="identifier" type="string"/>
                <element name="collectionDate" type="string"/>
                <element name="collectedBy" type="string"/>
                <element name="restrictionsOnUse"
                    ⌐type="string"/>
            </sequence>
        </extension>
    </complexContent>
</complexType>
```

Note that the metadata package is simply a list of metadata properties, that is, properties that describe the data in the GML feature or feature collection.

11.2.3 Rules for `metadata` properties

All GML objects have a generic `metaDataProperty` element defined as part of their content model. You can use this property to include metadata, or you can create custom metadata property elements that only allow for specific metadata packages. For example, you can define a specialized `bridgeInfo` property that encapsulates the `BridgeMetaData` package instead of `_MetaData`, as shown below:

```
<element name="bridgeInfo"
    ⌐type="app:BridgeMetaDataPropertyType"
    ⌐substitutionGroup="gml:metaDataProperty"/>
<complexType name="BridgeMetaDataPropertyType">
    <complexContent>
        <restriction
            ⌐base="gml:AbstractMetaDataPropertyType">
            <sequence>
                <element ref="app:BridgeMetaData"/>
            </sequence>
        </restriction>
    </complexContent>
</complexType>
```

It is also possible to use derivation by restriction so that all objects that derive from `AbstractBridgeType` can have the `bridgeInfo` property instead of the generic `metaDataProperty`. For example, you can specify the restriction in the `Bridge` feature's content model, as follows:

```
<element name="_Bridge" type="app:AbstractBridgeType"
    ⌐substitutionGroup="gml:_Feature"/>

<complexType name="AbstractBridgeType">
    <complexContent>
        <restriction base="gml:AbstractFeatureType">
            <sequence>
                <element ref="app:bridgeInfo" minOccurs="0"/>
                <element ref="gml:boundedBy"/>
            </sequence>
        </restriction>
    </complexContent>
</complexType>

<element name="Bridge" type="app:BridgeType"
    ⌐substitutionGroup="_Feature"/>

<complexType name="BridgeType">
    <complexContent>
        <extension base="app:AbstractBridgeType">
            <sequence>
                <element name="span" type="integer"/>
                <element name="height" type="integer"/>
                <element ref="gml:centerLineOf"/>
                <element ref="app:mobility"/>
            </sequence>
        </extension>
    </complexContent>
<complexType>
```

In the above example, the associated content model for the `Bridge` is `BridgeType`, which is derived by extension from `AbstractBridgeType`.

The content model for `AbstractBridgeType` is derived by restriction from `AbstractFeatureType`, and it contains the `bridgeInfo` metadata property.

11.2.4   Using the 'about' attribute with metadata

All properties of `MetaDataPropertyType` inherit an optional `about` attribute that can be used in GML instances to apply metadata to other GML objects outside of the object that a metadata property is attached to. For example, in a GML instance, the `bridgeInfo` property can have an `about` attribute that points to a `specific Bridge` feature, as shown in the following example. This applies the metadata in the `BridgeMetadata` package to the `Bridge` feature with the `Bridge001` id.

```
<app:BridgeDistrict gml:id="BridgeCollection001">
    ...
    <app:bridgeInfo gml:about="#Bridge001"/>
        <app:BridgeMetadata gml:id="Meta001">
            ...
        </app:BridgeMetadata>
    </app:bridgeInfo>
    <app:bridgeInfo gml:about="app:Bridge/
        ⌐app:mobility='DrawBridge'"
        ⌐xlink:href="#Meta001"/>
    <app:bridgeMember>
        <app:Bridge gml:id="Bridge001">
            <gml:centerLineOf>...</gml:centerLineOf>
            ...
        </app:Bridge>
    </app:bridgeMember>
    ...
    <app:bridgeMember>
    ...
    </app:bridgeMember>
    <app:area>500</app:area>
    <app:numBridges>73</app:numBridges>
</app:BridgeDistrict>
```

In this example, the first `bridgeInfo` property associates the values of the metadata instance `Meta001` with `Bridge001`. The second `bridgeInfo` property associates (by remote-property reference) the values of the same metadata instance to all `Bridge` features in the district that have a `mobility` property whose value is `DrawBridge`.

If the `about` attribute is not used, then the metadata only applies to the GML object that it is attached to. For example, if a `BridgeDistrict` feature collection has the `bridgeInfo` property that does not have an `about` attribute, then the metadata contained within or pointed to by the `bridgeInfo` property is only applicable to the `BridgeDistrict` instance that contains the `bridgeInfo` property.

11.3   How do I select the appropriate GML core schemas for my application domain?

Most GML applications will make use of only a subset of the GML core schemas described in this guide. As a starting point, you need the following:

- The `feature.xsd` if you are modelling geographic features.
- The `valueObjects.xsd` if your features have properties that make use of units of measure. You do not need the `measures.xsd` schema, unless you are defining units of measure, such as those that appear in a units of measure dictionary.
- The `geometryBasic0d1d.xsd` and `geometryBasic2d.xsd` schemas if you only need simple one- or two-dimensional geometries. You need the other geometry schemas only if you require support for complex, three-dimensional or non-linear geometry objects. Chapter 12 covers all of the geometry schemas.
- The `topology.xsd` if your features have topology properties.
- The CRS schemas if you are constructing CRS definitions for CRSs or support components, such as Coordinate Systems and Datums.
- The temporal schemas, `temporal.xsd` and `dynamicFeature.xsd`, if you are concerned with time-dependent feature properties or dynamic features.
- The coverage schemas, `coverage.xsd` and `grid.xsd`, if you are constructing coverages, such as remotely sensed images, aerial photographs, soil distribution or Digital Elevation Models (DEMs).
- The `valueObjects.xsd` and `observation.xsd` schemas if you are concerned with sensor data.

To create GML application schemas, you need to use XML Schema, import the required GML core schemas and follow a few simple GML rules. The GML core schemas have been modularized so that application schemas can import just a subset of GML schemas that are required. For example, if you are simply migrating a GML 2.0 application schema to GML 3.0 without adding any new definitions, the application schema can continue to import `feature.xsd`.

Figure 2 shows the dependencies of the GML core schemas. Note that the gray schemas represent GML 3.0 schemas, while the white schemas are external schemas that belong to a different namespace. All of the dashed arrows indicate dependency of one schema on another, in that the schema at the tail of the arrow depends upon the schema at the head of the arrow. For example, the feature schema depends on `geometryComplexes.xsd` and `temporal.xsd`, because these are the schemas that it includes.

The following top-level schemas are the roots of hierarchies of GML schemas:

- `coordinateReferenceSystem.xsd`
- `topology.xsd`
- `coverage.xsd`
- `dynamicFeature.xsd`
- `observation.xsd`
- `defaultStyle.xsd`.

If you import any of these schemas, you will also automatically import all of the schemas that they depend on. For example, the topology schema includes the

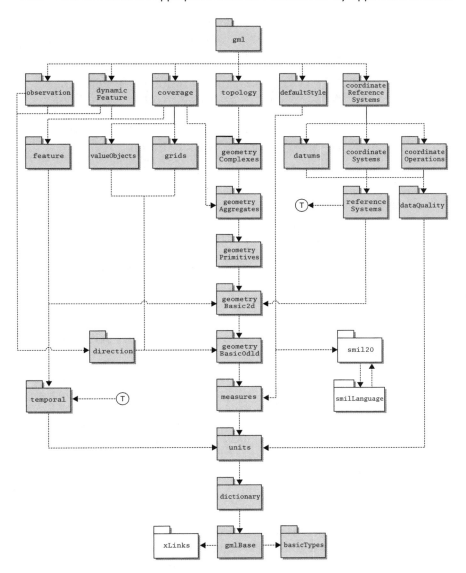

Figure 2 GML 3.0 core schema dependencies.

geometry complexes schema, which includes all of the other geometry schemas and the measures, units, temporal, GML base, basic types and xlink schemas.

Note that when some XML Schema parsers process import statements, they look for the first occurrence of an imported namespace and ignore subsequent occurrences of the same namespace, even if the import statements are for different schemas from the same namespace. XML Schema parsers cannot recognize that the application schema is actually importing two separate sets of schemas. Consequently, an application schema should not have more than one import statement for schemas from the GML namespace. For example, if your application schema has an import statement for `coordinateReferenceSystem.xsd`, you should not have another import statement for `defaultStyle.xsd`.

The `gml.xsd` schema was created to address this issue by providing a 'wrapper' for all of the GML core schemas. That is, if your application schema imports the `gml.xsd` schema, all of the GML core schemas are imported into the application schema. You can also create a custom profile of the GML core schema that only includes certain schema components. Profiles are covered in Clause 7.16 of the *GML Version 3.00 OpenGIS® Implementation Specification* (http://www.opengis.org/docs/02-023r4.pdf).

You can also use the schema-packaging tool (an XSLT application) that is contained in Annex F of the *GML Version 3.00 OpenGIS® Implementation Specification* (http://www.opengis.org/docs/02-023r4.pdf). This tool analyses the GML core schemas and works out the dependencies between the schemas. You then provide the tool with a list of the GML components of interest – for example, `Point`, `LineString` and `_Feature` – and it creates a derived stub schema – similar to `gml.xsd` but typically much smaller – that contains all the requested schema components and the components on which they depend. It is a recommended practise to use the subsetting tool, or a similar mechanism, to create a tailored schema.

11.4   How do I find and use existing application schemas?

GML application schemas can be shared on the Internet and can import or include one another. GML is designed specifically to support the development of interconnected and distributed schemas of geographic objects. The OGC is responsible for the GML core schemas. In addition to developing and maintaining the schemas, the OGC also provides them online 24 hours a day, seven days a week. As new GML application schemas are created, a similar level of service will be required for different domains. Eventually, there will be rich networks of GML application schemas covering many application domains.

Before you create application schemas, you need to determine where they will reside on your system and how software developers, database administrators and data modellers can access them. This is not necessarily an issue if you are deploying GML from a single geospatial web service. Most applications, however, require several geospatial web services. You should consider deploying the GML application schemas on a web site that is visible to your data users or through a schema registry.

11.5   Chapter summary

In GML 3.0 there are many different kinds of possible instance documents, including feature collections, dictionaries and topological complexes. Application schemas define the types that can be instantiated in these instance documents. To create application schemas, you need to follow a number of rules for each application schema.

Application schemas must declare a target namespace and import the appropriate core schemas from GML 3.0. All GML objects must derive, directly or indirectly, from an abstract GML type, and all object collections must derive from an abstract collection type. Unlike objects, GML properties are not required to derive from an abstract type, but they should follow the `AssociationType` pattern if they need to allow for remote values. All objects must conform to the rules respecting those objects. In this guide, these rules are discussed in the appropriate chapters. For example, the rules for creating user-defined geometry objects are discussed in Chapter 12.

If your application schema will contain features (and geometries), import `feature.xsd` (for dynamic features, you need to import `dynamicFeature.xsd`). To define a feature, its content model must derive from `AbstractFeatureType`, and feature collection content models must derive from `AbstractFeatureCollectionType`. If a feature collection's content model does not have any user-added content, you can use the built-in `FeatureCollectionType` as the value of the `type` attribute in the feature collection's element declaration.

Features (and feature collections) can have in-line and remote properties, which can be declared either as global elements within the application schema or locally as part of the feature's content model. Properties can be simple or complex. Simple properties can only have a simple value, which reflects simple data type, such as string or integer. Complex properties can have child elements and attributes, and these properties can be spatial, non-spatial, remote or expressions of relationships between features.

If you want to define new properties for your application schema, they can derive from `AssociationType` or follow the `AssociationType` pattern by including `AssociationAttributeGroup`. If the property needs to reference remote values, then you should use one of these methods. With GML, it is possible to have properties that can have xlink attributes (for referencing remote values) or in-line content. You can use a formal notation, such as Schematron (*ISO 19757-3*), to constrain the property type so that it can only have one kind of value.

To include custom metadata properties in your application, you can create a separate metadata schema that defines at least one metadata package. Each package contains a set of metadata properties. Values of the metadata package – that is, the values of the package's properties – can be either contained in-line within the `metaDataProperty` or pointed to using the remote property mechanism. To associate a GML object with a metadata package, you can either use the built-in `metaDataProperty` that is inherited by all GML objects, or you can create custom metadata properties.

When you import the GML core schemas into your application schema, you might only need a subset of the schemas. There are six separate 'modules' that include many of the other schemas and can be used independently. These 'modules' are `coordinateReferenceSystems.xsd`, `topology.xsd`, `coverage.xsd`, `dynamicFeature.xsd`, `observation.xsd` and `defaultStyle.xsd`.

If you decide to import one of these 'modules', XML Schema only recognizes that 'module'. In other words, schema imports are not transitive. If you need to import more than one of these 'modules', import `gml.xsd`, which is a wrapper that includes all of the GML core schemas, or use a GML profile. A schema-packaging tool is also provided with the GML 3.0 specification and it can be used to create a custom GML 'core' schema similar to `gml.xsd` that contains only the required schema components and their dependencies.

Currently, if your application domain wants to use GML application schemas, you need to make your own. In the next year, however, schema registries will become available. These registries will store application schema components that can be used by various application domains.

References

http://www.opengis.org/docs/02-023r4.pdf (October 15, 2003).

http://www.faqs.org/rfcs/rfc2141.html (October 17, 2003).

http://www.purl.org/ (October 17, 2003).

http://www.geoconnections.org/developersCorner/devCorner_devNetwork/components/GML_bpv1.3_E.pdf (October 22, 2003) p. 50.

http://www.w3.org/TR/xmlschema-0/ (September 20, 2003).

http://www.ascc.net/xml/resource/schematron/Schematron2000.html (October 16, 2003).

http://www.w3.org./TR/xptr/ (September 20, 2003).

Additional reference

http://schemas.opengis.net/gml/3.0.1/base/ (October 15, 2003).

Chapter 12

GML geometry

GML 3.0 provides support for modelling the geometric aspects of geographic features using points, curves, surfaces and solids. For an introduction to some of the concepts of geometry modelling, please refer to http://www.cs.mtu.edu/~shene/COURSES/cs3621/NOTES/notes.html. The GML 3.0 Geometry schemas are an implementation of a subset of *ISO DIS 19107 (ISO, 2000)*.

GML 2 has only one Geometry schema, `geometry.xsd`, however, in GML 3.0, the geometry model is more extensive, and the different types and elements are grouped in the following five schemas, which are available online at http://schemas.opengis.net/gml/3.0.1/base/:

- `geometryBasic0d1d.xsd`
- `geometryBasic2d.xsd`
- `geometryAggregates.xsd`
- `geometryPrimitives.xsd`
- `geometryComplex.xsd`

The first three schemas contain the types and elements that are required for backwards compatibility with GML 2. The last two Geometry schemas, `geometryPrimitives.xsd` and `geometryComplex.xsd`, comprise entirely new types and elements.

12.1 What if my geometry needs are very simple?

GML 3.0 contains geometry elements and types that are not required in many applications. Many applications will only require elements and types from one or two of the Geometry schemas.

12.1.1 Points and lines

If you are interested in using only points and lines to model the geometric properties of your geographic objects, you can use just `geometryBasic0d1D.xsd` and ignore the rest of the GML 3.0 Geometry schemas. The concrete geometry primitives defined in `geometryBasic0d1d.xsd` are `Point` and `LineString`.

Geography Mark-up Language (GML). R. Lake, D. S. Burggraf, M. Trninić, L. Rae © 2004 Galdos Systems Inc.
Published by John Wiley & Sons, Ltd ISBNs: 0-470-87153-9 (HB); 0-470-87154-7 (PB)

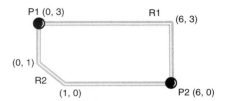

Figure 1 Simple highway network geometry.

These primitives can be used to model, for example, the simple geometry of the highway network shown in Figure 1. This figure shows two roads, R1 and R2, between two different points, P1 and P2. The point P1 can be encoded in GML 3.0 as shown in the following:

```
<gml:Point gml:id="P1" srsName="..">
   <gml:pos>0 3</gml:pos>
</gml:Point>
```

The following example shows how R1 and R2 can be encoded as LineString objects:

```
<gml:LineString gml:id="R1" srsName="...">
   <gml:coordinates>0,3 6,3 6,0</gml:coordinates>
</gml:LineString>
<gml:LineString gml:id="R2" srsName="..">
   <gml:posList dimension="2">0 3 0 1 1 0 6 0
      ⌣</gml:posList>
</gml:LineString>
```

In the above examples, Point has a pos property, while the two LineString objects have coordinates or posList properties. Note that posList will be introduced in GML 3.1. In GML 3.0, coordinates can be encoded using one of these different properties. In previous versions of GML, the coord and coordinates properties were used to encode coordinates. The coord property has been deprecated in GML 3.0. Note that the posList property was not finalized at the time of publication, so the exact syntax may differ from the one shown. You can also use the coordinates property, whose value is a whitespace-separated list of coordinate tuples. Both of these coordinate forms are described in this chapter.

Note that the LineString may also be defined using a combination of pos and coordinates properties, or pos and posList properties. Point elements can have identity (for example, <gml:Point gml:id="e22">), whereas pos, posList and coordinates designate coordinate tuples. GML thus distinguishes a Point object from the coordinate tuple that describes that Point in some Coordinate Reference System (CRS).

12.1.1.1 About the pos property

The pos property is known as a *direct position*, and its values represent coordinates relative to some CRS. Two optional attributes can be included with the pos property: dimension and srsName. The dimension attribute represents the number of coordinate entries contained in the pos property. The value of this attribute must match the number of axes in the CRS referenced by the srsName attribute.

The `srsName` attribute is used to reference a CRS that is used to interpret the coordinates contained within the `pos` property. Chapter 15 covers the rules for using this attribute. Note that the `srsName` attribute is not mandatory, because, in many cases, the `pos` property is contained within a geometry object that has an `srsName` attribute. The CRS value of the `srsName` attribute for the enclosing geometry also applies to the coordinates contained within the `pos` property.

12.1.1.2   About the `coordinates` property

The `coordinates` property provides a compact way for specifying the coordinates of a geometry object. Unlike `pos`, the `coordinates` property does not have an optional `srsName` attribute. The CRS must, therefore, be referenced from one of the parent geometry objects. The values of the property are encoded as a text string. For `LineString` objects, the `coordinates` property provides a much more compact encoding than the `pos` property. On the other hand, the coordinate values in the `pos` property are visible in XML, while with the `coordinates` property, only the entire coordinate string is visible in XML.

12.1.1.3   About the `posList` property

The `posList` property contains a list of direct positions. Because there is no tuple separator, the value of the `dimension` attribute tells you how to group the tuples in the list to determine the tuple boundaries. Consider, for example, the following encoding of a `LineString` instance:

```
<gml:LineString srsName="#myrefsys">
   <gml:posList dimension="2">10.1 20.1 10.6 24.2 13.3
      ⌣26.1</gml:posList>
</gml:LineString>
```

In this example, there are six coordinates in the list. The value of the `dimension` attribute is 2, which means that each tuple consists of two coordinates, and therefore, this example has three tuples – (10.1, 20.1), (10.6, 24.2), (13.3, 26.1) – each of which represents a point.

Note that the ordinates in the list are XML Schema double types. This form is to be used for all `double`, `integer`, `decimal` or `float` coordinates for which a precision of XML Schema double is acceptable, that is, in most cases. The `coordinates` property can be used for those cases in which string-valued coordinates are appropriate, such as D:M:S. Once `posList` is finalized, it is recommended that you use it in most cases instead of `coordinates`.

12.1.1.4   About the `pointRep` property

There are circumstances in which the points on a `LineString` (or other curves) can be defined elsewhere within the same instance or in another document. In these circumstances, the `pointRep` property can be used.

12.1.2   Points, lines and surfaces

If you are interested in only points, lines and surfaces, you can use geometry2D.xsd, which includes geometryBasic0d1d.xsd. The `Polygon`

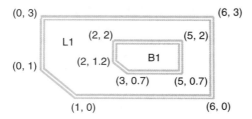

Figure 2 Simple land parcel geometry.

object is the main geometric primitive defined in `geometryBasic2d.xsd`. This primitive – together with the primitives from `geometryBasic0d1d.xsd` – can be used to represent the extent of the land parcel, L1, shown in Figure 2.

The following example shows how to encode the extent of this simple land parcel in GML, using `Polygon` and `LinearRing` elements:

```
<gml:Polygon gml:id="L1">
   <gml:exterior>
      <gml:LinearRing>
         <gml:coordinates>0,1 1,0 6,0 6,3 0,3 0,1
            ↳</gml:coordinates>
      </gml:LinearRing>
   </gml:exterior>
   <gml:interior>
      <gml:LinearRing>
         <gml:coordinates>2,1.2 2,2 5,2 5,0.7 3,0.7 2,1.2
            ↳</gml:coordinates>
      </gml:LinearRing>
   </gml:interior>
</gml:Polygon>
```

Note that the land parcel, L1, is represented by a `Polygon`, which has two boundary components: `exterior` and `interior`. Each boundary component is contained in either the `exterior` or `interior` property of the `Polygon` and is encoded using a `LinearRing`, which contains a `coordinates` property. Note that a `LinearRing` is a kind of `LineString`. The first and last points of the `LinearRing` must be identical, to ensure that it forms a closed loop.

12.2　Why do I need the extra geometries in GML 3.0?

You only need to consider the other three Geometry schemas if you are interested in non-linear and three-dimensional geometries or if you wish to construct curves and surfaces from other curve and surface components. These schemas contain additional geometric primitives, plus geometric aggregates and geometric composites. The extra geometries provide a richer and more descriptive geometry encoding. Note, however, that they can be more expensive to code, and computation can be more time-consuming and resource-intensive than with linear geometries.

12.2.1　Creating a composite curve

There are many cases – even with linear geometries – in which it is useful to construct a `Curve` or `Ring` from several line string segments. Consider the land

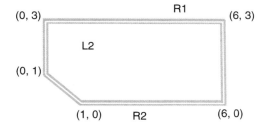

Figure 3 Land parcel as a `Polygon` bounded by two curve segments.

parcel shown in Figure 3, whose extent is a variation of the polygon L1 from Figure 2.

In GML 2, a `Polygon` can only have `outerBoundaryIs` and `inner-BoundaryIs` properties, whose values are `LinearRing` objects. The following example shows how the outer boundary of the `Polygon` can be encoded in GML 2:

```
<gml:Polygon gml:id="L2">
   <gml:outerBoundaryIs>
      <gml:LinearRing>
         <gml:coordinates>0,1 1,0 6,0 6,3 0,3 0,1
            ␣</gml:coordinates>
      </gml:LinearRing>
   </gml:outerBoundaryIs>
</gml:Polygon>
```

The `Polygon` can be encoded in GML 3.0 as follows:

```
<gml:Polygon gml:id="L2">
   <gml:exterior>
      <gml:LinearRing>
         <gml:coordinates>0,1 1,0 6,0 6,3 0,3 0,1
            ␣</gml:coordinates>
      </gml:LinearRing>
   </gml:exterior>
</gml:Polygon>
```

Note that since the `outerBoundaryIs` and `innerBoundaryIs` properties from GML 2 are deprecated in GML 3.0, the `exterior` and `interior` properties should be used instead.

In GML 2, the interior and exterior boundaries of a geometry object can only consist of `LinearRing` objects. This restriction has been relaxed in GML 3.0, as shown in the following encoding of a `Polygon` bounded by a `Ring` composed of curve members:

```
<gml:Polygon gml:id="L1">
   <gml:exterior>
      <gml:Ring>
         <gml:curveMember xlink:href="#R1"/>
         <gml:curveMember>
            <gml:OrientableCurve orientation="-">
               <gml:baseCurve xlink:href="#R2"/>
            </gml:OrientableCurve>
         </gml:curveMember>
```

```
                </gml:Ring>
              </gml:exterior>
           </gml:Polygon>
```

The exterior boundary of L1 first traverses R1, which is referenced using `xlink:href="#R1"`. Then it traverses R2 backwards, that is, with the opposite orientation of R2, which is encoded as an `OrientableCurve` with '-' orientation.

The `orientation` attribute is used to assign the orientation to the `OrientableCurve`. Every curve is inherently oriented by its coordinate description, that is, the first coordinate is the beginning and the last coordinate is the end. When a curve has a positive orientation (+), it follows the inherent orientation from the beginning to the end. If a curve has a negative orientation (−), the curve is navigated in the opposite direction. Orientation is provided in geometry to mirror the topology model covered in Chapter 13.

`Ring`, `OrientableCurve` and `curveMember` are discussed in the following section on geometry primitives. At least 30 different geometries, segments and patches are defined in the five Geometry schemas, and consequently, it is not possible to describe all of the GML 3.0 geometries in this chapter. For detailed information about all of the GML 3.0 geometries, see Clause 7.5 of the *GML Version 3.00 OpenGIS® Implementation Specification* (http://www.opengis.org/docs/02-023r4.pdf).

12.3   What are the geometry primitives?

The `geometryPrimitives.xsd` specifies geometric primitives on top of which other geometric constructs in GML are built. Table 1 lists the different primitives, their properties and the values of these properties.

All of these primitives are connected and continuous. A `Curve` can be oriented positively or negatively using `OrientableCurve`, and a `Surface` can be oriented using `OrientableSurface`. Both elements have an orientation attribute that has a value of '+' or '-'. `OrientableCurve` has a `baseCurve` property that can either contain or reference any object that is substitutable for `_Curve`. `OrientableSurface` follows a similar pattern, except that it has a `baseSurface` property, which contains or references an object from the `_Surface` substitution group.

Table 1   Geometry primitives from `geometryPrimitives.xsd`

Primitive	Property	Value
Curve	segments	Any object that is substitutable for `CurveSegment`, such as `LineStringSegment` and `Arc`.
Surface	patches	Any object that is substitutable for `_SurfacePatch`, such as `PolygonPatch`.
Solid	exterior interior	Any object that is substitutable for `_Surface`.

12.3.1 Encoding curve segments

A Curve is composed of one or more curve segments, which can be one of the following elements from geometryPrimitives.xsd:

- LineStringSegment
- Arc
- ArcString
- Circle
- ArcStringByBulge
- ArcByBulge
- ArcByCenterPoint
- CircleByCenterPoint
- CubicSpline
- Bspline
- Bezier.

Note that a curve segment (which is not itself a curve) is a portion of a curve in which a single interpolation method is used. For example, only linear interpolation is used within a LineStringSegment, while only circular interpolation is used within an Arc segment.

LineStringSegment is similar to LineString, but is substitutable for _CurveSegment, not _Curve. Because LineString is a geometry element from GML 2, it is not a curve segment and is used only in circumstances that require simple geometry. For more complex curves – for example, a combination of straight-line segments and arcs – use LineStringSegment instead.

Of the remaining curve segments, Arc and CubicSpline are the most likely to be used in GML 3.0. Arcs are the fundamental building blocks for encoding circles in GML 3.0. Figure 4 shows a simple Arc with three control points. The following example shows a how this Arc can be encoded in GML 3.0:

```
<gml:Arc>
    <gml:coordinates>2,0 0,2 -2,0</gml:coordinates>
</gml:Arc>
```

An Arc in GML is just a segment of a circle that is encoded using circular interpolation between three control points. An ArcString consists of one or more arcs that are contiguous; that is, the arcs are connected at endpoints. In Figure 5, for example, the two arcs in the ArcString are connected at $(-2, 0)$.

Figure 4 Arc with three control points.

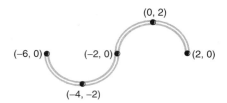

Figure 5 `ArcString` with five control points.

This `ArcString` can be encoded in GML 3.0 as follows:

```
<gml:ArcString numArc="2">
    <gml:coordinates>2,0 0,2 -2,0 -4,-2 -6,0•
      ⌣</gml:coordinates>
</gml:ArcString>
```

Note that the `ArcString` has a `numArc` attribute whose value is 2, thus indicating that the `ArcString` comprises two arcs. In contrast, an `Arc` is an `ArcString` consisting of a single arc. With `Arc` objects, the `numArc` attribute is fixed to '1', and therefore, the attribute does not need to be included in `Arc` instances.

A `Circle` is an `Arc` with an identical start and end. Note that in order for the `Circle` to be unambiguously defined, three distinct non-colinear control points are required. To form a complete `Circle`, the arc is extended past the third control point until the first control point is encountered. Figure 6 shows a circle whose control points are $(2, 0)$, $(0, 2)$ and $(-2, 0)$, where the first control point $(2, 0)$ represents the start and end of the circle.

The circle from Figure 1 can be encoded as

```
<gml:Circle>
    <gml:coordinates>2,0 0,2 -2,0</gml:coordinates>
</gml:Circle>
```

Figure 7 shows an example of another curve segment, `ArcByCenterPoint`, where the dimensions of the arc are determined by the arc's center point, in this case $(0, 0)$.

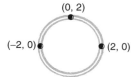

Figure 6 `Circle`.

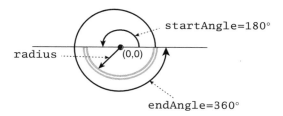

Figure 7 `ArcByCenterPoint`.

This arc can be encoded as follows:

```
<gml:ArcByCenterPoint>
    <gml:pos>0 0</gml:pos>
    <gml:radius uom="..">2</gml:radius>
    <gml:startAngle uom="..#degrees">180</gml:startAngle>
    <gml:endAngle uom="..#degrees">360</gml:endAngle>
</gml:ArcByCenterPoint>
```

In GML 3.0 instances, the `ArcByCenterPoint` object has a `pos` property – which defines the coordinates of the center – plus `radius`, `startAngle` and `endAngle` properties. The `startAngle` and `endAngle` values are angles measured counter-clockwise from the first axis in the referenced CRS. GML 3.0 also provides the `CircleByCenterPoint` arc, which is a circle whose dimensions are determined by its centre point. The values of the `startAngle` and `endAngle` properties must be the same.

Figure 8 shows an example of another curve segment, `ArcByBulge` that can be used to encode arcs in GML 3.0. The following example shows how this arc can be encoded:

```
<gml:ArcByBulge>
    <gml:coordinates>2,0 -2,0</gml:coordinates>
    <gml:bulge>2</gml:bulge>
    <gml:normal>-1</gml:normal>
</gml:ArcByBulge>
```

The arc has `coordinates` that represent the start and end points of the arc, and the height of the arc indicated by the `bulge` and `normal` properties. Another curve segment, `ArcStringByBulge`, serves a similar function as `ArcString`, except that it has `bulge` and `normal` properties instead of middle control points.

The `CubicSpline`, `BezierCurve` and `Bspline` elements are all examples of curve segments that can be parameterized piecewise by rational functions. That is, these three curve segments can be decomposed into pieces, each of which is described by a polynomial or, more generally, a rational function.

Like all curve segments, these segments can be composed together to form curves. Figure 9 shows an example of the `Arc` from Figure 4 with a `CubicSpline` that starts at the end point of the `Arc`.

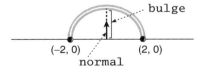

Figure 8 `ArcByBulge`.

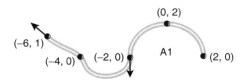

Figure 9 `Arc` with a `CubicSpline`.

The `CubicSpline` can be encoded as shown below:

```
<gml:CubicSpline>
    <gml:coordinates>-2,0 -4,0 -6,1</gml:coordinates>
    <gml:vectorAtStart>0 -1</gml:vectorAtStart>
    <gml:vectorAtEnd>-1 1</gml:vectorAtEnd>
<gml:CubicSpline>
```

In addition to providing the control points of the `CubicSpline`, the vectors in the `vectorAtStart` and `vectorAtEnd` properties uniquely determine the `CubicSpline` going through the given set of control points. In GML, arcs and cubic splines are in the substitution group _CurveSegment, and consequently, they need to be contained in the `segments` property of a Curve. The following example shows how the `Arc` and `CubicSpline` from Figure 9 can be encoded as curve segments inside of a `Curve`.

```
<gml:Curve gml:id="C11">
    <gml:segments>
        <gml:Arc>
            <gml:coordinates>2,0 0,2 -2,0</gml:coordinates>
        </gml:Arc>
        <gml:CubicSpline>
            <gml:vectorAtStart>0 -1</gml:vectorAtStart>
            <gml:vectorAtEnd>-1 1</gml:vectorAtEnd>
            <gml:coordinates>-2,0 -4,0 -6,1</gml:coordinates>
        <gml:CubicSpline>
    </gml:segments>
</gml:Curve>
```

Note that curve segments are not geometries and they cannot have unique identifiers. Instead, a unique identifier must be assigned to the geometry object that contains the curve segments. In the above example, the `Curve` has a unique identifier of C11.

12.3.2   Encoding surface patches

As shown in Table 1, `Surface` objects have a `patches` property, which serves a function similar to that of the `segments` property for Curve objects. Instead of containing curve segments, the `segments` property contains one or more surface patches, which are typically polygon patches. Note that `PolygonPatch`, not `Polygon`, can be contained within a `patches` property, because `Polygon` is a geometry object from GML 2 and is not substitutable for _SurfacePatch (`Polygon` is substitutable for _Surface).

The following example shows how a `PolygonPatch` can be encoded in GML 3:

```
<gml:PolygonPatch>
    <gml:exterior>
        <gml:Ring>
            <gml:curveMember>
                <gml:Curve gml:id="C1">
                    <gml:segments>
                        <gml:Circle>
```

```
            <gml:coordinates>1,0,0 -1,0,0
                 ⌣0,-1,0</gml:coordinates>
          </gml:Circle>
        </gml:segments>
      </gml:Curve>
    </gml:curveMember>
  </gml:Ring>
 </gml:exterior>
</gml:PolygonPatch>
```

In the above example, the exterior of the `PolygonPatch` is bounded by a `Ring`, which represents a single connected component of a surface boundary. A `Ring` can have one or more `curveMember` properties, each of which contains an object that is substitutable for `_Curve`. In the above example, the `Ring` only has one `curveMember` property whose value is a `Circle`. If the `Ring` comprises more than one curve, the curves must be contiguous and form a closed loop. If required, a `PolygonPatch` can also have an `interior` property.

The `geometryPrimitives.xsd` schema also provides the following simple surface patches that can also be contained within the `patches` property: `Triangle` and `Rectangle`, as shown in Figure 10. A `Triangle` can be encoded as follows:

```
<gml:Triangle>
   <gml:exterior>
      <gml:LinearRing>
         <gml:coordinates>0,0 2,0 1,1 0,0
             ⌣</gml:coordinates>
      </gml:LinearRing>
   </gml:exterior>
</gml:Triangle>
```

A `Rectangle` can be encoded as

```
<gml:Rectangle>
   <gml:exterior>
      <gml:LinearRing>
         <gml:coordinates>0,0 2,0 2,1 0,1 0,0
             ⌣</gml:coordinates>
      </gml:LinearRing>
   </gml:exterior>
</gml:Rectangle>
```

For both of these objects, the first and last coordinates are the same. Note that, unlike the `PolygonPatch`, the `Triangle` and `Rectangle` objects can only have an `exterior` property. To understand how surface patches are used to model the content of a `Surface`, consider the example of a tetrahedron shown in Figure 11.

Figure 10 `Triangle` and `Rectangle`.

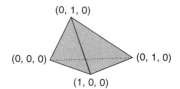

Figure 11 Tetrahedron surface with four `Triangle` patches.

The tetrahedron T1 comprises four `Triangle` patches, as shown in the following GML 3 encoding:

```
<gml:Surface gml:id="T1">
  <gml:patches>
    <gml:Triangle>
      <gml:exterior>
        <gml:LinearRing>
          <gml:coordinates>0,0,0 1,0,0 0,1,0 0,0,0
             ↳</gml:coordinates>
        </gml:LinearRing>
      </gml:exterior>
    </gml:Triangle>
    <gml:Triangle>
      <gml:exterior>
        <gml:LinearRing>
          <gml:coordinates>0,0,0 1,0,0 0,0,1 0,0,0
             ↳</gml:coordinates>
        </gml:LinearRing>
      </gml:exterior>
    </gml:Triangle>
    <gml:Triangle>
      <gml:exterior>
        <gml:LinearRing>
          <gml:coordinates>0,0,0 0,1,0 0,0,1 0,0,0
             ↳</gml:coordinates>
        </gml:LinearRing>
      </gml:exterior>
    </gml:Triangle>
    <gml:Triangle>
      <gml:exterior>
        <gml:LinearRing>
          <gml:coordinates>0,0,1 1,0,0 0,1,0 0,0,1
             ↳</gml:coordinates>
        </gml:LinearRing>
      </gml:exterior>
    </gml:Triangle>
  </gml:patches>
</gml:Surface>
```

The `patches` property is used to encapsulate the four `Triangle` patches.

12.3.2.1 Solids

GML 3 supports three-dimensional geometry by providing the three-dimensional geometry primitive, `Solid`. An example of a `Solid` with `CompositeSurface` content is provided in the following section.

Table 2 Complex geometries from `geometryComplexes.xsd`

Composite or Complex	Property	Value
`CompositeCurve`	`curveMember` (unbounded)	Any object that is substitutable for `_Curve`, such as `Curve`.
`CompositeSurface`	`surfaceMember` (unbounded)	Any object that is substitutable for `_Surface`.
`CompositeSolid`	`solidMember` (unbounded)	Any object that is substitutable for `_Solid`.
`GeometricComplex`	`element` (unbounded)	Any object that is substitutable for `_GeometryPrimitive`.

Note that all of the lower-dimensional geometry primitives (`Surface`, `Curve` and `Point`) can be embedded in a three-dimensional space using a three-dimensional CRS to describe the coordinates. Chapter 15 provides an example of a three-dimensional CRS.

12.4 What are the new complex geometries in GML 3.0?

The `geometryComplexes.xsd` includes geometric composites for `Curve`, `Surface` and `Solid` geometries, plus a `GeometricComplex`. Table 2 lists the different composites and complexes, their properties and the values of these properties. Note that all of the complex geometries can have unbounded occurrences of their properties.

A `GeometricComplex` is a collection of geoprimitives, whose interiors are mutually disjoint (that is, they do not intersect). A `GeometricComplex` contains all of the boundaries for all of the geometry primitives included in the complex.

12.4.1 Encoding a composite curve

A composite curve is composed of two or more contiguous curves; that is, the end point of one curve coincides with the start point of the next curve. Note that different curve elements that are substitutable for `_Curve` can be used to define each composite curve.

> *Note:* Based on ISO/TC 211 19107, there is a distinction between a composite curve, which is composed of curves, and a curve, which itself may be built from curve segments.

Figure 12 shows a variation of the simple highway network shown in Figure 1. In this new highway network, the `LineString`, R1, is replaced by a `CompositeCurve` consisting of two `curveMembers`: a `CompositeCurve` and a `Curve` consisting of a single `LineStringSegment`.

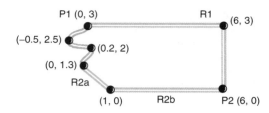

Figure 12 Highway network with a CompositeCurve.

The following encoding shows how to encode the whole network as a CompositeCurve (CC1) that contains the CompositeCurve R1 and the LineString R1:

```
<gml:CompositeCurve gml:id="CC1">
    <gml:curveMember>
        <gml:CompositeCurve gml:id="R2">
            <gml:curveMember>
                <gml:Curve gml:id="R2a">
                    <gml:segments>
                        <gml:CubicSpline>
                            <gml:coordinates>0,3 -0.5,2.5
                                ⌣0.2,2 0,1.3 1,0
                                ⌣</gml:coordinates>
                            <gml:vectorAtStart
                                ⌣srsName="..">0 -1
                                ⌣</gml:vectorAtStart>
                            <gml:vectorAtEnd
                                ⌣srsName="..">1 -1
                                ⌣</gml:vectorAtEnd>
                        </gml:CubicSpline>
                    </gml:segments>
                </gml:Curve>
            </gml:curveMember>
            <gml:curveMember>
                <gml:Curve gml:id="R2b">
                    <gml:segments>
                        <gml:LineStringSegment>
                            <gml:coordinates>1,0 6,0
                                ⌣</gml:coordinates>
                        </gml:LineStringSegment>
                    </gml:segments>
                </gml:Curve>
            </gml:curveMember>
        </gml:CompositeCurve>
    </gml:curveMember>
    <gml:curveMember>
        <gml:OrientableCurve orientation="-">
            <gml:baseCurve xlink:href="#R1"/>
        </gml:OrientableCurve>
    </gml:curveMember>
</gml:CompositeCurve>
```

As shown in the above example, the members of a CompositeCurve can also be CompositeCurve objects. The CompositeCurve CC1 represents the entire highway network, and its members are another CompositeCurve (R2) and an OrientableCurve (R1). The members of the

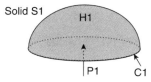

Figure 13 A Solid bounded by a CompositeSurface.

CompositeCurve R2 are two curves, one composed of a CubicSpline, and the other of a LineStringSegment. The CubicSpline begins at (0, 3) and ends at (1, 0). Note that the other curveMember of the Composite-Curve CC1, OrientableCurve, contains a baseCurve that references a LineString, which is defined outside of the CC1 instance.

12.4.2   Encoding a composite surface

A composite surface consists of two or more surface members, which are connected along their boundary curves. Figure 13 provides an example of a composite surface that can be encoded in GML. The exterior boundary surface of the solid S1 can be encoded as a composite surface whose surface members enclose S1 without overlaps or holes. In this figure, H1 represents the surface of the hemisphere, while P1 is a surface that represents a plane. C1 represents the Circle that forms the exterior of the plane P1. The surface H1 can be encoded in GML as follows:

```
<gml:Surface gml:id="H1">
  <gml:patches>
    <gml:Triangle>
      <gml:exterior>
        <gml:LinearRing>
          <gml:coordinates>1,0,0 0.99,0.14,0
            ⌐0.99,0,0.14 1,0,0</gml:coordinates>
        </gml:LinearRing>
      </gml:exterior>
    </gml:Triangle>
    <gml:Triangle>
    ...
    </gml:Triangle>
    <gml:Triangle>
    ...
    </gml:Triangle>
    ...
  </gml:patches>
</gml:Surface>
```

In this example, the hemisphere surface is essentially a geodesic dome composed of a collection of surface patches, each of which is a triangle. Figure 14 shows a portion of the triangle patches that make up the hemisphere surface. The triangle with darker shading represents the Triangle from the above encoding example.

Note that surface patches are only one of the possible approaches to encoding a hemisphere. For example, you could also model the hemisphere using a spherical surface interpolation, an example of which is provided in Section 12.7.

Figure 14 `Triangle` patches that approximate a hemisphere surface.

The plane P1 can be encoded in GML 3.0 as shown below. Note that in real world data instances, coordinates take up most of the 'volume' of the data.

```
<gml:Surface gml:id="P1">
    <gml:patches>
        <gml:PolygonPatch>
            <gml:exterior>
                <gml:LinearRing>
                    <gml:coordinates>
                        ⌣0.95,0.31,0 0.81,0.58,0 0.60,0.80,
                        ⌣0 0.36,0.93,0 0,1,0 -0.36,0.93,
                        ⌣0 -0.60,0.80,0 -0.81,0.58,0 -0.95,
                        ⌣0.31,0 -1,0,0 -0.95,-0.31,0 -0.81,
                        ⌣-0.58,0 -0.60,-0.80,0 -0.36,-0.93,
                        ⌣0 0,-1,0 0.36,-0.93,0 0.60,-0.80,
                        ⌣0 0.81,-0.58,0 0.95,-0.31,0 1,0,0
                        ⌣0.95,0.31,0</gml:coordinates>
                </gml:LinearRing>
            </gml:exterior>
        </gml:PolygonPatch>
    </gml:patches>
</gml:Surface>
```

The following example shows how these two surfaces can be encoded as part of a `CompositeSurface` that forms the exterior of the `Solid  S1` from Figure 13:

```
<gml:Solid gml:id="S1">
    <gml:exterior>
        <gml:CompositeSurface>
            <gml:surfaceMember>
                <gml:Surface gml:id="H1">
                    <gml:patches>
                        <gml:Triangle>
                            <gml:exterior>
                                <gml:LinearRing>
                                    <gml:coordinates>1,0,0
                                    ⌣0.95,0.31,0.14 0.95,
                                    ⌣0,0.14 1,0,0
                                    ⌣</gml:coordinates>
                                </gml:LinearRing>
                            </gml:exterior>
                        </gml:Triangle>
                        <gml:Triangle>
                        ...
                        </gml:Triangle>
                        ...
                    </gml:patches>
                </gml:Surface>
            </gml:surfaceMember>
            <gml:surfaceMember>
```

```
<gml:Surface gml:id="P1">
    <gml:patches>
        <gml:PolygonPatch>
            <gml:exterior>
                <gml:LinearRing>
                    <gml:coordinates>
                        ⌐0.95,0.31,0 0.81,0.58,
                        ⌐0 0.60,0.80,0 0.36,
                        ⌐0.93,0 0,1,0 -0.36,
                        ⌐0.93,0 -0.60, 0.80,0
                        ⌐-0.81,0.58,0 -0.95,
                        ⌐0.31,0 -1,0,0 -0.95,
                        ⌐-0.31,0 -0.81, -0.58,0
                        ⌐-0.60,-0.80,0 -0.36,
                        ⌐-0.93,0 0,-1,0 0.36,
                        ⌐-.93,0 0.60,-0.80,0
                        ⌐0.81,-0.58,0 0.95,
                        ⌐-0.31,0 1,0,0 0.95,0.31,0
                        ⌐</gml:coordinates>
                </gml:LinearRing>
            </gml:exterior>
        </gml:PolygonPatch>
    </gml:patches>
</gml:Surface>
            </gml:surfaceMember>
        </gml:CompositeSurface>
    </gml:exterior>
</gml:Solid>
```

The Solid has an exterior property, which in turn contains a CompositeSurface object with two surfaceMember properties, each of which contains a Surface object.

12.5   What are the geometry aggregates?

Table 3 lists the GML 3.0 geometry aggregates, their properties and the values of these properties. Unlike the composite geometries defined in geometryComplexes.xsd, the aggregate geometries do not need to be contiguous. For example, MultiCurve can be used to encode various disconnected rivers within a county, or MultiSurface to encode a series of lakes within a national park. The different aggregate geometries can also be used in coverages, and GML 3.0 already provides a MultiPointCoverage that uses the MultiPoint aggregate geometry. Chapter 17 and the *Worked Examples CD* contain examples of MultiPointCoverage encodings.

12.6   How do I express a feature's geometry?

Geometry-valued properties must be used to express a feature's geometry. GML 3.0 provides various convenience properties for linking a feature with geometry objects that express the geometric characteristics of the feature. As discussed in Chapter 9, GML provides predefined geometry properties for associating features with geometry objects. These properties are shown in Table 4.

Table 3   Geometry aggregates from `geometryAggregates.xsd`

Aggregate	Property	Value
MultiPoint	pointMember (unbounded)	A Point object.
	pointMembers	An array of Point objects.
MultiCurve	curveMember (unbounded)	Any object that is substitutable for _Curve, such as Curve.
	curveMembers	An array of objects that are substitutable for _Curve.
MultiSurface	surfaceMember (unbounded)	Any object that is substitutable for _Surface.
	surfaceMembers	An array of objects that are substitutable for _Surface.
MultiSolid	solidMember (unbounded)	Any object that is substitutable for _Solid.
	solidMembers	An array of objects that are substitutable for _Solid.
MultiGeometry	geometryMember (unbounded)	Any object that is substitutable for _Geometry.
	geometryMembers	An array of objects that are substitutable for _Geometry.

Table 4   Convenience properties from `feature.xsd`

Property	Value
boundedBy	An Envelope or a Null object. The Envelope can contain pos, posList or coordinates properties whose values describe the spatial boundaries of a feature. This property is mandatory for feature collections.
location	Any object that is substitutable for _Geometry, a location keyword or a location string. Chapter 9 contains a detailed discussion about the location property.
centerOf position	A geometry object that is substitutable for Point.
edgeOf centerLineOf	A geometry object that is substitutable for _Curve, such as LineString or Curve.
extentOf coverage	A geometry object that is substitutable for _Surface, such as Polygon or Surface.

The convenience geometry-valued properties are used to relate a feature to a geometry object, as shown in the following example, in which the `centerLineOf` property contains a `LineString` object:

```
<app:Bridge gml:id="..">
   ...
   <gml:centerLineOf>
      <gml:LineString gml:id="L1"  srsName="..">
         <gml:coordinates>...</gml:coordinates>
      </gml:LineString>
   </gml:centerLineOf>
</app:Bridge>
```

When features are defined in GML application schemas, they must derive from `AbstractFeatureType`, as discussed in Chapter 11. All features whose content models derive from this type inherit the `location` property, however, they do not automatically inherit the other convenience properties. The following schema fragment shows how a GML geometry property should be included in a feature's content model:

```
<element name="Bridge" type="BridgeType"
   ⌐substitutionGroup="_Feature">

<complexType name="BridgeType">
   <complexContent>
      <extension base="gml:AbstractFeatureType">
         <sequence>
            ...
            <element ref="gml:centerLineOf"/>
         </sequence>
      </extension>
   </complexContent>
<complexType>
```

Table 5 lists the additional convenience properties provided by `geometryAggregates.xsd`. In addition to these properties, another property, `solidProperty` – which is defined in `geometryPrimitives.xsd` – can be added to application schema definitions of features that need to have a property with a value that is substitutable for `_Solid`.

Table 5   Convenience properties from `geometryAggregates.xsd`

Property	Value
multiCenterOf multiPosition	An array of geometry objects that are substitutable for `Point`.
multiCenterLineOf multiEdgeOf	An array of geometry objects that are substitutable for `_Curve`, such as `LineString` or `Curve`.
multiExtentOf multiCoverage	An array of geometry objects that are substitutable for `_Surface`, such as `Polygon` or `Surface`.

> *Note:* Geometry properties describe the role of the geometry in relation to the referencing feature. It is recommended that users avoid meaningless properties, such as `pointProperty`.

12.6.1 Geometry complexes and feature collections

In many applications in which complex geographic objects are required, data modellers are usually faced with the following choice:

- Capture the object as a single feature but with a complex geometry (for example, the extent of the object is a kind of geometry complex such as a `CompositeSurface`), or
- Capture the object as a collection of features that are geometrically connected in the manner of the geometry complex (for example, each feature has in-line geometry content, such as a `Road` with a `centerLineOf` property that encapsulates a `LineString` object). In this case, no geometric complex is actually employed in the encoding.

In general there is no right or wrong answer. It depends very much on the application to which the data is being put. If the identities and properties of the individual features are paramount, then the second choice is preferred. On the other hand, if only the overall geometry of the single complex feature is of interest, the first choice is more appropriate.

It is also possible to take a middle-ground approach, in which the single feature is a feature collection with a geometry property whose value is a geometric complex, and the member features of the feature collection have geometry properties that point to specific geometric elements in the geometric complex that applies to the whole complex feature. This last approach is illustrated in Figure 15.

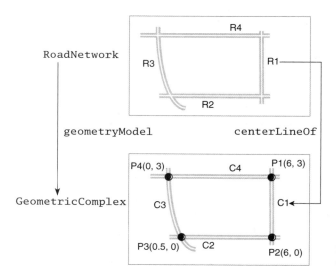

Figure 15 `RoadNetwork` feature collection with a `GeometricComplex`.

In this figure, a `RoadNetwork` feature collection contains four `Road` feature members (R1, R2, R3 and R4) and a user-defined `geometryModel` property whose value is a `GeometricComplex` object that contains the feature collection's underlying geometry model. The Road features can also have `centerLineOf` properties that point to the appropriate curves in the underlying geometry model, for example the Road R1 has a `centerLineOf` property that points to C1.

The `RoadNetwork` feature collection can be encoded as follows:

```
<app:RoadNetwork gml:id="RN1">
   <gml:boundedBy>
      ...
   </gml:boundedBy>
   <gml:featureMembers>
      <app:Road gml:id="R1">
         <gml:centerLineOf xlink:href="#C1"/>
      </app:Road>
      <app:Road gml:id="R2">
         <gml:centerLineOf xlink:href="#C2"/>
      </app:Road>
      <app:Road gml:id="R3">
         <gml:centerLineOf xlink:href="#C3"/>
      </app:Road>
      <app:Road gml:id="R4">
         <gml:centerLineOf xlink:href="#C4"/>
      </app:Road>
   </gml:featureMembers>
   <app:geometryModel>
      <gml:GeometricComplex>
         <gml:element>
            <gml:LineString gml:id="C1">
               <gml:coordinates">6,3 6,0
                  ⌣</gml:coordinates>
            </gml:LineString>
         </gml:element>
         <gml:element>
            <gml:LineString gml:id="C2">
               <gml:coordinates>6,0 0.5,0
                  ⌣</gml:coordinates>
            </gml:LineString>
         </gml:element>
         <gml:element>
            <gml:Curve gml:id="C3">
               <gml:segments>
                  <gml:CubicSpline>
                     ...
                  </gml:CubicSpline>
               </gml:segments>
            </gml:Curve>
         </gml:element>
         <gml:element>
            <gml:LineString gml:id="C4">
               ...
            </gml:LineString>
         </gml:element>
         <gml:element>
            <gml:Point gml:id="P1">
```

```
            . . .
        </gml:Point>
      </gml:element>
            . . .
      </gml:GeometricComplex>
    </app:geometryModel>
  </app:RoadNetwork>
```

Note that the GeometricComplex should also contain or reference (via the element property) all of the boundary primitives in the underlying model. For example, the geometry model for Figure 15 also consists of four Point objects. A more complete version of the above encoding is available in RoadNetwork.xml document on the *Worked Examples CD*.

Although GML does not provide a convenience property for encapsulating GeometricComplex objects, GeometricComplexPropertyType can be used to create user-defined properties that encapsulate GeometricComplex, as shown in the following schema fragment for a RoadNetwork feature collection:

```
<element name="RoadNetwork"
  ⌐type="app:RoadNetworkCollectionType"/>

<complexType name="RoadNetworkCollectionType">
  <complexContent>
    <extension base="gml:AbstractFeatureCollectionType">
      <sequence>
        <element name="geometryModel"
          ⌐type="app:GeometricComplexPropertyType"
          ⌐minOccurs="0"/>
      </sequence>
    </extension>
  </complexContent>
</complexType>
```

Note that a property of GeometricComplexPropertyType can contain a choice of one of the following complex geometry objects: GeometricComplex, CompositeCurve, CompositeSurface and CompositeSolid.

12.6.2   Creating a user-defined geometry-valued property

All new geometry-valued properties should follow the pattern used by GeometryPropertyType; that is, they should contain the AssociationAttributeGroup attribute. The following schema fragment shows an element declaration and type definition for a user-defined expanseOf element:

```
<element name="expanseOf"
  ⌐type="gml:SurfaceOrSolidPropertyType"/>

<complexType name="SurfaceOrSolidPropertyType ">
  <choice minOccurs="0">
    <element ref="gml:_Surface"/>
    <element ref="gml:_Solid"/>
  </choice>
  <attributeGroup ref="gml:AssociationAttributeGroup"/>
</complexType>
```

The value of this property can be an object that is substitutable for `_Surface` or `_Solid`. In a GML application schema, this property can added to the content model of features that contain either a solid or surface.

12.7   How do I add new geometry objects in an application schema?

If you need a geometry element that is not provided in GML 3.0, you can create it in an application schema. The geometry element's content model must derive, directly or indirectly, from `AbstractGeometryType`. Figure 16 shows two examples of a user-defined geometry called `SphericalCap` that has three-dimensional coordinates and four control points.

These two spherical caps can be encoded as follows:

```
<app:SphericalCap gml:id="SC1">
   <gml:coordinates>0,-2,0 2,0,0 0,2,0 0,0,2
      ↲</gml:coordinates>
</app:SphericalCap>

<app:SphericalCap gml:id="SC2">
   <gml:coordinates>0,-1,0 1,0,0 0,1,0 0,1,1
      ↲</gml:coordinates>
</app:SphericalCap>
```

The following schema fragment shows a possible definition of `SphericalCap`:

```
<element name="SphericalCap" type="app:SphericalCapType"
   ↲substitutionGroup="gml:_Surface"/>

<complexType name="SphericalCapType">
   <complexContent>
      <restriction base="gml:AbstractSurfaceType">
         <sequence>
            <choice>
               <choice minOccurs="4" maxOccurs="4">
                  <element ref="gml:pos"/>
                  <element ref="gml:pointRep"/>
               </choice>
               <element ref="gml:coordinates"/>
            </choice>
         </sequence>
         <attribute name="interpolation"
            ↲type="gml:SurfaceInterpolationType"
            ↲fixed="spherical">
      </restriction>
   </complexContent>
</complexType>
```

The `SphericalCap` can have four `pos` or `pointRep` properties, or a `posList` or `coordinates` property. The `interpolation` attribute specifies that the interpolation mechanism for this surface is 'spherical'. The interpolation method returns a subset of control points on a sphere. The first three

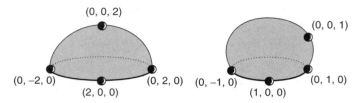

Figure 16 `SphericalCap`.

control points define the latitudinal circle that forms the boundary of the spherical cap. Circular interpolation is used for these three points.

The fourth point is the last control point needed to completely determine the surface of the spherical cap, and spherical interpolation through the four points is used to complete the spherical cap. Note that the first three points cannot be colinear; that is, they cannot lie on the same line. The fourth point cannot lie on the plane containing the other three points.

When creating user-defined geometry elements and types, always use the most specific abstract or concrete type from the GML Geometry schemas as the starting point for any user-defined geometry. For example, if you are creating a new type of curve, extend `AbstractCurveType` instead of `AbstractGeometryType`.

Note: You cannot expect general GML-aware software to be able to process custom geometry objects.

12.8   Chapter summary

The GML geometry model supports a wide range of approaches to modelling the geometric properties of geographic objects. There are four basic categories of geometric primitives: points, curves, surfaces and solids. GML 3.0 provides a number of different geometry types that derive from one of these four primitives. GML 2 supported some of these primitives; however, the GML 3.0 geometry model is far more comprehensive, and as a result comprises five separate schemas.

The coordinate values of all geometries are expressed, directly or indirectly, as coordinates via a `coordinates`, `posList` or `pos` property. Another property, `pointRep`, can be used in cases that require a reference to coordinate values that are defined elsewhere. The deprecated GML 2 property, `coord`, is also supported, but if you plan to use GML 3.0 to model the geometry of your geographic objects, it is recommended that you avoid using this property. In most cases, it is recommended that you use the `pos` or `posList` properties. The `coordinates` property can be used where string-valued coordinates are required.

Data modellers who prefer to use the linear geometry types to model their geographic objects can use the first two schemas – `geometryBasic0d1d.xsd` and `geometryBasic2d.xsd` – to model points, line strings and polygons. Note that these are all linear geometries that were provided in GML 2. These linear geometries can be used to model the geometry of various geographic objects, such as highway networks and land parcels.

The additional geometries in GML 3.0 provide a more descriptive encoding and the ability to model three-dimensional geometries. By using some of the new geometries, such as `OrientableCurve`, it is possible to include attributes to express the orientation of the geometries. This is not possible with the linear geometries from GML 2.

The remaining GML 3.0 Geometry schemas are `geometry-Primitives.xsd`, `geometryComplexes.xsd` and `geometryAggre-gates.xsd`. The `geometryPrimitives.xsd` schema defines various geometry primitives for modelling complex geometries. These primitives include `Curve`, `Surface` and `Solid`.

A `Curve` has a `segments` property that can contain one or more objects from the `_CurveSegment` substitution group, including `LineStringSegment`, `Arc`, `ArcString` or `CubicSpline`. Arcs are the fundamental building blocks for encoding circles in GML 3.0.

The `geometryComplexes.xsd` schema defines types that are used to encode collections of geometric primitives, including `CompositeCurve`, `CompositeSurface`, `CompositeSolid` and `GeometryComplex`. Each of these composite geometries can contain the geometry objects implied by their name. For example, via the `curveMember` property, a `CompositeCurve` can comprise one or more geometries that are substitutable for `_Curve`. The `GeometryComplex` has an `element` property that can encapsulate any geometry primitives that are substitutable for `_GeometryPrimitive`, such as `Curve` or `Surface`.

The `geometryAggregates.xsd` schema defines aggregate geometries that, unlike complex geometries, contain members that do not need to be connected. The aggregates are `MultiPoint`, `MultiCurve`, `MultiSurface`, `MultiSolid` and `MultiGeometry`.

Geometry properties are used to associate a feature (or other geographic object) with its geometric description. The `feature.xsd` schema defines the following simple geometry properties: `location`, `centerOf`, `position`, `edgeOf`, `centerLineOf`, `extentOf` and `coverage`. Additional properties – `multiCenterOf`, `multiPosition`, `multiCenterLineOf`, `multi-EdgeOf`, `multiExtentOf`, `multiExtentOf` and `multiCoverage` – are defined in `geometryAggregates.xsd`.

Aside from `location`, the remaining properties can only be used in GML instances if they are added to the content model of a feature in a GML application schema. GML 3.0 provides the same geometry properties as GML 2, but these are extended so that, for example, `centerLineOf` can be a curve and not simply a `LineString` as in GML 2. If you are modelling more complex geometry, it is also possible to organize the geometry model in a `Geometric-Complex` that can be contained within or referenced by a user-defined property of `GeometricComplexPropertyType`.

User-defined geometry types and properties can also be created in GML application schemas. All user-defined properties should follow the same pattern as `GeometryPropertyType` (from `geometryBasic0d1d.xsd`), in that they contain the `AssociationAttributeGroup` attribute. All user-defined geometry types must derive, directly or indirectly, from `AbstractGeometryType`. Always use the most specific abstract or concrete type from the GML Geometry schemas as the starting point for any new geometry added in an application schema.

References

http://www.cs.mtu.edu/~shene/COURSES/cs3621/ NOTES/notes.html (October 15, 2003).

International Organization for Standardization. (2000) ISO Technical Committee 211 Geographic Information/Geomatics. *ISO DIS 19107, Geographic Information – Spatial Schema.*

http://schemas.opengis.net/gml/3.0.1/base/ (October 15, 2003).

http://www.opengis.org/docs/02-023r4.pdf (October 15, 2003).

Additional references

MARSH, D. (1999) *Applied Geometry for Computer Graphics and CAD.* Springer-Verlag, London.

MOLENAAR, M. (1998) *An Introduction to the Theory of Spatial Object Modelling for GIS.* Taylor & Francis, London.

SCHNEIDER, P.J. and EBERLY, D.H. (2002) *Geometric Tools for Computer Graphics.* Morgan Kaufmann, San Francisco, CA.

http://mathworld.wolfram.com/ (October 13, 2003).

http://velab.cau.ac.kr/lecture/Bspline.ppt (October 13, 2003).

http://www.vrac.iastate.edu/~carolina/519/notes/ 13.bspline.pdf (October 13, 2003).

http://w3imagis.imag.fr/~Brian.Wyvill/course/notes/ bspline.pdf (October 13, 2003).

http://schemas.opengis.net/gml/3.0.1/base/geometry-Basic0d1d.xsd (October 15, 2003).

http://schemas.opengis.net/gml/3.0.1/base/geometry-Basic2d.xsd (October 15, 2003).

http://schemas.opengis.net/gml/3.0.1/base/geometry-Aggregates.xsd (October 15, 2003).

http://schemas.opengis.net/gml/3.0.1/base/geometry-Primitives.xsd (October 15, 2003).

http://schemas.opengis.net/gml/3.0.1/base/geometry-Complex.xsd (October 15, 2003).

Chapter 13

GML topology

Topology is a branch of pure mathematics that is concerned with the study of properties of objects, which are preserved through deformations, twistings and stretchings (that is, bicontinuous, one-to-one transformations). Topology has many branches including algebraic, combinatorial and point-set topology. Algebraic and combinatorial topology are the branches that are most applicable to describing geographic objects. For additional information about topological concepts in relation to geographic information, please refer to http://mathworld.wolfram.com/.

> *Note:* The discipline of algebraic topology is popularly known as 'rubbersheet geometry', and provides the mathematical machinery for studying different kinds of hole structures. The term 'algebraic' is used because many hole structures are represented best by algebraic objects such as groups and rings.

The GML 3 topology model is based on a subset of the topological types defined in *ISO TC 211/DIS 19107* (ISO, 2000) and comprises four topological primitives and a number of properties. These primitives and properties are also discussed in Clause 7.7 of the *GML Version 3.00 OpenGIS® Implementation Specification* (http://www.opengis.org/docs/02-023r4.pdf). The topology.xsd schema is available online at http://schemas.opengis.net/gml/3.0.1/base/topology.xsd.

Table 1 lists the different topological primitives in GML 3 and the corresponding geometry objects that realize them. Note that Chapter 12 covers all of the geometry objects listed in this table.

Consider the example of a simple highway network from Chapter 12 shown in Figure 1. The geometry objects represented in this figure are `Point` and `LineString`. In the GML topology model, the road intersection, whose position was modelled by the `Point P1`, can be topologically modelled as a `Node`, as shown below:

```
<gml:Node gml:id="n1"/>
```

Geography Mark-up Language (GML). R. Lake, D. S. Burggraf, M. Trninić, L. Rae © 2004 Galdos Systems Inc.
Published by John Wiley & Sons, Ltd   ISBNs: 0-470-87153-9 (HB); 0-470-87154-7 (PB)

Table 1   Topological primitives and their related geometries

Topological Primitive	Corresponds to...
Node	Point
Edge	Curve
Face	Surface
TopoSolid	Solid

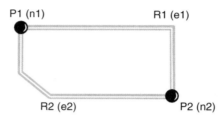

Figure 1 Simple highway network geometry.

The gml:id attribute identifies the Node as n1. A curve such as a Line-String can be topologically modelled as an Edge, as follows:

```
<gml:Edge gml:id="e1">
    <gml:directedNode orientation="-" xlink:href="#n1"/>
    <gml:directedNode orientation="+" xlink:href="#n2"/>
</gml:Edge>
```

In this example, the xlink:href attribute is used to reference nodes n1 and n2. The orientation attribute is used to assign a negative orientation (-) to signify that n1 is the start of e1, whereas a positive orientation (+) signifies that n2 is the end of e1.

Figure 2 shows the topology model that corresponds to the geometry in Figure 1. Topological primitives can be 'realized' by corresponding geometric primitives. For example, a Node is realized as a Point. This is covered in Section 13.3.

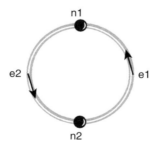

Figure 2 Topology model corresponding to Figure 1.

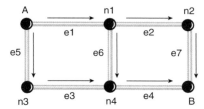

Figure 3 Road network topology.

13.1   How do I encode a simple network with GML topology?

Suppose that the road network of a city is stored in a geographic database and that you have access to this database via a Route Service that supports the following query:

'Which route from point A to point B has the fewest intersections along the way?'

The answer is naturally found by looking at the topology of the road network rather than the geometry. The topology model of the road network encodes intersections as nodes, road segments as edges and the connective relationships between the edges and nodes.

Figure 3 shows an example of a topology model for a road network. This example has six nodes (A, n1, n2, n3, n4 and B) and seven edges (e1, e2, e3, e4, e5, e6 and e7). The distances between nodes and the directions of the edges – which are normally captured by the coordinate values of the geometry of the road network – are completely ignored in the topology model. One simple way to answer the query about the optimal route from point A to B is to look at the possible paths from A to B along edges in the topology model and count the number of nodes traversed along each path.

Note that each Edge is directed by its start and end node and can be traversed in two ways: positively (forwards) or negatively (backwards). A route through the road network is determined by a sequence of directed edges. One of the possible routes from A to B can be expressed as the following sequence of directed edges {+e1, +e2, +e7}. The following example shows an encoding of all of the nodes and edges from Figure 3:

```
<gml:Node gml:id="A"/>
<gml:Node gml:id="n1"/>
<gml:Node gml:id="n2"/>
<gml:Node gml:id="n3"/>
<gml:Node gml:id="n4"/>
<gml:Node gml:id="B"/>
<gml:Edge gml:id="e1">
  <gml:directedNode orientation="-" xlink:href="#A"/>
  <gml:directedNode orientation="+" xlink:href="#n1"/>
</gml:Edge>
<gml:Edge gml:id="e2">
  <gml:directedNode orientation="-" xlink:href="#n1"/>
  <gml:directedNode orientation="+" xlink:href="#n2"/>
</gml:Edge>
<gml:Edge gml:id="e3">
  <gml:directedNode orientation="-" xlink:href="#n3"/>
  <gml:directedNode orientation="+" xlink:href="#n4"/>
```

```
  </gml:Edge>
  <gml:Edge gml:id="e4">
    <gml:directedNode orientation="-" xlink:href="#n4"/>
    <gml:directedNode orientation="+" xlink:href="#B"/>
  </gml:Edge>
  <gml:Edge gml:id="e5">
    <gml:directedNode orientation="-" xlink:href="#A"/>
    <gml:directedNode orientation="+" xlink:href="#n3"/>
  </gml:Edge>
  <gml:Edge gml:id="e6">
    <gml:directedNode orientation="-" xlink:href="#n1"/>
    <gml:directedNode orientation="+" xlink:href="#n4"/>
  </gml:Edge>
  <gml:Edge gml:id="e7">
    <gml:directedNode orientation="-" xlink:href="#n2"/>
    <gml:directedNode orientation="+" xlink:href="#B"/>
  </gml:Edge>
```

Note that all of the edges contain `directedNode` properties that have values expressed by the `orientation` and `xlink:href` attributes. The `xlink:href` attribute points to the Node, which is a value for the `directedNode` property.

13.2 How do I encode parcel networks?

Topological characteristics – such as the adjacency of one parcel to another – are often used for modelling land parcel networks. A simple two-dimensional topology model for a land parcel network can be encoded using faces, edges and nodes, as shown in Figure 4.

The two land parcels are represented topologically by the faces `f1` and `f2`. In this topology model, each face is assigned a counterclockwise orientation and has a boundary that consists of a list of directed edges. The orientation of each face induces a relative orientation on each face's boundary edges. The relative orientation is positive if the orientation of the face agrees with the inherent direction of the edge; it is negative if it disagrees.

> *Note:* The assignment of the orientation to the faces is completely arbitrary and can be clockwise or counterclockwise.

For example, the boundary of the Face labelled `f1`, when traversed counterclockwise, corresponds to the directed edges in the set {-e6, -e1, +e5, +e3} and is encoded in GML 3 as follows:

```
<gml:Face gml:id="f1">
  <gml:directedEdge orientation="-" xlink:href="#e6"/>
  <gml:directedEdge orientation="-" xlink:href="#e1"/>
  <gml:directedEdge orientation="+" xlink:href="#e5"/>
  <gml:directedEdge orientation="+" xlink:href="#e3"/>
</gml:Face>
```

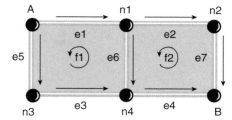

Figure 4 Land parcel network topology.

Note that the Face contains four directedEdge properties whose values are edges, which are referenced remotely with the xlink:href attribute. Each directedEdge property also has an orientation attribute. An Edge is inherently directed by its start and end node, and can be traversed in two ways: positively or negatively. For example, the directed edge -e4 corresponds to traversing the path from B to n4, and the directed edge +e6 traverses the path from n1 to n4.

In Figure 4, e6 is the only edge that has a face on either side of it, that is, both faces f1 and f2 contain the edge e6 – with either a positive or a negative orientation – in their boundary lists. In this case, the faces f1 and f2 are said to be in the coboundary of e6, each with the same orientation that the directed edge e6 has in the boundary lists of f1 and f2. In GML 3.0, the Edge primitive has an optional property called directedFace, whose value is a Face that is in the coboundary of the Edge. For example, the Edge e6 can be encoded as

```
<gml:Edge gml:id="e6">
   <gml:directedNode orientation="-" xlink:href="#n1"/>
   <gml:directedNode orientation="+" xlink:href="#n4"/>
   <gml:directedFace orientation="-" xlink:href="#f1"/>
   <gml:directedFace orientation="+" xlink:href="#f2"/>
</gml:Edge>
```

Similarly, each Node can be encoded with a coboundary list of directed edges to represent the edges that are incident upon the Node. For example, the Node n4 from Figure 4 can be encoded as

```
<gml:Node gml:id="n4">
   <gml:directedEdge orientation="-" xlink:href="#e4"/>
   <gml:directedEdge orientation="+" xlink:href="#e6"/>
   <gml:directedEdge orientation="+" xlink:href="#e3"/>
</gml:Node>
```

Note that for the directed edges that form the coboundary of a node, the assignment of orientation does not follow the same pattern as for edges that form the boundary of a face. That is, the orientation of an edge in relation to a node is based on whether or not it is pointing towards or away from the node. For example, +e6 in relation to n4 means that e6 is pointing towards n4, while +e6 in relation to the Face f2 means that the direction of e6 agrees with the counterclockwise orientation of f2. The cobounding edges of the Node n4 are shown in Figure 5.

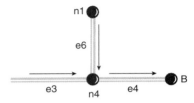

Figure 5 Coboundary of a Node.

Table 2   Topological primitives and their geometric realizations

Topological Primitive	Geometry-Valued Property	Geometry Value
Node	pointProperty	A Point
Edge	curveProperty	An object that is substitutable for _Curve
Face	surfaceProperty	An object that is substitutable for _Surface

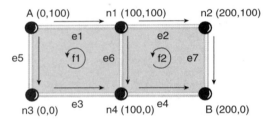

Figure 6 Land parcel network topology with geometry.

13.3   How is topology connected to geometry?

GML supports independent geometric and topological descriptions of geographic features. A geographic feature can be described in GML by its geometry-valued properties and, separately, by its topology-valued properties. These descriptions may or may not be connected. A geographic feature may have a topological description, a geometric (coordinate) description or both.

To relate the topology and geometry model, you explicitly map each topology primitive in the model to a corresponding geometry primitive. The corresponding geometry primitives are referred to as geometric realizations, and they are expressed by using the properties listed in Table 2.

If you assume that the land parcels from Figure 4 lie in a coordinate plane, then coordinates can be assigned to each node, as shown in Figure 6. For this example, the geometric realization of Node A can be encoded in GML as follows:

```
<gml:Node gml:id="A">
   <gml:pointProperty>
      <gml:Point>
```

```
        <gml:pos>0 100</gml:pos>
    </gml:Point>
  </gml:pointProperty>
</gml:Node>
```

Note that the `Point` represents the geometric realization of Node A, and the optional coboundary encoding is not included in the above example.

The edges in Figure 6 are designated as e1, e2 and so on. For example, the Edge `e1` starts at Node A and ends at Node n1, as shown in the following instance of an Edge that was discussed in Section 13.2.

```
<gml:Edge gml:id="e6">
    <gml:directedNode orientation="-"  xlink:href="#n1"/>
    <gml:directedNode orientation="+"  xlink:href="#n4"/>
    ...
</gml:Edge>
```

You can use any curve, such as `LineString`, from `geometryBasic0d1d.xsd` to realize these edges, as shown in the following GML instance of Edge e1:

```
<gml:Edge gml:id="e1">
    <gml:directedNode orientation="-"  xlink:href="#A"/>
    <gml:directedNode orientation="+"  xlink:href="#n1"/>
    <gml:curveProperty>
        <gml:LineString>
            <gml:coordinates>0,100 100,100</gml:coordinates>
        </gml:LineString>
    </gml:curveProperty>
</gml:Edge>
```

The Face `f1` can be realized by any surface, such as a `Polygon` object, as shown in the following GML encoding of `f1`:

```
<gml:Face gml:id="f1">
    <gml:directedEdge orientation="-"  xlink:href="#e6"/>
    <gml:directedEdge orientation="-"  xlink:href="#e1"/>
    <gml:directedEdge orientation="+"  xlink:href="#e5"/>
    <gml:directedEdge orientation="+"  xlink:href="#e3"/>
    <gml:surfaceProperty>
        <gml:Polygon>
            <gml:exterior>
                <gml:LinearRing>
                    <gml:coordinates>0,0 100,0 100,100 0,100
                       ⤶0,0</gml:coordinates>
                </gml:LinearRing>
            </gml:exterior>
        </gml:Polygon>
    </gml:surfaceProperty>
</gml:Face>
```

13.4   How do I encode faces with interior boundary rings?

Figure 7 shows a more complex use of two-dimensional topology. The Face f1 has an exterior boundary Edge and two interior boundary rings (expressed as a set of directed edges) that intersect at the common Node n2. There is also a dangling Edge, e4, and an isolated Node, n3, in the interior of the face.

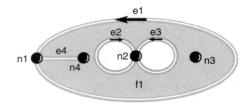

Figure 7 A more complex boundary
configuration of a Face.

The Face f1 can be encoded in GML 3.0 as follows:

```
<gml:Face gml:id="f1">
    <gml:isolated>
        <gml:Node gml:id="n3"/>
    </gml:isolated>
    <gml:directedEdge orientation="+">
        <gml:Edge gml:id="e1">
            <gml:directedNode orientation="-">
                <gml:Node gml:id="n1"/>
            </gml:directedNode>
            <gml:directedNode orientation="+"
                ⌣xlink:href="#n1"/>
        </gml:Edge>
    </gml:directedEdge>
    <gml:directedEdge orientation="+">
        <gml:Edge gml:id="e2">
            <gml:directedNode orientation="-">
                <gml:Node gml:id="n2"/>
            </gml:directedNode>
            <gml:directedNode orientation="+"
                ⌣xlink:href="#n2"/>
        </gml:Edge>
    </gml:directedEdge>
    <gml:directedEdge orientation="-">
        <gml:Edge gml:id="e3">
            <gml:directedNode orientation="-"
                ⌣xlink:href="#n2"/>
            <gml:directedNode orientation="+"
                ⌣xlink:href="#n2"/>
        </gml:Edge>
    </gml:directedEdge>
    <gml:directedEdge orientation="-">
        <gml:Edge gml:id="e4">
            <gml:directedNode orientation="-">
                <gml:Node gml:id="n4"/>
            </gml:directedNode>
            <gml:directedNode orientation="+"
                ⌣xlink:href="#n1"/>
        </gml:Edge>
    </gml:directedEdge>
    <gml:directedEdge orientation="+" xlink:href="#e4"/>
</gml:Face>
```

The isolated property is used to encode the isolated Node n3 that is not
bounded by any of the edges in Figure 7. Note also that, unlike geometry, exterior and interior properties are not required for encoding the boundaries

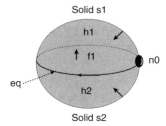

Solid s1

Solid s2

Figure 8 Two TopoSolids in the coboundary of a face.

of topological objects. In the above example, although the edges e2 and e3 are both interior boundary rings for the Face f1, the interior relationship is not explicitly encoded in GML.

Note also that to ensure that a GML instance document with topology content is valid, an even number of directed edges must be incident to each node that is incident to a boundary edge of a face. With the above encoding, you can verify this inherent rule: n1 is incident to the directed edges in the list {+e1, -e1}, and n2 is incident to the directed edges in the list {+e2, -e2, +e3, -e3}.

13.5   How do I encode a three-dimensional solid in topology?

Faces can also be encoded with a coboundary list of directed TopoSolids. For example, consider the Face f1 that represents the equatorial plane of the solid ball in Figure 8. The ball is composed of two topo solids: s1 and s2. The boundary of s1 consists of two faces, represented by upper hemisphere, h1, and the equatorial plane, f1. The solid, s2, also has two faces in its boundary, f1 (with the opposite orientation as with s1) and h2, represented by the lower hemisphere.

The Face f1 can be encoded as follows:

```
<gml:Face gml:id="f1">
   <gml:directedEdge orientation="-">
      <gml:Edge gml:id="eq">
         <gml:directedNode orientation="-">
            <gml:Node gml:id="n0"/>
         </gml:directedNode>
         <gml:directedNode orientation="+"
            ⌣xlink:href="#n0"/>
      </gml:Edge>
   </gml:directedEdge>
   <gml:directedTopoSolid orientation="+"
      ⌣xlink:href="#s1"/>
   <gml:directedTopoSolid orientation="-"
      ⌣xlink:href="#s2"/>
</gml:Face>
```

Note that the Face f1 has two directedTopoSolid properties that reference the s1 and s2 solids defined elsewhere. The solids s1 and s2 can be encoded as follows:

```
<gml:TopoSolid gml:id="s1">
   <gml:directedFace orientation="+" xlink:href="#h1"/>
   <gml:directedFace orientation="+" xlink:href="#f1"/>
```

```
    </gml:TopoSolid>
    <gml:TopoSolid gml:id="s2">
        <gml:directedFace orientation="+" xlink:href="#h2"/>
        <gml:directedFace orientation="-" xlink:href="#f1"/>
    </gml:TopoSolid>
```

Note that s1 has two directedFace properties with positive orientation, while s2 has one with negative orientation. With a GML 3.0 TopoSolid, the normal vector of each boundary face points into the solid (note that the normal vector points outward in GML 3.1). In the above example, h1 and h2 have positive orientation because their normal vectors point into the solids. The Face f1 has positive orientation in the boundary of s1, because it points upwards into s1, but it has negative orientation as a boundary of s2, because it points outward, away from s2.

13.6   How do I encode the topological properties of a feature?

In GML 3.0, topological relationships with a feature can be encoded in either of the following two ways:

- Via a topology-primitive-valued property – such as one of the topology-valued properties (for example, directedNode) or a user-defined property.
- As part of a topology-aggregate- or topology-composite-valued property.

The location of a BusStop feature, for example, can be encoded topologically, using a user-defined property, as shown in the following:

```
<app:BusStop gml:id="stopA">
    <app:position xlink:href="#A"/>
</app:BusStop>
```

The xlink:href attribute is used to associate the BusStop with the topology primitive Node A, which is defined elsewhere in the same instance document. This approach is useful when a feature's topology is quite simple, however, aggregate topologies should be used for more complex topology models. Table 3 lists the aggregate and composite topology objects and their constituent topology members.

Note that TopoPoint is an aggregate object and TopoCurve, Topo-Surface and TopoVolume are composite objects because the values of their constituent members must be connected. For example, a TopoCurve must have directedEdge properties whose values are edges that are connected. For the first four objects in this table, the properties have an unbounded maximum occurrence. The TopoPoint, for example, can have an unlimited number of directedNode properties. The properties of the TopoComplex object are covered in Section 13.8.

GML 3 provides the following built-in properties that encapsulate the topology objects listed in Table 3:

- topoPointProperty
- topoCurveProperty

Table 3   Topology aggregate and composite objects

Object	Property	Value
TopoPoint	directedNode	Node
TopoCurve	directedEdge	Edge
TopoSurface	directedFace	Face
TopoVolume	directedTopoSolid	TopoSolid
TopoComplex	maximalComplex	TopoComplex
	superComplex	TopoComplex
	subComplex	TopoComplex
	topoPrimitiveMember	_TopoPrimitive
	topoPrimitiveMembers	An unbounded list of _TopoPrimitive objects

- topoSurfaceProperty
- topoVolumeProperty
- topoComplexProperty.

In the following example, a BusRoute feature has a topoCurveProperty whose value is a TopoCurve, which represents the path of the bus route:

```
<app:BusRoute gml:id="rt44">
    <gml:topoCurveProperty>
        <gml:TopoCurve>
            <gml:directedEdge orientation="+"
                ⌐xlink:href="#e5"/>
            <gml:directedEdge orientation="+"
                ⌐xlink:href="#e3"/>
            <gml:directedEdge orientation="-"
                ⌐xlink:href="#e6"/>
            <gml:directedEdge orientation="+"
                ⌐xlink:href="#e2"/>
            <gml:directedEdge orientation="+"
                ⌐xlink:href="#e7"/>
        </gml:TopoCurve>
    </gml:topoCurveProperty>
</app:BusRoute>
```

There are circumstances in which data modellers will want to create user-defined topology-valued properties to connect features to an underlying topology model. This is covered in Section 13.9.

13.7   How do I relate the topology of a feature collection to the topology of its members?

The members of a feature collection can use remote properties to reference the feature collection's topology model. For example, Figure 9 shows a Transit

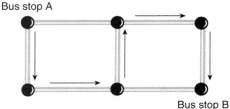

Figure 9 `Transit` feature collection with bus stop and bus route members.

feature collection with an underlying topology model. A bus stop is associated with the nodes A and B, and a bus route is associated with a path from A to B.

The following example shows how to attach the underlying topology model to the `Transit` feature collection:

```
<app:Transit gml:id="..">
   <gml:topoComplexProperty>
      <gml:TopoComplex>
         <gml:topoPrimitiveMembers>
            <gml:Node gml:id="A"/>
            ...
            <gml:Node gml:id="B"/>
            ...
         </gml:topoPrimitiveMembers>
      </gml:TopoComplex>
   </gml:topoComplexProperty>
   ...
</app:Transit>
```

Note that the `topoComplexProperty` associates the `Transit` feature collection with a `TopoComplex`, which represents the topology model. In addition to containing a topology model, the `Transit` feature collection also contains feature members that are related to the feature collection's underlying topology model. For example, one of the feature members could be a `BusStop` feature, as shown in the following:

```
<gml:featureMember>
   <app:BusStop gml:id="stopA">
      <app:position xlink:href="#A"/>
   </app:BusStop>
</gml:featureMember>
```

The `xlink:href` attribute is used to associate the `BusStop` with `Node` A from the feature collection's topology model, which is defined in the topology complex discussed above. Note that the `BusStop` feature is only one of the members of the `Transit` feature collection. The *Worked Examples CD* contains the complete version of the `Transit` instance document. For more information about associations between features and other objects, see the discussion in Chapter 12 regarding the association of features and geometry objects.

13.8   How is a topology complex encoded?

Topology complexes form the foundation of a feature collection's underlying topology model. The topological model of a road segment as a `directedEdge` is only meaningful if the underlying `Edge` is an element in a larger complex of edges and nodes. In other words, it is not sufficient to simply associate a feature member with a series of `Edge` objects without providing their context within the topology model of the feature collection.

A `TopoComplex` can be used to encode the underlying topology model of a feature collection, such as a land subdivision. The land subdivision's `TopoComplex` can then contain a hierarchy of subcomplexes, each of which represents a different subset of the underlying topology model.

The definition of a topology complex in the GIS world is somewhat vague; however, a cellular complex such as a CW-complex can be used in most cases. For more information about CW-complexes, please see 'Combinatorial Homotopy I' by J.H.C. Whitehead (*Bull. Amer. Math. Soc.*, 1949). To visualize a CW-complex, consider the following example:

1. Start with a discrete collection of nodes (or 0-cells) called $X0$.
2. Attach one-dimensional edges (or 1-cells) so that each end of every 1-cell is incident to a 0-cell. Let X1 be the collection of all 1-cells and 0-cells.
3. Attach two-dimensional faces (or 2-cells) to the 1-cells along their boundaries and write X2 for the new space consisting of 2-cells, 1-cells and 0-cells.
4. Attach three-dimensional topological solids (or 3-cells) to the 2-cells along their boundaries and let X3 be the space that comprises all 3-cells, 2-cells, 1-cells and 0-cells.

A CW-complex is any space that has this sort of decomposition into subspaces built up in a hierarchical fashion. This process is illustrated in Figure 10. Note that in Figure 10, the `Face f6` is a hemispherical surface and the topological `Solid s1` is represented by a hemispherical solid. The `Solid s1` is bounded by `f6` and `f5` represented by a planar disk. Note further that the CW-complex has a hierarchical structure with cells in dimensions 0 through 3; that is, the boundary of each cell is also contained in the complex.

In GML, the cell complex is an aggregation of cells of different dimensions. The different spaces shown in Figure 10 can be encoded as a graded sequence of topological complexes, each of which represents one of the diagrams shown in the figure.

The zero-dimensional skeleton (the first diagram in Figure 10) of the cell complex X3 can be encoded as `TopoComplex X0`, as shown in the following:

```
<gml:TopoComplex gml:id="X0">
    <gml:maximalComplex xlink:href="#X3"/>
    <gml:superComplex xlink:href="#X1"/>
    <gml:superComplex xlink:href="#X2"/>
    <gml:topoPrimitiveMembers>
        <gml:Node gml:id="n1"/>
        <gml:Node gml:id="n2"/>
        <gml:Node gml:id="n3"/>
        <gml:Node gml:id="n4"/>
        <gml:Node gml:id="n5"/>
        <gml:Node gml:id="n6"/>
        <gml:Node gml:id="n7"/>
```

```
        <gml:Node gml:id="n8"/>
        <gml:Node gml:id="n9"/>
    </gml:topoPrimitiveMembers>
</gml:TopoComplex>
```

Note: A topology complex must have at least one node.

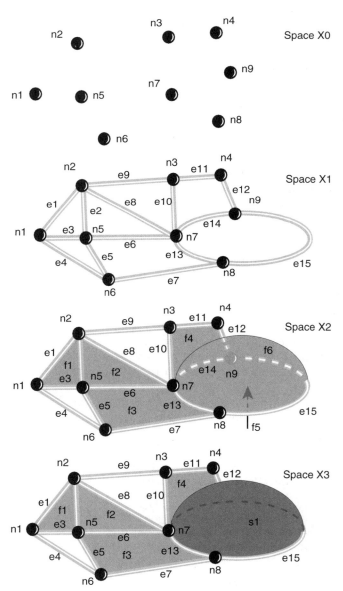

Figure 10 Construction of a CW-complex.

The set of nodes and edges from the second diagram in Figure 10 can form a one-dimensional skeleton of X3 encoded as `TopoComplex X1`. The subcomplex X1 can be encoded in GML as follows:

```
<gml:TopoComplex gml:id="X1">
   <gml:maximalComplex xlink:href="#X3"/>
   <gml:superComplex xlink:href="#X2"/>
   <gml:subComplex xlink:href="#X0"/>
   <gml:topoPrimitiveMember xlink:href="n1"/>
   <gml:topoPrimitiveMember xlink:href="n2"/>
   <gml:topoPrimitiveMember xlink:href="n3"/>
   ...
   <gml:topoPrimitiveMember xlink:href="n9"/>
   <gml:topoPrimitiveMembers>
      <gml:Edge gml:id="e1">
         <gml:directedNode orientation="-"
            ⌣xlink:href="#n1"/>
         <gml:directedNode orientation="+"
            ⌣xlink:href="n2"/>
      </gml:Edge>
      <gml:Edge gml:id="e2">...</gml:Edge>
      <gml:Edge gml:id="e3">...</gml:Edge>
      ...
      <gml:Edge gml:id="e15">...</gml:Edge>
   </gml:topoPrimitiveMembers>
</gml:TopoComplex>
```

As shown in the above encoding examples, the `maximalComplex` of X1 and X0 is X3. Both the X0 and X1 complexes also have `superComplex` properties that reference the `TopoComplex` X2, which corresponds to the third diagram in Figure 10. Although an encoding of this `TopoComplex` is not provided here, it should follow the same pattern as X0 and X1, and both X0 and X1 should be encoded as subcomplexes.

The `TopoComplex` X3 – which corresponds to the fourth diagram in Figure 10 – can be encoded as follows:

```
<gml:TopoComplex gml:id="X3" isMaximal="true">
   <gml:maximalComplex xlink:href="#X3"/>
   <gml:topoPrimitiveMember xlink:href="#n1"/>
   <gml:topoPrimitiveMember xlink:href="#n2"/>
   ...
   <gml:topoPrimitiveMember xlink:href="#n9"/>
   <gml:topoPrimitiveMember xlink:href="#e1"/>
   <gml:topoPrimitiveMember xlink:href="#e2"/>
   ...
   <gml:topoPrimitiveMember xlink:href="#e15"/>
   <gml:topoPrimitiveMember xlink:href="#f1"/>
   <gml:topoPrimitiveMember xlink:href="#f2"/>
   ...
   <gml:topoPrimitiveMember xlink:href="#f5"/>
   <gml:topoPrimitiveMember xlink:href="#f6"/>
   <gml:topoPrimitiveMembers>
      <gml:TopoSolid gml:id="s1">
         <gml:directedFace orientation="+"
            ⌣xlink:href="#f5"/>
         <gml:directedEdge orientation="+"
            ⌣xlink:href="#f6"/>
```

```
        </gml:TopoSolid>
      </gml:topoPrimitiveMembers>
  </gml:TopoComplex>
```

The TopoComplex X3 has a set of topoPrimitiveMember properties, each of which references one of the topological primitives contained within the X3 complex. Although it is possible to encode each primitive in-line, in this example, the topoPrimitiveMember properties point to instances that are encoded in subcomplexes in the same document. The TopoComplex also has a topoComplexMembers property with in-line content, in this case the s1 TopoSolid with two directedFace properties.

The isMaximal attribute indicates that the TopoComplex X3 is a maximal complex, which is a topological complex that is not contained by any other topological complex. The maximal complex is the top-level topological complex in a hierarchy of topological complexes, and the hierarchy can only have one maximal complex. Note that the TopoComplex X3 also has a maximalComplex property that points to itself as the maximal complex. This property is required for all TopoComplex objects, including maximal complexes. Note that it is also possible to include subcomplex properties for all of the topological complexes that are subcomplexes of the maximal complex.

13.9   How do I add new topology properties to an application schema?

Application developers are not restricted to using predefined topology properties. They can add topology properties in application schemas. For example, a topoModelProperty property – which has the type TopoComplexMemberType – can be added to a Transit schema, as shown below:

```
<element name="Transit" type="app:TransitType"
  ⌐substitutionGroup="_FeatureCollection"/>

<complexType name="TransitType">
  <complexContent>
    <extension base="gml:AbstractFeatureCollectionType">
      <sequence>
        <element name="topoModelProperty"
            ⌐type="gml:TopoComplexMemberType"/>
      </sequence>
    </extension>
  </complexContent>
</complexType>
```

Note that a complex type definition is not required for the topoComplexProperty because TopoComplexMemberType is defined in GML. Consider the following schema definition in a Transit feature collection with a user-defined topoSurfaceAndVolumeProperty property:

```
<element name="Transit" type="app:TransitType"
  ⌐substitutionGroup="_FeatureCollection"/>

<complexType name="TransitType">
  <complexContent>
    <extension base="gml:AbstractFeatureCollectionType">
```

```
        <sequence>
          <element
             ⌐ref="app:topoSurfaceAndVolumeProperty"/>
        </sequence>
      </extension>
    </complexContent>
  </complexType>
```

The `topoSurfaceAndVolumeProperty` property can be defined as follows:

```
<element name="topoSurfaceAndVolumeProperty"
   ⌐type="app:TopoFaceAndVolumeMemberType"/>

<complexType name="TopoSurfaceAndVolumeMemberType">
  <choice minOccurs="0">
    <element ref="gml:TopoSurface"/>
    <element ref="gml:TopoVolume"/>
  </choice>
  <attributeGroup ref="gml:AssociationAttributeGroup"/>
</complexType>
```

This property can contain a `TopoSurface` and a `TopoVolume` aggregate topology object and follows the standard pattern for defining properties by including `AssociationAttributeGroup`. Note that you will probably not need to create new topology primitives, but if it is required, the new content model should ultimately derive from `AbstractTopologyType`.

In general, the following should be considered a best practice for creating new topology properties and objects:

1. Create a new GML application schema element and type only as a last resort when existing elements and types do not meet your requirements.
2. Derive the content model from the narrowest type possible. For example, if you want to create a special kind of Node, derive the new Node content model from `NodeType` rather than from `AbstractTopologyType`.

13.10 What about arbitrary relationships, which may or may not be topological?

There are certain relationships that cannot be modelled with GML 3 Topology. For example, Figure 11 shows a courtyard as a region that can be perceived to be 'inside' a hedge boundary. Topologically speaking, the courtyard is neither inside nor outside the hedge because the hedge does not form a closed loop. As a result, you cannot use GML 3 Topology to model the containment of the courtyard.

The perception that the courtyard is 'inside' of the hedge, however, can be modelled geometrically in various ways. If the curve that describes the hedge is assigned coordinates, then it essentially becomes 'fixed' in some coordinate system – unlike topological edges and nodes, which are allowed to be continuously deformed. Once the hedge is fixed in this way – which is the way it tends to be perceived – it is possible to define the 'inside' of the hedge. You can do this by specifying the following: for any two points on the hedge, all points on the straight-line segment that joins the two points are 'inside' the hedge. In

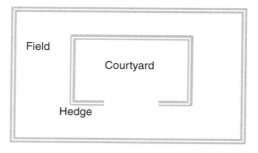

Figure 11 Courtyard with Hedge boundary.

mathematics, this definition of the 'inside' of the hedge is known as the 'convex hull' of the hedge.

A still more qualitative relationship can simply be asserted in GML by using a user-defined relationship property. This can be written as

```
<app:CourtYard gml:id="f12">
    <gml:name>La Notre</gml:name>
    <app:boundedBy xlink:href="#hedge1"/>
</app:CourtYard>

<app:Hedge gml:id="hedge1">
    <gml:name>Strange Cedar Hedge</gml:name>
    <gml:centerLineOf>
        <gml:LineString srsName="#myrefsys">
            <gml:coordinates>40,40 40,60 60,60 60,
                40</gml:coordinates>
        </gml:LineString>
    </gml:centerLineOf>
    <app:bounds xlink:href="f12"/>
</app:Hedge>
```

In this case, the application schema modeller is responsible for defining the semantics of the boundedBy and bounds properties.

13.11   Chapter summary

As opposed to the geometry model – which requires the encoding of coordinates – GML Topology provides a model for encoding the structural relationships of abstract objects. The GML 3 Topology model is based on a subset of *ISO 19107*. Although future versions of GML will likely contain more of the topology types defined in the ISO specification, the current model was designed to include the topology types that are most commonly used to model geographic features. The GML 3.0 Topology model is also covered in Clause 7.7 of *GML Version 3.00 OpenGIS® Implementation Specification* (http://www.opengis.org/docs/02-023r4.pdf).

GML 3.0 provides the following four topological primitives: Node, Edge, Face and TopoSolid, each of which is realized by different geometry objects. That is, a Node is realized by a Point object; an Edge by any object that is substitutable for _Curve; a Face by a Surface; and TopoSolid by a Solid.

Nodes and edges can be used to encode the connectivity of different physical networks, such as a road network in which the nodes represent intersections, and the edges represent road segments. Each Node and Edge has two possible orientations, negative and positive. For example, an edge with two nodes has two directedNode properties, each of which can either contain or reference a node with an orientation attribute. The node with negative (–) orientation represents the start point of the edge, while the node with positive orientation represents the endpoint. The inherent direction of an edge is from the start Node to the end Node.

A Face can have a set of directedEdge properties, each of which contains or references an Edge with a specific orientation. In other words, each Face has a boundary that consists of a list of directed edges, and each boundary is typically navigated in a counterclockwise direction.

Geometric realizations can be used to encode the geometric objects that realize topology objects. For example, to realize the geometry of a face, the surfaceProperty can be used to encapsulate or reference a Surface object, such as Polygon.

In GML 3, the TopoSolid primitive can be used to encode three-dimensional solids. The boundary of a TopoSolid contains a set of directedFace properties. Because the coboundary of a TopoSolid is empty, it does not require representation.

Properties can be used to describe the topological aspects of a feature. For example, a BusStop feature can have a position property that topologically models the 'location' of the feature. If the underlying feature model is more complex, however, aggregate or composite topology objects can also be used. GML 3 provides a topology aggregate (TopoPoint) and composites (TopoCurve, TopoSurface and TopoVolume), each of which can contain unbounded lists of one of the directed topology properties. For example, TopoPoint can contain a list of directedNode properties.

The TopoComplex aggregate can be used to contain all of the different topology primitives. Although built-in topology-valued properties, such as topoPointProperty, can be used to associate features with aggregate and composite topology objects, it is also possible to create user-defined properties that can contain one or more of the aggregates or composites.

A feature collection can comprise a series of features and a topology model that can be referenced from the different features within the collection. The xlink:href attribute can be used to reference any topology object that is included within a feature collection's topology model, represented by a topology complex.

Although GML 3 defines a number of convenience topology properties for associating features with a topology model, these properties are not part of the definition of the abstract feature type. Therefore, if you want a feature to have topological properties, they must be added to the feature's content model in a GML application schema. You can also create user-defined properties to connect a feature to an underlying topology model. If these properties have a user-defined complex type, the content model should follow the standard pattern for defining properties in GML application schemas. That is, they should include the AssociationAttributeGroup.

GML 3 Topology cannot be used to model all relationships. For example, a courtyard inside an incomplete hedge cannot be modelled topologically, because the hedge does not form a closed loop. It is possible, however, to model this relationship geometrically.

References

http://mathworld.wolfram.com/ (October 13, 2003).

International Organization for Standardization. (2000) ISO Technical Committee 211 Geographic Information/Geomatics. ISO DIS 19107, *Geographic Information – Spatial Schema.*

http://www.opengis.org/docs/02-023r4.pdf (October 15, 2003).

Whitehead, J.H.C. (1949) "Combinatorial Homotopy I." *Bull. Amer. Math. Soc.*, **55**, 213–245.

Additional references

MOLENAAR, M. (1998) *An Introduction to the Theory of Spatial Object Modelling for GIS.* Taylor & Francis, London.

http://www.math.toronto.edu/~drorbn/People/Eldar/thesis/index.html (October 13, 2003).

http://www.colorado.edu/geography/courses/geog_5003_s03/lecnotes/Modeling%20Our%20World.pdf (October 13, 2003).

http://www.cs.albany.edu/~amit/tutijcai.html (October 13, 2003).

http://schemas.opengis.net/gml/3.0.1/base/topology.xsd (October 15, 2003).

Chapter 14

GML temporal elements and dynamic features

Almost all geographic phenomena involve change. For example, flood waters rise and recede, persons and vehicles move around in a city, electoral boundaries are defined and reshaped, and land parcels are divided and merged. In many cases, these changes are simple, involving only the change in a few characteristics of a feature or feature collection. In other cases, the change is profound and complex, and the evolution of the type and identity of the features may not be clear. The inability to deal with such changes has long been a serious shortcoming of modern GIS technology. To remedy this shortcoming, GML 3 provides components for modelling time-dependent phenomena.

In GML 3, the `dynamicFeatures.xsd` and `temporal.xsd` schemas are used to describe dynamic features and the temporal characteristics of geographic data. The `temporal.xsd` schema provides primitives and properties for representing temporal instants and periods. It also provides types and elements for defining various kinds of temporal Coordinate Reference Systems (CRSs). The `dynamicFeature.xsd` schema provides elements and types for representing dynamic features, that is, features that change over time. All of the constructs discussed in this chapter are covered in Clause 7.8 of the *GML Version 3.00 OpenGIS® Implementation Specification* (http://www.opengis.org/docs/02-023r4.pdf). The schemas are available online at http://schemas.opengis.net/gml/3.0.1/base/.

Two geometric time primitives are provided in GML 3.0, namely `TimeInstant` and `TimePeriod`. `TimeInstant` is used to represent positions in time, while `TimePeriod` represents temporal length or duration. In GML instances, these two primitives typically occur as the value of a temporal property that provides a time stamp for some aspect of a geographic feature.

Using GML 3, it is possible to model a dynamic feature such as a moving object or a feature that has a subset of its properties that change over time. For example, forest-fire or flood boundaries, moving vehicles and evolving disaster situations are some of the things that can be represented. Note, however, that GML 3.0 does not provide support for modelling temporal topological constructs or features that change their identities over time. GML 3.1 will introduce temporal topology. It is also likely that future versions of GML will support feature succession.

Geography Mark-up Language (GML). R. Lake, D. S. Burggraf, M. Trninić, L. Rae © 2004 Galdos Systems Inc.
Published by John Wiley & Sons, Ltd   ISBNs: 0-470-87153-9 (HB); 0-470-87154-7 (PB)

The conceptual model underlying the representation of temporal objects in GML constitutes a profile of the conceptual schema described in *ISO DIS 19108* (ISO, 2002). Temporal topology objects and temporal feature relationships are not included in the current versions of the temporal schemas.

Two other ISO standards are relevant to describing temporal objects: *ISO 8601* and *ISO 11404* (ISO, 2001 and ISO, 1996). ISO 8601 describes encodings for time instants and time periods, as text strings using the specified syntax. ISO 11404 provides a detailed description of time intervals as part of a general discussion of language independent datatypes.

14.1   How do I specify which temporal reference system is being used?

Temporal reference systems, including calendars, provide the basis for time measurements. The default temporal reference system in GML is the same as the one defined in *ISO 8601* – the Gregorian calendar with Coordinated Universal Time (UTC). Other reference systems – such as the Global Positioning System (GPS) calendar and Julian Day Numbers – can also be used.

To indicate the specific reference system for a time measurement, you need to include the `frame` attribute with one of the temporal objects discussed above. The following example shows a `TimeInstant` that includes a URI-valued reference to a definition of the ISO 8601 reference system (Gregorian calendar with UTC time):

```
<gml:TimeInstant>
  <gml:timePosition frame="urn:opengis:temporal:ISO-8601">
    ⌣2003-02-13T12:28-08:00</gml:timePosition>
</gml:TimeInstant>
```

This example uses a location-independent identifier – a URN – to indicate the relevant temporal reference system. Here it is assumed that the actual GML 3 definition for the temporal reference system has been registered and can be retrieved from a Catalog Service or a registry using the specified identifier. Note that registration is not a requirement and the data consumer may decide to interpret the URN as if it were a well-known identifier for the frame. Chapter 15 contains a more detailed discussion of URNs in GML. For additional information about URNs, please see http://www.faqs.org/rfcs/rfc2141.html.

If the `timePosition` property does not have a `frame` attribute, then the default reference system, `ISO-8601`, is assumed. Note that temporal reference systems can include temporal ordinal reference systems (for example, a stratigraphic sequence) and temporal coordinate reference systems (for example, Julian Day Numbers).

14.2   How do I define an ordinal temporal reference system?

With GML, you can define ordinal temporal reference systems to express the relative position of events in time. An ordinal reference system is commonly used when dealing with prehistoric or archaeological events that cannot be precisely dated, such as those that comprise the geological time scale. An example of an

Table 1   Fragment of the geological time scale – an ordinal temporal
reference system

Eon	Era	Period
Phanerozoic	Cenozoic	Quaternary
		Tertiary
	Mesozoic	Cretaceous
		Jurasic
		Triasic
	Paleozoic	Permian
		Carboniferous
		Devonian
		Silurian
		Ordovician
		Cambrian

ordinal temporal reference system – a fragment of the geological time scale – is
shown in Table 1.

GML provides `TimeOrdinalReferenceSystem` and `TimeOrdi-`
`nalEra` constructs for defining ordinal temporal reference systems. The `com-`
`ponent` property of an ordinal reference system is used to encapsulate each of
the component eras that make up the reference system. This is shown in the
following stratigraphic sequence defined for the relative dating of archaeological
excavations in a region of Turkey:

```
<gml:TimeOrdinalReferenceSystem gml:id="YHSS">
    <gml:name>Yassihoyuk Stratigraphic Sequence</gml:name>
    <gml:domainOfValidity>Yassihöyük,Turkey
        ⌣</gml:domainOfValidity>
    <gml:component>
        <gml:TimeOrdinalEra gml:id="YHSS.1">
            <gml:description>10th-12th century AD
                ⌣</gml:description>
            <gml:name>Medieval</gml:name>
        </gml:TimeOrdinalEra>
    </gml:component>
    <gml:component>
        <gml:TimeOrdinalEra gml:id="YHSS.2">
            <gml:description>1st-3rd century AD
                ⌣</gml:description>
            <gml:name>Roman</gml:name>
        </gml:TimeOrdinalEra>
    </gml:component>
    <gml:component>
        <gml:TimeOrdinalEra gml:id="YHSS.3">
            <gml:name>Hellenistic periods</gml:name>
            <gml:member>
                <gml:TimeOrdinalEra gml:id="YHSS.3A">
                    <gml:description>early 3rd century-150 BC
                        ⌣</gml:description>
                    <gml:name>Hellenistic-Galatian</gml:name>
```

```
                        </gml:TimeOrdinalEra>
                    </gml:member>
                    <gml:member>
                        <gml:TimeOrdinalEra gml:id="YHSS.3B">
                            <gml:description>late 4th-early 3rd century
                                ⌣BC</gml:description>
                            <gml:name>Hellenistic</gml:name>
                        </gml:TimeOrdinalEra>
                    </gml:member>
                </gml:TimeOrdinalEra>
            </gml:component>
            <gml:component>
                <gml:TimeOrdinalEra gml:id="YHSS.4">
                    <gml:description>550-330 BC</gml:description>
                    <gml:name>Late Phrygian/Persian</gml:name>
                </gml:TimeOrdinalEra>
            </gml:component>
            <gml:component>
                <gml:TimeOrdinalEra gml:id="YHSS.5">
                    <gml:description>700-550 BC</gml:description>
                    <gml:name>Middle Phrygian</gml:name>
                </gml:TimeOrdinalEra>
            </gml:component>
        </gml:TimeOrdinalReferenceSystem>
```

Note that the unique identifier for the TimeOrdinalReferenceSystem in this example is YHSS. This reference system contains a series of component properties, each of which encapsulates a TimeOrdinalEra instance that has its own unique identifier, such as YHSS.1 and YHSS.2. Note that it is also possible to define a TimeOrdinalEra instance as an aggregate that contains multiple TimeOrdinalEra instances as components. For example, the YHSS.3 TimeOrdinalEra contains two TimeOrdinalEra members: YHSS.3a and YHSS.3b.

14.3   How do I define a temporal coordinate reference system?

The temporal schema also provides the TimeCoordinateSystem element, which can be used to define reference systems for cardinal temporal coordinates. Temporal CRSs are based on continuous interval scales that are defined in terms of single time intervals. As a result, the calculation of temporal length generally involves simple arithmetic operations.

A temporal CRS is more precise than an ordinal system and can be used for prehistoric times, where only rough estimates of the boundaries between the periods are available. For example, you can use a temporal CRS to represent the geological time scale, where geological events can be approximately dated by radiometric means.

The following example defines a temporal CRS that measures the number of days elapsed since – or preceding, if the value is negative – midnight on January 1, 4713 BC:

```
<gml:TimeCoordinateSystem gml:id="CJD"
    ⌣xmlns="http://www.opengis.net/gml">
    <gml:description>Number of days elapsed since
        ⌣midnight, January 1st, 4713 B.C.</gml:description>
    <gml:name>Chronological Julian Days</gml:name>
```

```
    <gml:domainOfValidity>global</gml:domainOfValidity>
    <gml:origin>4712-01-01</gml:origin>
    <gml:interval>d</gml:interval>
</gml:TimeCoordinateSystem>
```

The `TimeCoordinateSystem` content model adds two more properties to those inherited from `TimeReferenceSystemType`: `origin` and `interval`. The `origin` property provides the start date of the temporal CRS. The `interval` property is a unit of measure that specifies the granularity of the time values. The `frame` attribute in the following `TimeInstant` instance references the CJD temporal coordinate system defined above:

```
<gml:TimeInstant>
    <gml:timePosition
        ⌣frame="CJD">2452623</gml:timePosition>
</gml:TimeInstant>
```

The number 2452623 indicates the number of days that have elapsed since January 1, 4713 BC. This value corresponds to the date 2002-12-14 in the Gregorian calendar. Note that in the above example, the `TimeInstant` and the `TimeCoordinateSystem` are assumed to be in the same XML document, given that the `frame` attribute value (#CJD) contains only a fragment identifier. Note further that although this example describes a temporal CRS with an interval of days, other temporal reference systems can be defined with intervals of minutes, seconds or hours, depending on the requirements.

14.4 How do I define a dynamic feature in an application schema?

GML provides several convenience types for defining dynamic features and feature collections. Both `DynamicFeatureType` and `DynamicFeatureCollectionType` contain a property group called `dynamicProperties` group, which includes `timeStamp`, `history` and `dataSource` properties. Note that it is also possible to define your own custom properties for a dynamic feature, however, this is not covered in this book.

Note: The `timestamp` property is likely to be replaced by the `validTime` property in GML 3.1.

To define a dynamic feature in an application schema, the feature's content model must derive from `DynamicFeatureType`, as shown in the following element declaration and type definition for a `Boat` feature:

```
<element name="Boat" type="app:BoatType"
    ⌣substitutionGroup="gml:_Feature"/>

<complexType name="app:BoatType">
    <complexContent>
        <extension base="gml:DynamicFeatureType">
            <sequence>
                <element name="loa" type="integer"/>
```

```
            <element name="sailArea" type="decimal"/>
        </sequence>
    </extension>
  </complexContent>
</complexType>
```

In addition to deriving from `DynamicFeatureType`, the `BoatType` also specifies additional properties to describe a `Boat` instance. An example of a `Boat` instance is provided later in this chapter.

Note that a feature collection does not have to be dynamic in order to contain dynamic feature members. If the feature collection has its own temporal properties, however, the feature collection must derive from `DynamicFeatureCollectionType`, as shown in the following element declaration and type definition for a `SchoolDistrict` feature collection:

```
<element name="SchoolDistrict"
    ⌐type="app:SchoolDistrictType"
    ⌐substitutionGroup="gml:_FeatureCollection"/>

<complexType name="SchoolDistrictType">
    <complexContent>
        <extension base="gml:DynamicFeatureCollectionType">
            <sequence>
                <element ref="gml:extentOf"/>
            </sequence>
        </extension>
    </complexContent>
</complexType>
```

14.5   What is a 'snapshot' and how is it used in GML?

A dynamic feature can have a `timeStamp` property, which generates a 'snapshot' of a feature. A 'snapshot' is a representation of the state of a feature at a point in time or during some interval. The concept of a 'snapshot' comes from the Snapshot Model for temporal GIS, as discussed in *Temporal GIS and Spatio-Temporal Modeling* (http://www.ncgia.ucsb.edu/conf/SANTA_FE_CD-ROM/sf_papers/yuan_may/may.html). With a 'snapshot', the time-varying properties of a feature are expressed at the feature level. Snapshots also include time-invariant properties, and, therefore, successive snapshots may contain much redundant information if only a few of the feature's properties are dynamic.

To illustrate how snapshots are expressed in GML 3.0, consider the following example of a school district with two different time stamps. Each snapshot represents the same school district – identified as SD28 – but at different points in time.

14.5.1   First snapshot

In a GML instance document, the first dynamic `SchoolDistrict` feature collection called 'District 28' can be encoded as follows:

```
<app:SchoolDistrict gml:id="SD28">
    <gml:name>District 28</gml:name>
    <gml:boundedBy>
```

```
        <gml:EnvelopeWithTimePeriod srsName="..">
           ...
        </gml:EnvelopeWithTimePeriod>
     </gml:boundedBy>
     <gml:featureMember>
        <app:School>
           <gml:name>Alpha</gml:name>
           <app:address>100 Cypress Ave.</app:address>
           <gml:location>
              ...
           </gml:location>
        </app:School>
     </gml:featureMember>
     <gml:featureMember>
        <app:School>
           <gml:name>Beta</gml:name>
           <app:address>1673 Balsam St.</app:address>
           <gml:location>
              ...
           </gml:location>
        </app:School>
     </gml:featureMember>
     <gml:extentOf>
        ...
     </gml:extentOf>
     <gml:timeStamp>
        <gml:TimeInstant>
           <gml:timePosition>1965-01-01</gml:timePosition>
        </gml:TimeInstant>
     </gml:timeStamp>
  </app:SchoolDistrict>
```

Note that, in addition to the name, boundedBy and extentOf properties, District 28 has two school members – Alpha and Beta – and a timeStamp property. The value of the timeStamp property is a TimeInstant element with a temporal position of 1965-01-01 (January 1st, 1965).

14.5.2   Second snapshot

In the second snapshot, District 28 has only one school member, Beta, and the timePosition property has a value of 1975-01-01.

```
     <app:SchoolDistrict gml:id="SD28">
        <gml:name>District 28</gml:name>
        <gml:boundedBy>
           <gml:EnvelopeWithTimePeriod srsName="..">
              ...
           </gml:EnvelopeWithTimePeriod>
        </gml:boundedBy>
        <gml:featureMember>
           <app:School>
              <gml:name>Beta</gml:name>
              <app:address>1673 Balsam St.</app:address>
              <gml:location>
                 ...
              </gml:location>
           </app:School>
```

```
    </gml:featureMember>
    <gml:extentOf>
    ...
    </gml:extentOf>
    <gml:timeStamp>
        <gml:TimeInstant>
            <gml:timePosition>1975-01-01</gml:timePosition>
        </gml:TimeInstant>
    </gml:timeStamp>
</app:SchoolDistrict>
```

The two snapshots of the school district are illustrated in Figure 1. These two instances represent two different time-stamped versions of the same feature – they can exist as independent representations. Note that snapshots are typically used to update a feature or feature collection that has already been defined. Because snapshots of a feature or feature collection have the same id, they cannot co-exist in the same instance document. In many circumstances, time slices are used instead of snapshots to avoid this potential id conflict and also for improved efficiency. Time slices are discussed later in this chapter.

14.6   How is the boundedBy property encoded for dynamic features?

As discussed in Chapter 9, the boundedBy property is required for feature collections. Given that DynamicFeatureCollectionType derives from FeatureCollectionType, dynamic feature collections must also have a boundedBy property. With dynamic feature collections, however, the property can contain an EnvelopeWithTimePeriod object, which is an Envelope with two timePosition properties. The two timePosition properties specify the temporal extent of the feature collection.

Consider again the example of the SchoolDistrict feature collection, from Section 14.4, whose time stamp is January 1, 1965. The values of the time-Position properties, in the EnvelopeWithTimePeriod, can represent the

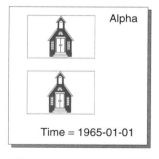

Alpha

District 28: Snapshot 1

Time = 1965-01-01

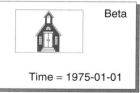

Beta

District 28: Snapshot 2

Time = 1975-01-01

Figure 1 Snapshots in GML.

beginning and end of the time period that contains 1965, for example January 1, 1960 to January 1, 1970. This is shown in the following example:

```
<app:SchoolDistrict gml:id="SD100">
    <gml:name>District 28</gml:name>
    <gml:boundedBy>
        <gml:EnvelopeWithTimePeriod srsName="..">
            <gml:timePosition>1960-01-01
                ⌣</gml:timePosition>
            <gml:timePosition>1970-01-01
                ⌣</gml:timePosition>
            <gml:pos>...</gml:pos>
            <gml:pos>...</gml:pos>
        </gml:EnvelopeWithTimePeriod>
    </gml:boundedBy>
    ...
</app:SchoolDistrict>
```

The beginning and end of the time period coincide with significant change events – for example, adding or removing schools from the `SchoolDistrict`. This allows you to deduce the composition of the school district at different times within the interval, which is 1960 to 1970 in the above example.

A dynamic feature can also have a `boundedBy` property with an `EnvelopeWithTimePeriod`, but this is not mandatory. Note that if a dynamic feature or feature collection only has one time stamp, then it is possible for the two `timePosition` properties to have the same value. A more detailed discussion of the `boundedBy` property is provided in Chapter 9.

14.7   What does the `history` property represent?

A dynamic feature can have a `history` property that may be used to express its historical development. The `history` property of a dynamic feature associates a feature instance with a sequence of time slices that encapsulate the evolution of the feature over time. Each time slice includes a subset of feature properties – those properties that change over time. With time slices, a feature's time-variable properties are expressed at the property level, as opposed to a snapshot, where the time-variable properties are expressed at the feature level. This approach often leads to a more economical representation of a dynamic feature.

The concepts of history and time slices are represented in Figure 2. The `DynamicFeature` has time-invariant and time-varying properties. The time-varying properties are encapsulated within the `history` property whose value is a series of objects that are substitutable for the abstract `_TimeSlice` object. Note that it is possible to define application-specific time slices and history properties. This is covered in Section 14.8.

14.7.1   About `track` and `MovingObjectStatus`

The `track` property is a specific kind of `history`, defined in GML 3, for modelling dynamic features that represent moving rigid objects, such as a person, ground vehicle, a boat or an aircraft. A dynamic feature with a `track` property corresponds to a G-XML Mover – a concept defined in the `Mover.xsd` schema in G-XML. A Mover is a dynamic feature with a series of `MovingObjectStatus`

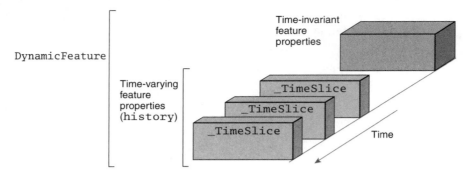

Figure 2 The history property and time slices.

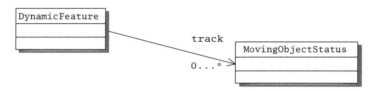

Figure 3 Mover diagram.

instances, which are specialized time slices that show the progress of a moving object over time. Figure 3 illustrates the essential concept of a Mover.

The MovingObjectStatus includes properties for expressing the location, speed and direction of a moving object, as shown in the following example:

```
<gml:MovingObjectStatus gml:id="wp-0">
   <gml:timeStamp>
      <gml:TimeInstant>
         <gml:timePosition
            ⌐frame="urn:opengis:temporal:GPS">
            ⌐995738736</gml:timePosition>
      </gml:TimeInstant>
   </gml:timeStamp>
   <gml:location>
      <gml:Point srsName="..">
         <gml:pos>
         ...
         </gml:pos>
      </gml:Point>
   </gml:location>
   <gml:speed>4.2</gml:speed>
   <gml:direction>268</gml:direction>
</gml:MovingObjectStatus>
```

The following example shows how a series of MovingObjectStatus instances are encoded for the Boat feature that was defined in Section 14.4:

```
<app:Boat gml:id="G1251">
   <gml:track>
```

```
      <gml:MovingObjectStatus gml:id="wp-0">
         <gml:timeStamp>
            <gml:TimeInstant>
               <gml:timePosition
                  ↳frame="urn:opengis:temporal:GPS">
                  ↳995738736</gml:timePosition>
            </gml:TimeInstant>
         </gml:timeStamp>
         <gml:location>
            <gml:Point srsName="..">
               <gml:pos>
               ...
               </gml:pos>
            </gml:Point>
         </gml:location>
         <gml:speed>4.2</gml:speed>
         <gml:bearing>268</gml:bearing>
      </gml:MovingObjectStatus>
      <gml:MovingObjectStatus gml:id="wp-1">
         <gml:timeStamp>
            <gml:TimeInstant>
               <gml:timePosition
                  ↳frame="urn:opengis:temporal:GPS">
                  ↳995738801</gml:timePosition>
            </gml:TimeInstant>
         </gml:timeStamp>
         <gml:location>
            <gml:Point srsName="..">
               <gml:pos>
               ...
               </gml:pos>
            </gml:Point>
         </gml:location>
         <gml:speed>6.7</gml:speed>
         <gml:bearing>348</gml:bearing>
      </gml:MovingObjectStatus>
      <gml:MovingObjectStatus gml:id="wp-2">
         <gml:timeStamp>
            <gml:TimeInstant>
               <gml:timePosition
                  ↳frame="urn:opengis:temporal:GPS">
                  ↳995738967</gml:timePosition>
            </gml:TimeInstant>
         </gml:timeStamp>
         <gml:location>
            <gml:Point srsName="..">
               <gml:pos>
               ...
               </gml:pos>
            </gml:Point>
         </gml:location>
         <gml:speed>5.3</gml:speed>
         <gml:bearing>260</gml:bearing>
      </gml:MovingObjectStatus>
   </gml:track>
   <app:loa>228</app:loa>
   <app:sailArea>200</app:sailArea>
</app:Boat>
```

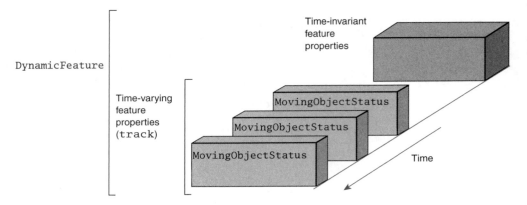

Figure 4 The track property and MovingObjectStatus.

The concepts of track and MovingObjectStatus are depicted in Figure 4. In the above example, the track property provides a relationship between the Boat feature and different MovingObjectStatus instances. The static properties of the boat are recorded only once. Note also that the frame attribute indicates that these values are recorded in GPS time, instead of in units associated with the default ISO 8601 reference system.

Note that with the track property, time slices are always encoded as MovingObjectStatus instances. The track property, however, is only one kind of history property. It is also possible to create user-defined history properties and time slices. The GML history property exemplifies the basic pattern for defining other history properties.

14.8   How do I create user-defined history properties and time slices?

To create user-defined history properties and time slices, you need to define the new history property and time slice object in an application schema. Consider the following example of a schema for archaeological excavations that includes a user-defined time slice called OccupationInterval. The element declaration and type definition appear below:

```
<element name="OccupationInterval"
  ⌐type="app:OccupationIntervalType"
  ⌐substitutionGroup="gml:TimeSlice"/>

<complexType name="OccupationIntervalType">
  <complexContent>
    <extension base="gml:AbstractTimeSliceType">
      <sequence>
        <element name="settlementAreas"
          ⌐type="gml:MultiPolygonPropertyType"/>
        <element ref="gml:description" minOccurs="0"/>
      </sequence>
    </extension>
  </complexContent>
</complexType>
```

Note that the content model for `OccupationInterval` derives by extension from `AbstractTimeSliceType`, and that an `OccupationInterval` instance is in the `_TimeSlice` substitution group – that is, it may appear wherever a `_TimeSlice` element is expected. This derivation and the associated substitution group are important as they are what enables GML-aware software to detect that the `OccupationInterval` is a GML time slice.

Before you can use `OccupationInterval` in a GML instance, you need to create a user-defined history property that can link a dynamic feature with the time slice. The following schema fragment shows how to define a `settlementHistory` property that does this.

```
<element name="settlementHistory"
   ⌐type="app:SettlementHistoryPropertyType"
   ⌐substitutionGroup="gml:history"/>

<complexType name="SettlementHistoryPropertyType">
   <complexContent>
      <restriction base="gml:HistoryPropertyType">
         <sequence>
            <element ref="app:OccupationInterval"
               ⌐maxOccurs="unbounded"/>
         </sequence>
         <attributeGroup
            ⌐ref="gml:AssociationAttributeGroup"/>
      </restriction>
   </complexContent>
</complexType>
```

The `settlementHistory` property must be substitutable for `history`, and hence its content model must derive from `HistoryPropertyType`.

The following fragment shows how the `OccupationInterval` time slice and the `settlementHistory` property can be encoded in a GML instance. The `Gordian` feature is a dynamic feature, an archaeological site, that is defined using the same pattern employed to define the `Boat` feature discussed earlier in this chapter.

```
<app:Gordian gml:id="..">
   ...
   <gml:settlementHistory>
      <app:OccupationInterval>
         <gml:timeStamp>
            <gml:TimeInstant>
               <gml:timePosition frame="#YHSS">#YHSS.3B
                  ⌐</gml:timePosition>
            </gml:TimeInstant>
         </gml:timeStamp>
         <app:settlementAreas>
            <gml:MultiSurface srsName="..">
               ...
            </gml:MultiSurface>
         </app:settlementAreas>
         <gml:description>Domestic architecture mainly
            ⌐mudbrick with stone foundations.
            ⌐</gml:description>
      </app:OccupationInterval>
```

```
<app:OccupationInterval>
    <gml:timeStamp>
        <gml:TimeInstant>
            <gml:timePosition frame="YHSS">#YHSS.4
                ⌐</gml:timePosition>
        </gml:TimeInstant>
    </gml:timeStamp>
    <app:settlementAreas>
        <gml:MultiSurface srsName="..">
        ...
        </gml:MultiSurface>
    </app:settlementAreas>
    <gml:description>Extensive reconstruction
        ⌐following fire.</gml:description>
</app:OccupationInterval>
</app:settlementHistory>
<app:chronology>
    <gml:TimeOrdinalReferenceSystem gml:id="YHSS">
    ...
    </gml:TimeOrdinalReferenceSystem>
</app:chronology>
</app:Gordian>
```

Note that the frame attribute is used to reference the temporal ordinal reference system (identified as YHSS) that defines a stratigraphic sequence for relative dating. The time values are fragment identifiers that specify particular eras in the sequence – for example, #YHSS.3B indicates the Hellenistic era, late fourth to early third century BC. This example is illustrated in Figure 5.

Note further that in the above example, the ordinal temporal reference system is located in the same document and attached to the Gordian feature via a user-defined chronology property. In many cases, the reference system is located outside of the instance document and is simply referenced from a frame attribute. The *Worked Examples CD* contains an example of a user-defined chronology property.

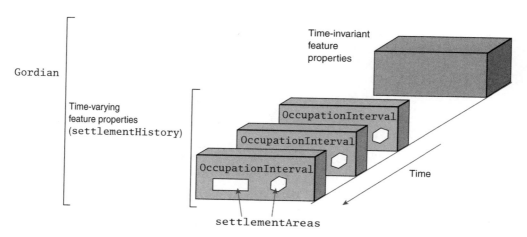

Figure 5 Gordian feature with user-defined time slices.

14.9   How is the `TimePeriod` primitive encoded in GML?

The `TimePeriod` primitive is typically used to encode an interval of time between two time instants. In GML instances, a `TimePeriod` must have a `begin` or an `end` property. Each property has a child `TimeInstant` element that provides the time values for the beginning or end of the period. If only one of these properties is provided, then a `duration` must also be included to implicitly specify the position of the other one. For example, an interval that begins at 2003-02-12 and has a `duration` of five days ends at 2003-02-17.

To illustrate how `TimePeriod` is encoded in GML, consider again the `Gordian` feature from the previous section. The following example shows how to use the `TimePeriod` primitive to include temporal values for the beginning and end of a period of time:

```
<app:Gordian gml:id="..">
   ...
   <gml:settlementHistory>
      <app:OccupationInterval>
         <gml:timeStamp>
            <gml:TimePeriod>
               <gml:begin>
                  <gml:TimeInstant>
                     <gml:timePosition
                        frame="#YHSS">#YHSS.3A
                        </gml:timePosition>
                  </gml:TimeInstant>
               </gml:begin>
               <gml:end>
                  <gml:TimeInstant>
                     <gml:timePosition
                        frame="#YHSS">#YHSS.3B
                        </gml:timePosition>
                  </gml:TimeInstant>
               </gml:end>
            </gml:TimePeriod>
         </gml:timeStamp>
         <app:settlementAreas>
            <gml:MultiSurface srsName="..">
            ...
            </gml:MultiSurface>
         </app:settlementAreas>
         <gml:description>Domestic architecture mainly
            mudbrick with stone foundations.
            </gml:description>
      </app:OccupationInterval>
      <app:OccupationInterval>
      ...
      </app:OccupationInterval>
   </app:settlementHistory>
   <gml:chronology>
      <gml:TimeOrdinalReferenceSystem gml:id="YHSS">
      ...
      </gml:TimeOrdinalReferenceSystem>
   </gml:chronology>
</app:Gordian>
```

14.10 Chapter summary

GML provides support for modelling dynamic features, which are features with time-varying properties. It also includes constructs for creating and referencing temporal reference systems. The spatio-temporal model in GML 3.0 is based on the conceptual model described in *ISO 19108*. GML does not currently support all of the temporal types defined in *ISO 19108*. Time primitives in GML are also based on ISO 8601, and time intervals are partially based on the structures defined in ISO 11404. Two geometric time primitives are defined in GML: `TimeInstant` and `TimePeriod`. GML 3.0 does not currently provide any support for modelling temporal topology or feature succession, however, temporal topology will be introduced in GML 3.1.

The `temporal.xsd` schema provides support for three kinds of temporal reference systems: temporal, ordinal temporal and temporal coordinate. The `frame` attribute – which qualifies the `timePosition` property – is used in GML instances to reference all of these kinds of reference systems. It is not likely that you will need to use GML to create new temporal reference systems, given that standardized calendars are already available. However, GML 3 may be used to represent existing standard temporal reference systems or to define localized ordinal reference systems.

GML provides `TimeOrdinalReferenceSystem`, `component` and `TimeOrdinalEra` for creating temporal ordinal reference systems. In GML instances, the `component` property encapsulates a series of `TimeOrdinalEra` instances. The `TimeCoordinateSystem` element is used to instantiate temporal CRSs, such as Julian Day Numbers.

Features can only have temporal properties if they are defined as dynamic features in application schemas. You can designate a feature as dynamic by having its content model derive from `DynamicFeatureType`. Feature collections are made dynamic by having their content models derive from `DynamicFeatureCollectionType`. These types provide features and feature collections with dynamic properties that contain time primitives, such as `TimeInstant` and `TimePeriod`, or time slices.

One of the dynamic properties is called `timeStamp`. When a dynamic feature in a GML instance has a `timeStamp` property, this property serves to define a 'snapshot' of the feature that represents the feature's state at some instant or during some interval. You can create multiple time-stamped versions of the same feature, however, to include these different versions in the same instance – for example, as part of a feature collection – they need to be included through a history property. Note that in GML 3.1, the `timeStamp` property is expected to be renamed as `validTime`.

A `history` is another dynamic property that links dynamic features with a series of time slices, each of which includes a subset of feature properties that change over time; static properties are not included. GML 3 provides a convenience property called `track`, which is a kind of `history` property. The `track` property contains a GML-defined time slice called `MovingObjectStatus`, which has properties that describe the state of a moving object at a point in time.

The `history` property defines a basic pattern for creating user-defined history properties. When you create a user-defined history property, it should be substitutable for the `history` property and derive by restriction from `HistoryryPropertyType`. The user-defined `history` property must also contain a user-defined time slice that derives by extension from `AbstractTimeSliceType`.

Of the two geometric time primitives provided by GML, `TimeInstant` is used most often. `TimePeriod` is used to express duration – the temporal distance between two different time positions. In GML instances, `TimePeriod` objects are related to `TimeInstant` objects that indicate the beginning and ending of the period. It is not necessary to explicitly specify both the start and end points of a period. If one 'boundary' is omitted, then a `duration` property must be included to implicitly specify the other one.

References

http://www.opengis.org/docs/02-023r4.pdf (October 15, 2003).

http://schemas.opengis.net/gml/3.0.1/base/ (October 15, 2003).

International Organization for Standardization. (1996) ISO 11404, *Information Technology – Programming Languages, their Environments and System Software Interfaces – Language-Independent Datatypes.*

International Organization for Standardization. (2001) ISO 8601, *Data Elements and Interchange Formats – Information Interchange – Representation of Dates and Times.*

International Organization for Standardization. (2002) ISO Technical Committee 211 Geographic Information/Geomatics. ISO 19108, *Geographic Information – Temporal Schema.*

http://www.iso.ch/iso/en/ittf/PubliclyAvailableStandards/s019346_ISO_IEC_TR_11404_1996(E).zip (October 17, 2003).

http://www.faqs.org/rfcs/rfc2141.html (October 17, 2003).

http://www.ncgia.ucsb.edu/conf/SANTA_FE_CD-ROM/sf_papers/yuan_may/may.html (October 17, 2003).

Additional references

Langran, G. (1992) *Time in Geographic Information Systems.* Taylor & Francis, London.

http://schemas.opengis.net/gml/3.0.1/base/dynamicFeature.xsd (October 15, 2003).

http://schemas.opengis.net/gml/3.0.1/base/temporal.xsd (October 15, 2003).

Chapter 15

GML coordinate reference systems

This chapter describes the Coordinate Reference System (CRS) model in GML. To properly interpret the coordinates captured in GML geometry objects, you need to consider the CRS in which the coordinates are expressed. You only need to be familiar with the material in this chapter if

1. you are creating a CRS dictionary, or
2. you are a data provider and you need to understand CRS definitions.

Note that in the second case you only need a general 'reader' level of understanding. When more CRS dictionaries become available as online web services (for example, http://crs.opengis.org/crsportal), it is likely that you will need even less of a technical understanding of the contents of this chapter.

Coordinate Reference Systems are part of a more general model called Spatial Reference Systems (SRS), which includes referencing by coordinates and geographic identifiers. Currently, GML only provides support for referencing by coordinates, however, it is likely that future versions will support referencing by geographic identifiers.

With GML, CRS data is typically encoded in dictionaries that contain CRS and support component definitions. GML provides a general dictionary model that can be used for encoding CRS dictionaries. This chapter covers the rules for referencing and encoding CRS dictionaries and dictionary entries.

The GML implementation of CRS is based on the following OGC and ISO documents:

- *Recommended Definition Data for Coordinate Reference Systems and Coordinate Transformations* (http://member.opengis.org/tc/archive/ arch01/01-014r5.pdf)
- *OpenGIS Abstract Specification: Topic 2 – Spatial Referencing by Coordinates* (http://www.opengis.org/docs/02-102.pdf)
- *ISO DIS 19111 (ISO, 2000)*

The GML encoding of CRS elements and types is covered in *Recommended XML Encoding of Coordinate Reference System Definitions* (http://www.opengis. org/docs/03-010r7.pdf). This document was approved by the OGC Technical and

Geography Mark-up Language (GML).   R. Lake, D. S. Burggraf, M. Trninić, L. Rae © 2004 Galdos Systems Inc.
Published by John Wiley & Sons, Ltd   ISBNs: 0-470-87153-9 (HB); 0-470-87154-7 (PB)

Planning Committees. Before reading this chapter, you may wish to become familiar with basic CRS terminology as used by geodesists, GIS developers or mathematicians.

15.1   How are CRS dictionaries used in practice?

CRS dictionaries are XML documents – typically created by data administrators who specialize in reference systems – that contain a collection of CRS and support component definitions. In GML, CRS definitions are referenced from geometry objects using the `srsName` attribute, as shown in the following example of a `Tower` feature:

```
<app:Tower>
   <gml:centerOf>
      <gml:Point
         ⌣srsName="http://www.ukusa.org/
         ⌣CRSDictionary.xml#crs1111">
         <gml:pos dimension="2">-13411428  -35959999
            ⌣</gml:pos>
      </gml:Point>
   </gml:centerOf>
</app:Tower>
```

As discussed in Chapter 12, all GML 3.0 geometry objects have an optional `srsName` attribute, which is of `anyURI type`. The value of the attribute is a link that points to a CRS definition in a CRS dictionary. In the above example, the unique identifier of the CRS definition is `crs1111`.

The reference in the `srsName` attribute must be a valid URI; however, this constraint is fairly lax, since the `anyURI` type does not constrain the value. In the above example, the value of the `srsName` attribute is a URL. According to recommendations in *Developing and Managing GML Application Schemas: Best Practices* (http://www.geoconnections.org/developersCorner/devCorner_devNetwork/ components/GML_bpv1.3_E.pdf), the value of the `srsName` attribute should be a location independent identifier such as a URN. For example, the `srsName` reference can be encoded as follows:

```
"urn:NID:v1.0:crsdictionary:crs1111"
```

The different components of the above URN are listed in Table 1.

CRS definitions are not necessarily referenced on the basis of a file name. Consider the following example:

```
"urn:NID:v1.0:coordinatereferencesystem:crs1111"
```

The first three components of this URN are the same as those listed in Table 1, however, `coordinatereferencesystem` represents the resource category (in this case, the resource is a CRS) and `crs1111` represents the unique identifier of the CRS resource. You can find additional information about URNs at http://www.faqs.org/rfcs/rfc2141.html.

CRS dictionaries can be private to an individual or an organization. They can also be shared by a group of individuals or organizations, or be made available to anyone over the Internet. Note also that CRS dictionaries will be increasingly

Table 1   Content of a URN

Components	Represents
urn:	The standard and mandatory URN prefix
NID:	The namespace identifier of the authority that maintains the resource
v1.0:	The version of the resource. Note that the version is optional
crsdictionary	The parent resource (in this case, an xml file) containing the CRS definitions
crs1111	The unique identifier of the resource (in this case a CRS definition) in the file

deployed as web services, likely using the OGC Catalog Service (also known as WRS), an example of which can be seen at http://crs.opengis.org/crsportal.

Most data collectors who encode GML instances only need to know two things about CRS dictionaries: how to use the srsName attribute to reference a CRS definition and the list of CRS definitions that can be referenced. You do not need to read the rest of this chapter if you are not interested in learning how to create CRS dictionaries and schemas.

15.2 How do I provide local visibility for CRS information?

The idea of CRS dictionaries seems like a good one, but what if data users cannot access a dictionary? If you are a data provider, how can you communicate at least some of the CRS information, such as axis names, axis order and units of measure? GML provides optional attributes that can appear on any element that can have an srsName attribute. These attributes are illustrated in the following example:

```
<gml:Point srsName="#myrefsys" srsDimension="2"
    ⌐axisLabels="lat long" uomLabels="deg deg">
    <gml:pos>36.35 120.44</gml:pos>
</gml:Point>
```

Note that the exact grammar for the encoding of this local CRS information was not formalized at the time of publication and may differ from that shown in this section. Note further that local visibility is introduced in GML 3.1, and is not part of GML 3.0. Two of the local visibility attributes are axisLabels and uomLabels. Both attributes are whitespace-separated lists of tokens that must agree in name and order with the corresponding items defined in the CRS dictionary definition referenced through the srsName attribute. The value of the axisLabels attribute shows the axis abbreviations and the uomLabels value shows the units of measure catalog symbols defined in the CRS dictionary.

Note that the local CRS information is provided for visibility only and cannot override or conflict with the referenced CRS definition. It is up to a data provider to ensure that the referenced CRS is correct for the data provided. If a data provider elects to also provide local visibility for some of the CRS information, the data provider is responsible for seeing that this local information is completely consistent with the referenced CRS definition. This is, in effect, the contract between the data provider and the data consumer. If you are an intermediate data provider in a chain of providers and consumers, you need to guarantee that your data meets these conditions, even if you are only a 'pass through' for the data from another source.

15.3   How do I create a CRS dictionary?

The `dictionary.xsd` schema provides types for encoding different kinds of dictionaries, including CRS dictionaries. A GML dictionary should have a root-level `Dictionary` element that has `dictionaryEntry` properties, each of which contains a dictionary entry that is a `Definition` object. For CRS dictionaries, the dictionary entries are typically CRS definitions, such as `GeocentricCRS` or `ProjectedCRS`. These CRS definitions are defined in the `coordinateReferenceSystems.xsd` schema.

15.3.1   Encoding a CRS dictionary

A CRS dictionary can be encoded as shown in the following example of a simple dictionary with three CRS definitions:

```
<gml:Dictionary gml:id="CRSD1">
  <gml:name>...</gml:name>
  <gml:dictionaryEntry>
    <gml:GeocentricCRS gml:id="crs1111">
      <gml:srsName>...</gml:srsName>
      <gml:srsID>
        <gml:code>...</gml:code>
      </gml:srsID>
      <gml:usesCartesianCS
        ⌐xlink:href="CSDictionary.xml#cs2222"/>
      <gml:usesGeodeticDatum
        ⌐xlink:href="DatumDictionary.xml#dt4444"/>
    </gml:GeocentricCRS>
  </gml:dictionaryEntry>
  <gml:dictionaryEntry>
    <app:Geographic2DCRS gml:id="crs1112">
      <gml:srsName>Geographic Two-Dimensional
        ⌐Coordinate Reference System 1112
        ⌐</gml:srsName>
      <gml:srsID>
        <gml:code>1112</gml:code>
      </gml:srsID>
      <app:usesEllipsoidal2DCS
        ⌐xlink:href="CSDictionary.xml#cs2223"/>
      <gml:usesGeodeticDatum
        ⌐xlink:href="DatumDictionary.xml#dt4444"/>
    </app:Geographic2DCRS>
  </gml:dictionaryEntry>
  <gml:dictionaryEntry>
    <gml:ProjectedCRS gml:id="crs1113">
      <gml:srsName>...</gml:srsName>
      <gml:srsID>
        <gml:code>...</gml:code>
      </gml:srsID>
      <gml:baseCRS
        ⌐xlink:href="CRSDictionary.xml#crs1112"/>
```

```
            <gml:definedByConversion>
                ...</gml:definedByConversion>
            <gml:usesCartesianCS
                xlink:href="CSDictionary.xml#cs2222"/>
          </gml:ProjectedCRS>
       </gml:dictionaryEntry>
    </gml:Dictionary>
```

The root element of this dictionary is `Dictionary`, with a unique identifier, CRSD1, and it has a series of `dictionaryEntry` properties, each of which contains a CRS definition. Note that the unique identifier (`gml:id`) value of the first CRS definition, `GeocentricCRS`, is `crs1111`, which is the dictionary entry that is referenced from the `Point` geometry in the above example of a `Tower` feature. Every CRS definition has a unique identifier by which they can be referenced, and these definitions can also reference each other. For example, the `ProjectedCRS` has a `baseCRS` property that references the `Geographic2dCRS` definition with the unique identifier `crs1112`.

Of the three CRS definitions in the above example, `GeocentricCRS` and `ProjectedCRS` are concrete CRS elements from GML 3.0, while `Geographic2dCRS` is a user-defined element. In most cases, only the predefined GML 3.0 CRS elements are required, but there are circumstances in which new CRS elements need to be defined. The concrete CRS elements are listed in Section 15.8. The rules for creating these new elements are covered in Section 15.9.

As with other GML instances, the root element of a CRS dictionary must have a `schemaLocation` attribute that points to the location of the schema that supports the content of the dictionary. If the dictionary contains user-defined elements from the same application namespace, the `schemaLocation` should point to the schema (or schemas) that includes and imports all of the necessary elements and types. For example, the `Dictionary` element in the above example might have the following attributes:

```
    <gml:Dictionary gml:id="CRSD1"
        xmlns:app="http://www.ukusa.org/app"
        xmlns="http://www.ukusa.org/app"
        xmlns:gml="http://www.opengis.net/gml"
        xmlns:xlink="http://www.w3.org/1999/xlink"
        xmlns:xsi="http://www.w3.org/2001/XMLSchema-instance"
        xsi:schemaLocation="http://www.opengis.net/gml
        http://schemas.opengis.net/gml/3.0.1/base/
            coordinateReferenceSystems.xsd
        http://www.ukusa.org/app CRSDictionary.xsd">
        ...
    </gml:Dictionary>
```

In the above example, the `schemaLocation` attribute references the `gml.xsd` schema in the GML namespace, which contains the definition of `Dictionary`, and the `CRSDictionary.xsd` application schema that contains the definition of the user-defined `Geographic2dCRS`. If a CRS dictionary does not contain any user-defined types, the `schemaLocation` attribute should reference the appropriate GML schema(s), for example `coordinateReferenceSystems.xsd`. Note that a complete version of the above example of a CRS dictionary is available in the `CRSDictionary1.xml` document on the *Worked Examples CD*. The `CRSDictionary.xsd` schema is also on this CD.

15.3.2   Encoding CRS support component dictionaries

Each CRS definition has properties that reference or contain other definitions, in particular CS and Datum definitions. Note that these CS and Datum definitions are essentially 'building blocks' for encoding CRS dictionary entries. These 'building blocks' (also known as support components) can be encoded as part of the same dictionary or in separate dictionaries containing definitions that are referenced from the CRS dictionary.

For example, the `GeocentricCRS` definition in the above example has `usesCartesianCS` and `usesDatum` properties, both of which reference definitions in other dictionaries. The `CartesianCRS` definition, which is referenced from the `usesCartesianCS` property, can be encoded as part of a CS dictionary, as shown in the following example:

```
<gml:Dictionary gml:id="CSD1">
   <gml:dictionaryEntry>
      <gml:CartesianCS gml:id="cs2222">
         <gml:csName>3D Cartesian Coordinate System 2222
            ↳</gml:csName>
         <gml:csID>
            <gml:code>...</gml:code>
         </gml:csID>
         <gml:usesAxis>
            <gml:CoordinateSystemAxis gml:id="axis3333"
               ↳gml:uom="urn:x-epsg:v6.1:uom:degree">
               <gml:axisName>First Axis</gml:axisName>
               <gml:axisAbbrev>X</gml:axisAbbrev>
               <gml:axisDirection>east
                  ↳</gml:axisDirection>
            </gml:CoordinateSystemAxis>
         </gml:usesAxis>
         <gml:usesAxis>
            <gml:CoordinateSystemAxis gml:id="axis3334"
               ↳gml:uom="urn:x-epsg:v6.1:uom:degree">
               <gml:axisName>Second Axis</gml:axisName>
               <gml:axisAbbrev>Y</gml:axisAbbrev>
               <gml:axisDirection>north
                  ↳</gml:axisDirection>
            </gml:CoordinateSystemAxis>
         </gml:usesAxis>
         <gml:usesAxis>
            <gml:CoordinateSystemAxis gml:id="axis3335"
               ↳gml:uom="urn:x-epsg:v6.1:uom:degree">
               <gml:axisName>Third Axis</gml:axisName>
               <gml:axisAbbrev>Z</gml:axisAbbrev>
               <gml:axisDirection>up
                  ↳</gml:axisDirection>
            </gml:CoordinateSystemAxis>
         </gml:usesAxis>
      </gml:CartesianCS>
   </gml:dictionaryEntry>
</gml:Dictionary>
```

The `CartesianCS` has three `CoordinateSystemAxis` definitions, each of which has its own unique identifier. A `CoordinateSystemAxis` is a support component for each CS definition, which in turn, is a support component for CRS definitions. There are a number of different support components that can be used to create CRS dictionaries. These components are covered later in this chapter in Section 15.6.

Note that each coordinate axis has an associated direction property, `axisDirection`, whose value is a text description of the direction or orientation of the axis. This description provides some reference to the physical world. Many different descriptions are possible, such as `North` or `South`, up or down, or 'the axis is along the corner, where the two walls meet, and points upward'. The description of the axis and its direction should be unambiguous, and it must be consistent across all axes in the set of axes definitions.

Note: The `CoordinateSystemAxis` entries in the above examples have `gml:uom` attributes for referencing units of measure dictionaries. Chapter 16 covers the GML rules for referencing and encoding these dictionaries.

15.4   How do I create user-defined dictionary elements and types for CRS?

GML 3.0.1 provides support for encoding most kinds of CRSs and support component definitions, and therefore, in most circumstances additional user-defined dictionary elements do not need to be created. There are some circumstances, however, that might require user-defined dictionary elements. For example, you might want to create a CRS dictionary that can only contain CRS definitions instead of all kinds of definitions.

Consider the following variation of the CRSD1 dictionary from the previous section:

```
<app:CRSDictionary gml:id="CRSD2">
  <app:crsDictionaryEntries>
    <gml:GeocentricCRS gml:id="crs1111">
      . . .
    </gml:GeocentricCRS>
    <app:Geographic2DCRS gml:id="crs1112">
      . . .
    </app:Geographic2DCRS>
    <gml:ProjectedCRS gml:id="crs1113">
      . . .
    </gml:ProjectedCRS>
  </app:crsDictionaryEntries>
</app:CRSDictionary>
```

The root element of the dictionary is a user-defined `CRSDictionary` element whose content model derives from `DictionaryType`, as shown in the following schema fragment:

```
<element name="CRSDictionary"
  ⌐type="app:CRSDictionaryArrayType"
```

```
⌐substitutionGroup="gml:Dictionary"/>

<complexType name="CRSDictionaryArrayBaseType">
    <complexContent>
        <restriction base="gml:DictionaryType">
            <sequence>
                <element ref="gml:metaDataProperty"
                    ⌐minOccurs="0" maxOccurs="unbounded"/>
                <element ref="gml:description" minOccurs="0"/>
                <element ref="gml:name"
                    ⌐minOccurs="0"/>
            </sequence>
        </restriction>
    </complexContent>
</complexType>

<complexType name="CRSDictionaryArrayType">
    <complexContent>
        <extension base="app:CRSDictionaryArrayBaseType">
            <sequence>
                <element ref="app:crsDictionaryEntries"
                    ⌐maxOccurs="unbounded"/>
            </sequence>
        </extension>
    </complexContent>
</complexType>
```

Note that instead of dictionaryEntry, the CRSDictionary element
has a crsDictionaryEntries property that can contain an array of objects
from the _CRS substitution group. The following schema fragment shows how
the crsDictionaryEntries property can be defined in a GML application schema:

```
<element name="crsDictionaryEntries"
    ⌐type="app:CRSArrayRefType"/>

<complexType name="CRSArrayRefType">
    <annotation>
        <documentation>A container for CRSs.</documentation>
    </annotation>
    <sequence>
        <element ref="gml:_CRS" maxOccurs="unbounded"/>
    </sequence>
</complexType>
```

15.5   What is the difference between a CRS and CS?

A CRS uses a Datum to relate a CS to the earth or something in the real
world. A CS is a means of associating coordinates to points on what mathematicians call a manifold, which can be a curve (one-dimensional manifold), surface
(two-dimensional manifold) or solid (three-dimensional manifold). The CS is
determined by a coordinate mapping or function from the manifold to the CS.
The dimension of the manifold is determined by the number of axes used by the
CS. The coordinate mapping must be continuous and have a continuous inverse.

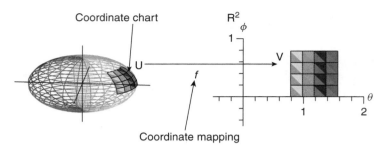

Figure 1 A coordinate mapping function. Reproduced by permission of the Open GIS Consortium.

Both the CRS and CS are illustrated in Figure 1. The CS is denoted by R^2 and consists of the θ and ϕ axes along with an assignment of units of measure (for example, degrees or radians) for these axes. A CRS that is valid in the region U consists of the set of points in U, the CS and the coordinate mapping f.

15.6   What are the CRS support components?

CRS definitions consist of a number of support components, each of which describes a particular aspect of a CRS definition. For example, in the CRSDictionary example discussed earlier in this chapter, most of the CRS definitions had CS and Datum support components. In GML, the following set of support components are provided to create CRS definitions:

- Coordinate Systems
- Coordinate System Axes
- Geodetic Datums
- Ellipsoids
- Prime Meridians
- Operations
- Operation Methods
- Operation Parameters.

The most important of the above-mentioned components are the CS and Datum. Most CRS definitions either reference or contain a CS and Datum definition. In some cases, CRS definitions also consist of a Coordinate Operation definition. Some datum definitions can consist of a Prime Meridian and an Ellipsoid definition, CS definitions have one or more Coordinate System Axis definitions, and a Coordinate Operation definition can have Operation Method and Operation Parameter definitions. Figure 2 shows the relationship between the different support components.

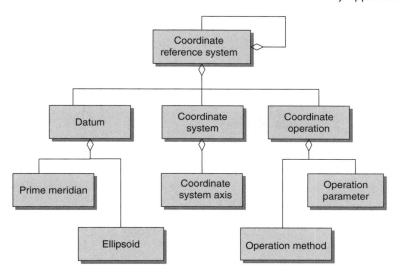

Figure 2 Coordinate reference system component relationships.

15.7   Which CRS schemas should I use in my application schema?

If a dictionary's schemaLocation attribute references an application schema and not a GML schema, the referenced application schema must import the appropriate GML schema. In the case of CRS dictionaries, the schema should either import the coordinateReferenceSystem.xsd schema or a schema that includes or imports coordinateReferenceSystem.xsd. All of the CRS schemas are available online at http://schemas.opengis.net/gml/3.0.1/base/.

With CRS, the definitions are organized in a similar fashion as the CRS components that are shown in Figure 2, however, there are not separate schemas for each kind of definition. Instead of nine schemas, the definitions are located in five schemas. For example, in GML, the datums.xsd schema defines more than just datums; it also has Ellipsoid and Prime Meridian definition elements and types.

These relationships between the schemas are important, given that when you create new CRS types, you need to know which GML schema contains the definition elements you require. Figure 3 shows the dependencies between all of the CRS-related schemas and other required GML schemas. Note that the darker schemas represent CRS schemas, while the lighter schemas are non-CRS schemas. All of the dashed arrows indicate dependency of one schema on another, in that the schema at the tail of the arrow depends upon the schema at the head of the arrow.

By importing the coordinateReferenceSystems.xsd schema, all of the other schemas in Figure 3 are also included. The temporal.xsd, geometryBasic2d.xsd and units.xsd schemas include additional GML base schemas, such as gmlBase.xsd and basicTypes.xsd. If the CRS definitions require GML types that are not included in these schemas – for example, complex geometry or topology types – the gml.xsd wrapper schema can be imported. All of the GML dependencies are covered in Chapter 11.

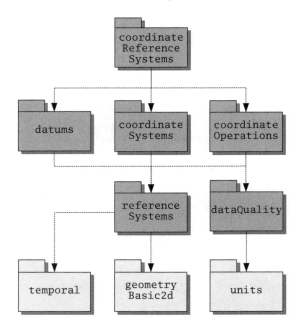

Figure 3 CRS dependencies
in GML.

15.8   What are the CRS definition elements and types provided by GML?

The GML 3.0 CRS schemas provide the concrete definition elements that are used to encode CRS dictionaries and abstract types that are used to extend the CRS model. The concrete elements are the most commonly used definitions for all of the different CRS components, and in many cases, you will not need any definition elements other than the ones already defined in these schemas.

Note that this section is intended as an overview of many of the definition elements and properties defined in the CRS schemas. For detailed descriptions of all CRS types in the GML 3.0 schemas, please refer to the *Recommended XML Encoding of Coordinate Reference System Definitions* (http://www.opengis.org/docs/03-010r7.pdf).

The CRS and support component definitions have certain common characteristics. First, they all derive indirectly from `DefinitionType` in the `dictionary.xsd` schema; that is, they are all definitions. Second, they all have a mandatory `gml:id` attribute, which is required to properly reference the different definitions.

15.8.1   Definition types from `referenceSystems.xsd`

The `referenceSystems.xsd` schema provides the abstract base type (`AbstractReferenceSystemBaseType`) for all of the CRS definition types, but not for the support component definition types. The abstract base types for the support components are defined in the `coordinateSystems.xsd`, `datums.xsd` and `coordinateOperations.xsd` schemas. The content model for `AbstractReferenceSystemBaseType` contains the optional `metaDataProperty` and `remarks` properties, and a required `srsName` property. Consequently, all CRS definitions must contain an `srsName` property, because they derive indirectly from `AbstractReferenceSystemBaseType`.

In addition to the abstract base type, the `referenceSystems.xsd` schema also provides the `_ReferenceSystem` element and its type, `AbstractReferenceSystemType`, from which the `_CRS` content model (`AbstractCRSType`) derives. The `AbstractReferenceSystemType` content model contains optional `srsID`, `validArea` and `scope` properties. The `srsID` property is a complex property with a required `code` element and optional `codeSpace`, `version`, and `remarks` elements. These elements are covered in detail in the Recommended XML Encoding of Coordinate Reference System Definitions (http://www.opengis.org/docs/03-010r7.pdf). The `validArea` property contains elements that describe the spatial, vertical and temporal extent of a reference system definition, and the value of `scope` is a string that describes the domain of usage for which a reference system definition is valid.

Table 2   Concrete elements for coordinate reference systems

Concrete Elements	Properties	Value
CompoundCRS	includesCRS (at least two occurrences)	A definition that is substitutable for _CoordinateReferenceSystem
GeographicCRS	usesEllipsoidalCS	An EllipsoidalCS definition
	usesGeodeticDatum	A GeodeticDatum definition
VerticalCRS	usesVerticalCS	A VerticalCS definition
	usesVerticalDatum	A VerticalDatum definition
GeocentricCRS	usesCartesianCS or usesSphericalCS	A CartesianCS or SphericalCS definition
	usesGeodeticDatum	A GeodeticDatum definition
ProjectedCRS	usesCartesianCS	A CartesianCS definition
	baseCRS	The CRS definition that is used by the ProjectedCRS; it can be any definition that is substitutable for _CoordinateReferenceSystem
	definedByConversion	The Conversion that is used to define the ProjectedCRS; it can be any definition that is substitutable for _GeneralConversion, such as Conversion
DerivedCRS	derivedCRSType	A description of the code of the derived CRS; this property also has a required codeSpace attribute
	usesCS	A definition that is substitutable for _CoordinateSystem

(*continued overleaf*)

Table 2   (*continued*)

Concrete Elements	Properties	Value
	baseCRS	The CRS definition that is used by the DerivedCRS; it can be any definition that is substitutable for _CoordinateReferenceSystem
	definedByConversion	The Conversion that is used to define the DerivedCRS; it can be any definition that is substitutable for _GeneralConversion
EngineeringCRS	usesCS	A definition that is substitutable for _CoordinateSystem
	usesEngineeringDatum	An EngineeringDatum definition
ImageCRS	usesCartesianCS or usesObliqueCartesianCS	A CartesianCS or ObliqueCartesianCS definition
	usesImageDatum	An ImageDatum definition
Temporal CRS	usesTemporalCS	A TemporalCS definition
	usesTemporalDatum	A TemporalDatum definition

15.8.2   Definition elements from coordinateReferenceSystems.xsd

The abstract element _CoordinateReferenceSystem, is defined in the coordinateReferenceSystems.xsd schema. The content model for this element derives from AbstractCRSType (the content model for _CRS). A number of concrete CRSs are defined in the coordinateReferenceSystems.xsd schema, as shown in Table 2.

All of the CRS elements listed in Table 2 also have as srsName property plus a number of other properties, including validArea, scope, srsID, remarks and version. These properties are defined in the referenceSystems.xsd schema. With the exception of CompoundCRS, which derives directly from AbstractCRSType, the remaining CRS types derive, directly or indirectly, from AbstractCoordinateReferenceSystemType. A CompoundCRS must have at least two includeCRS properties, each of which contains an element in the substitution group headed by _CoordinateReferenceSystem. ProjectedCRS and DerivedCRS derive from AbstractGeneralDerivedCRSType, which provides a baseCRS property and a definedByConversion property.

Note also that the coordinateReferenceSystem.xsd schema defines convenience properties for containing each CRS type. For example, the geocentricCRSRef property can only contain a GeocentricCRS definition. These properties are not likely to be required if you use the dictionary model for encoding CRSs, however, they can be used for user-defined CRSs that need to contain a specific kind of CRS definition.

15.8.3 Definition elements from `coordinateSystems.xsd`

The following CS definition elements are defined in the `coordinateSystems.xsd` schema:

- `EllipsoidalCS`
- `CartesianCS`
- `VerticalCS`
- `TemporalCS`
- `LinearCS`
- `UserDefinedCS`
- `SphericalCS`
- `PolarCS`
- `CylindricalCS`
- `ObliqueCartesianCS`.

All of the content models for these definition elements derive from `AbstractCoordinateSystemType` and are substitutable for `_CoordinateSystem`, and they all have an unbounded number of `usesAxis` properties, each of which contains the `CoordinateSystemAxis` element shown in Table 3. When they are instantiated in GML dictionaries, these CS definitions must have a `csName` property and can have an unbounded list of `csID` properties, plus `metaDataProperty` and `remarks` properties.

Note that `CoordinateSystemAxis` also has a `gml:uom` attribute for referencing a units of measure dictionary. All `CoordinateSystemAxis` definitions can also have optional `metaDataProperty` and `remarks` properties.

Table 3 Concrete definition elements for coordinate system axes

Concrete Elements	Properties	Value
CoordinateSystemAxis	axisID (optional)	Identification information for a CoordinateSystemAxis definition; CoordinateSystemAxis can have an unbounded list of axisID properties
	axisAbbrev	A CodeType string that indicates the abbreviation used for a CoordinateSystemAxis definition, such as X
	axisDirection	A CodeType string that indicates the direction, such as south, of a CoordinateSystemAxis definition
	axisName	A SimpleNameType string describing the name of the CoordinateSystemAxis definition

15.8.4   Definition elements from `datums.xsd`

In addition to deriving from `AbstractDatumType` and belonging to the `_Datum` substitution group, all of the concrete datums have a mandatory `datumName` property and optional `datumID`, `anchorPoint` and `realizationEpoch` properties. As with other identifier properties, the `datumID` property has unbounded cardinality, is of `IdentifierType` and must contain a `code` property, plus optional `codeSpace`, `version` and `remarks` properties.

The `anchorPoint` property describes the point or points at which the datum is anchored to the earth. The `realizationEpoch` property indicates the time period in which the datum is valid. Table 4 shows all of the concrete datum elements defined in the `datums.xsd` schema and any additional properties that they have. Note that `EngineeringDatum` does not have any additional properties, because it is intended mainly for use in a local context.

Table 5 lists the properties of `PrimeMeridian` and `Ellipsoid` definitions. Note that both `PrimeMeridian` and `Ellipsoid` definitions can also have optional `metaDataProperty` and `remarks` properties.

15.8.5   Definition elements from `coordinateOperations.xsd`

In addition to CSs and datums, CRS definitions can also contain coordinate operation support components. The GML constructs for encoding these components are defined in the `coordinateOperations.xsd` schema. These constructs are listed in Table 6. All of the coordinate operation definition types have a required `coordinateOperationName` property and optional `metaDataProperty`, `remarks`, `coordinateOperationID`, `validArea`, `scope` and `_positionalAccuracy` properties. With the exception of `Conversion`,

Table 4   Concrete elements for datums

Concrete Elements	Additional Properties	Value
EngineeringDatum	n/a	n/a
ImageDatum	pixelInCell	A string describing the association between the image grid and the image data; pixelInCell also has a codeSpace attribute
VerticalDatum	verticalDatumType	A string describing the code of the vertical datum; pixelInCell also has a codeSpace attribute
TemporalDatum	origin	A simple datetime value describing the data and time origin of the TemporalDatum definition
GeodeticDatum	usesPrimeMeridian	A PrimeMeridian definition (see Table 5)
	usesEllipsoid	An Ellipsoid definition(see Table 5)

Table 5 Concrete elements for prime meridians and ellipsoids

Concrete Elements	Properties	Value
PrimeMeridian	meridianID (optional)	Identification information about the PrimeMeridian definition; meridianID is of IdentifierType
	greenwichLongitude	The longitude of the prime meridian, as an angle or a dmsAngle, measured from the Greenwich meridian
	meridianName	A string (of SimpleNameType) describing the name of the PrimeMeridian definition
Ellipsoid	ellipsoidalID (optional)	Identification information about the Ellipsoid; ellipsoidID is of IdentifierType type
	semiMajorAxis	A LengthType describing of the semi-major axis of the ellipsoid
	SecondDefiningParameter	A definition – expressed as an inverseFlattening, semiMinorAxis or isSphere – of the second parameter that defines the shape of the ellipsoid
	meridianName	A string (of SimpleNameType) describing the name of the Ellipsoid definition

Table 6 Concrete elements for coordinate operations

Concrete Elements	Additional Properties	Value
ConcatenatedOperation	usesSingleOperation (at least two occurrences of this property)	A definition that is substitutable for _SingleOperation, such as Conversion
PassThroughOperation	modifiedCoordinate (unbounded)	An ordered sequence of positive integers that defines the coordinates affected by the pass-through operation
	usesOperation	An definition that is substitutable for _Operation
Conversion	usesMethod	An OperationMethod object that is used by the Conversion
	usesValue (optional and unbounded)	A choice of in-line or referenced parameter values, plus a valueOfParameter that provides an association to an OperationParameter
Transformation	usesMethod	An OperationMethod object that is used by the Transformation
	usesValue (optional and unbounded)	See Conversion

Table 7   Concrete elements for operation methods and operation parameters

Concrete Elements	Properties	Value
OperationMethod	usesParameter (optional and unbounded)	An OperationParameter definition
	methodID (optional and unbounded)	Identification information of the OperationMethod
	methodFormula	A string that lists the formula used by the operation method
	methodName	A SimpleNameType string describing the name of the OperationMethod definition
OperationParameter	parameterID (optional and unbounded)	Identification information of the OperationParameter
	minimumOccurs (optional)	A NonNegativeIntegerType that indicates the minimum times that the values for the OperationParameter are required; the default is one
	parameterName	A SimpleNameType string describing the name of the OperationParameter definition
OperationParameterGroup	groupID (optional and unbounded)	Identification information of the OperationParameterGroup
	GroupName	A SimpleNameType string describing the name of the OperationParameterGroup definition
	minimumOccurs (optional)	A nonNegativeIntegerType that indicates the minimum times that the values for the OperationParameterGroup are required; the default is one
	maximumOccurs (optional)	A positiveInteger that indicates the maximum number of times that the values for the OperationParameterGroup can be included; the default is one
	includesParameter (at least two occurrences of this property)	An association to an OperationParameter or OperationParameterGroup definition

the operation definitions also have optional `operationVersion`, `sourceCRS` and `targetCRS` properties.

The `validArea` property contains information about the extent of a coordinate operation definition, such as its valid spatial extent. The `scope` property contains the domain of usage for which the coordinate operation definition is valid. The `_positionalAccuracy` property is defined in the `dataQuality.xsd` schema, and the `operationVersion` property contains a string that indicates the version of the coordinate operation. The `sourceCRS` property contains a reference to the source CRS definition, and the `targetCRS` property contains a reference to the target CRS definition.

`Conversion` and `Transformation` definitions reference `Operation-Method` definitions via the `usesMethod` property. Table 7 lists the properties and property values of `OperationMethod`, `OperationParameter`, and `OperationParameterGroup`. Note that `OperationParameter` and `OperationParameterGroup` both have optional `metaDataProperty` and `remarks` properties.

15.9 How do I create user-defined CRS elements and types?

GML CRS schemas are created so that they are easily extensible. The schemas provide a number of abstract elements and types for extending the CRS model, including those listed in Table 8.

All of these abstract types are derived indirectly from `DefinitionType`, and consequently, all objects with content models that derive from these types can be encoded as definitions. `AbstractCRSType` and `AbstractCoordinateReferenceSystemType`, for example, both derive indirectly from `AbstractReferenceSystemBaseType`, which is derived from `DefinitionType`.

The `AbstractCoordinateReferenceSystemType` derives from `AbstractCRSType`. Most user-defined CRSs should derive from `AbstractCoordinateReferenceSystemType` unless they are compound CRSs, which should derive from `AbstractCRSType`.

GML CRS definitions follow the GML inheritance rules, which are covered in Chapters 10 and 11. The following example shows an element declaration and content model for a concrete `GeographicCRS`:

```
<element name="GeographicCRS" type="gml:GeographicCRSType"
  ↳substitutionGroup="gml:_CoordinateReferenceSystem"/>

<complexType name="GeographicCRSType">
  <annotation>
    <documentation>A coordinate reference system based
      ↳on an ellipsoidal approximation of the geoid;
      ↳this provides an accurate representation of the
      ↳geometry of geographic features for a large
      ↳portion of the earth's surface.</documentation>
  </annotation>
  <complexContent>
    <extension base=
      ↳"gml:AbstractCoordinateReferenceSystemType">
      <sequence>
        <element ref="gml:usesEllipsoidalCS"/>
```

```
              <element ref="gml:usesGeodeticDatum"/>
          </sequence>
      </extension>
  </complexContent>
</complexType>
```

As shown in the above schema fragment, the GeographicCRSType
content model extends, through derivation, the AbstractCoordinateReferenceSystemType content model. The same approach is used for creating
user-defined CRS elements and types, as shown in the following definition of
Geographic2DCRS from the CRS dictionary example discussed earlier in
this chapter.

```
<element name="Geographic2DCRS"
    ⌣type="app:Geographic2DCRSType"
    ⌣substitutionGroup="gml:_CoordinateReferenceSystem"/>

<complexType name="Geographic2DCRSType">
    <complexContent>
        <restriction base="gml:GeographicCRSType">
            <sequence>
                <element ref="gml:metaDataProperty"
                    ⌣minOccurs="0" maxOccurs="unbounded"/>
                <element ref="gml:remarks" minOccurs="0"/>
                <element ref="gml:srsName"/>
                <element ref="gml:srsID"
                    ⌣minOccurs="0" maxOccurs="unbounded"/>
                <element ref="gml:validArea" minOccurs="0"/>
                <element ref="gml:scope" minOccurs="0"/>
                <element ref="app:usesEllipsoidal2DCS"/>
                <element ref="gml:usesGeodeticDatum"/>
            </sequence>
            <attribute ref="gml:id" use="required"/>
        </restriction>
    </complexContent>
</complexType>
```

Note that Geographic2DCRSType derives by restriction from GeographicCRSType, which derives directly from AbstractCoordinateReferenceSystemType. This properly classifies the new CRS type as a CRS
definition type. The *Worked Examples CD* contains a complete example of
the CRSDictionary.xsd schema, which contains Geographic2DCRSType,
plus other user-defined types, such as Ellipsoidal2DCSRefType.

Table 8   Abstract CRS elements and types

Abstract Element	Abstract Type
_CRS	AbstractCRSType
_CoordinateReferenceSystem	AbstractCoordinateReferenceSystemType
_CoordinateSystem	AbstractCoordinateSystemType
_Datum	AbstractDatumType
_CoordinateOperation	AbstractCoordinateOperationType

Although GML allows for the creation of stand-alone support component definitions, they are not typically used independent of CRS definitions. It is anticipated that the support components – for example, Prime Meridians and Geodetic Datums – are sufficient for the definition of most CRSs and that user-defined support components will not normally be required.

If you need to create new definition types for any of the support components, such as a new CS or Datum definition type, the rules are similar to those for creating new CRS definitions, except that the new types must derive from the appropriate abstract type. For example, a new CS must derive, directly or indirectly, from `AbstractCoordinateSystemType`, and a Datum from `AbstractDatumType`.

15.10 How do I create a simple engineering CRS?

Figure 4 shows a simple engineering CRS that could be used in a construction project or for the position of objects relative to a moving object such as an airplane or an automobile. This engineering CRS uses the concrete elements and types for CRS definitions described in Section 15.8. Note that it is assumed that the engineering CRS is not related to any other CRS.

The CRS is determined by an origin point and three frame vectors as shown in the figure, which are encoded in GML as `CoordinateSystemAxis` definitions. The CRS is encoded in GML as follows:

```
<gml:EngineeringCRS gml:id="crs1123">
   <gml:srsName>BodyFixedFrame</gml:srsName>
   <gml:srsID>
      <gml:code>OXYZ</gml:code>
   </gml:srsID>
   <gml:usesCS>
      <gml:CartesianCS gml:id="c1">
         <gml:csName>3D Orthogonal Frame</gml:csName>
         <gml:usesAxis>
            <gml:CoordinateSystemAxis gml:id="x1"
               ⌐gml:uom="urn:x-si:v1999:uom:metre">
               <gml:remarks>X is along the longitudinal
                  ⌐reference line of the aircraft,
                  ⌐pointing forward. A positive rotation
                  ⌐about the X-axis corresponds to right
                  ⌐wing down.</gml:remarks>
               <gml:axisName>X-axis</gml:axisName>
               <gml:axisAbbrev>OX</gml:axisAbbrev>
               <gml:axisDirection>positive toward front of
                  ⌐aircraft</gml:axisDirection>
            </gml:CoordinateSystemAxis>
         </gml:usesAxis>
         <gml:usesAxis>
            <gml:CoordinateSystemAxis gml:id="x2"
               ⌐gml:uom="urn:x-si:v1999:uom:metre">
               <gml:remarks>Y is along the lateral
                  ⌐reference line of the aircraft,
                  ⌐pointing along the right wing. A
                  ⌐positive rotation about the Y-axis
                  ⌐corresponds to nose pitch
                  ⌐up.</gml:remarks>
```

```
            <gml:axisName>Y-axis</gml:axisName>
            <gml:axisAbbrev>OY</gml:axisAbbrev>
            <gml:axisDirection>positive toward the
                ⌐right wing</gml:axisDirection>
         </gml:CoordinateSystemAxis>
      </gml:usesAxis>
      <gml:usesAxis>
         <gml:CoordinateSystemAxis gml:id="x3"
             ⌐gml:uom="urn:x-si:v1999:uom:metre">
            <gml:remarks>Z is orthogonal to X and Y,
                ⌐pointing downward. A positive rotation
                ⌐about the Z-axis corresponds to a
                ⌐positive counter-clockwise rotation in
                ⌐yaw.</gml:remarks>
            <gml:axisName>Z-axis</gml:axisName>
            <gml:axisAbbrev>OZ</gml:axisAbbrev>
            <gml:axisDirection>Downward direction
                ⌐defined by cross product of X and
                ⌐Y</gml:axisDirection>
         </gml:CoordinateSystemAxis>
      </gml:usesAxis>
   </gml:CartesianCS>
</gml:usesCS>
<gml:usesEngineeringDatum>
   <gml:EngineeringDatum gml:id="O">
      <gml:remarks> Used to define points relative to
          ⌐the center of mass O of the aircraft
          ⌐</gml:remarks>
      <gml:datumName>Vehicle CenterOfMass
          ⌐</gml:datumName>
      <gml:anchorPoint>O</gml:anchorPoint>
       <gml:scope>Used to define points relative to the
          ⌐aircraft.</gml:scope>
   </gml:EngineeringDatum>
</gml:usesEngineeringDatum>
</gml:EngineeringCRS>
```

Note the following about this CRS definition:

1. The CS definition does not include the anchor point. The CS corresponds to the frame, and the `EngineeringDatum` provides the anchor point or the point where the frame is attached to the curve in Figure 4.
2. The CS axes are not defined relative to another CRS.

Note that this dictionary entry is also available in the `EngineeringCRS`. `xml` document on the *Worked Examples CD*.

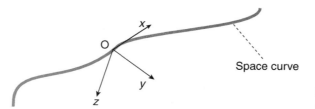

Space curve

Figure 4 CCRS1 – engineering coordinate reference system.

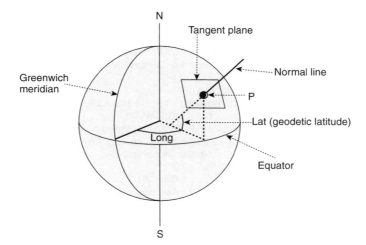

Figure 5 Geographic
CRS – illustration for
geodetic latitude
and longitude.

15.11 How do I create a simple geographic CRS?

Figure 5 shows a simple two-dimensional geographic CRS that expresses a location on the earth's surface using geodetic latitude and longitude. To encode this Geographic CRS definition in GML, you need to do the following:

1. Use the `GeographicCRS` element.
2. Specify that this `GeographicCRS` uses an `EllipsoidalCS`.
3. Specify the `GeodeticDatum` used by this `GeographicCRS`. This typically includes the referenced `Ellipsoid`, an anchor point and a prime meridian or azimuth line.

The `Ellipsoid` approximates the shape of the earth and is defined by a semi-major axis and one other parameter, which can be the inverse flattening or semi-minor axis. Note that all axes have required `gml:uom` attributes, which typically point to definitions in units of measure dictionaries. The `GeographicCRS` definition is encoded as follows:

```
<gml:GeographicCRS gml:id="EPSG62696405">
  <gml:srsName>NAD 83 (deg)</gml:srsName>
  <gml:srsID>
    <gml:code>62696405</gml:code>
    <gml:codeSpace>EPSG</gml:codeSpace>
    <gml:version>0.1</gml:version>
  </gml:srsID>
  <gml:validArea>
    <gml:description>North America</gml:description>
  </gml:validArea>
  <gml:usesEllipsoidalCS>
    <gml:EllipsoidalCS gml:id="EPSG6405">
      <gml:remarks>Axis order is by element order.
        ↵</gml:remarks>
```

```
        <gml:csName>ellipsoidal</gml:csName>
        <gml:csID>
           <gml:code>6405</gml:code>
           <gml:codeSpace>EPSG</gml:codeSpace>
           <gml:version>6.0</gml:version>
        </gml:csID>
        <gml:usesAxis>
           <gml:CoordinateSystemAxis
              ⌣gml:id="EPSG9901"
              ⌣gml:uom="urn:x-epsg:v0.1:uom:degree">
              <gml:axisName>Geodetic latitude
                 ⌣</gml:axisName>
              <gml:axisID>
                 <gml:code>9901</gml:code>
                 <gml:codeSpace>EPSG</gml:codeSpace>
                 <gml:version>6.0</gml:version>
              </gml:axisID>
              <gml:axisAbbrev>Lat</gml:axisAbbrev>
              <gml:axisDirection>north
                 ⌣</gml:axisDirection>
           </gml:CoordinateSystemAxis>
        </gml:usesAxis>
        <gml:usesAxis>
           <gml:CoordinateSystemAxis
              ⌣gml:id="EPSG9902"
              ⌣gml:uom="urn:x-epsg:v0.1:uom:degree">
              <gml:axisName>Geodetic longitude
                 ⌣</gml:axisName>
              <gml:axisID>
                 <gml:code>9902</gml:code>
                 <gml:codeSpace>EPSG</gml:codeSpace>
                 <gml:version>6.0</gml:version>
              </gml:axisID>
              <gml:axisAbbrev>Long</gml:axisAbbrev>
              <gml:axisDirection>east
                 ⌣</gml:axisDirection>
           </gml:CoordinateSystemAxis>
        </gml:usesAxis>
     </gml:EllipsoidalCS>
  </gml:usesEllipsoidalCS>
  <gml:usesGeodeticDatum>
     <gml:GeodeticDatum gml:id="EPSG6269">
        <gml:datumName>North American Datum 1983
           ⌣</gml:datumName>
        <gml:datumID>
           <gml:code>6269</gml:code>
           <gml:codeSpace>EPSG</gml:codeSpace>
           <gml:version>0.1</gml:version>
        </gml:datumID>
        <gml:usesPrimeMeridian>
           <gml:PrimeMeridian gml:id="EPSG8901">
              <gml:meridianName>Greenwich
                 ⌣</gml:meridianName>
              <gml:meridianID>
                 <gml:code>8901</gml:code>
                 <gml:codeSpace>EPSG</gml:codeSpace>
                 <gml:version>6.0</gml:version>
              </gml:meridianID>
```

```
            <gml:greenwichLongitude>
                <gml:angle
                    ⌐uom="urn:x-epsg:v0.1:uom:degree">
                    ⌐0</gml:angle>
            </gml:greenwichLongitude>
        </gml:PrimeMeridian>
    </gml:usesPrimeMeridian>
    <gml:usesEllipsoid>
        <gml:Ellipsoid gml:id="EPSG7019">
            <gml:ellipsoidName>GRS 1980
                ⌐</gml:ellipsoidName>
            <gml:ellipsoidID>
                <gml:code>7019</gml:code>
                <gml:codeSpace>EPSG
                    ⌐</gml:codeSpace>
                <gml:version>6.0</gml:version>
            </gml:ellipsoidID>
            <gml:semiMajorAxis
                ⌐uom="urn:x-si:v1999:uom:metre">
                ⌐6378137</gml:semiMajorAxis>
            <gml:secondDefiningParameter>
                <gml:inverseFlattening
                    ⌐uom="urn:x-bagug:v0.1:
                    ⌐dictionary:ifu">
                    ⌐298.257222101
                    ⌐</gml:inverseFlattening>
            </gml:secondDefiningParameter>
        </gml:Ellipsoid>
    </gml:usesEllipsoid>
        </gml:GeodeticDatum>
    </gml:usesGeodeticDatum>
</gml:GeographicCRS>
```

Note that this dictionary is also available in the `GeographicCRS.xml` document on the *Worked Examples CD*.

15.12   How do I handle a standard CRS as defined by EPSG?

There are a variety of formal and semi-formal 'dictionaries' of CRS that are in common usage. Organizations such as the USA National Imagery and Mapping Agency (NIMA), the United States Geological Survey (USGS), and the European Petroleum Survey Group (EPSG) maintain relatively comprehensive lists of CRS definitions. In addition, most national and regional mapping agencies such as Great Britain's Ordnance Survey or the Canadian Geological Survey maintain key CRS definitions of national interest. The content of the global or national CRS dictionaries can be completely represented in GML simply by creating a suitable CRS dictionary (for example, a NIMA CRS Dictionary or an EPSG CRS Dictionary). It is hoped that these agencies will put their existing CRS definitions online in machine readable form using GML in the near future. Note that this same principle applies to any organization and a CRS dictionary can be created solely for use within a single enterprise. Consider the EPSG here as just one example of 'standard' dictionary of CRS definitions.

The EPSG has developed a set of standard CRSs for use in the petroleum industry. This set comprises instances of common CRS types, such as Geographic

Table 9   EPSG component types and subtypes

Type	Subtype
CoordinateReferenceSystem	Geographic2D, Geographic3D, Projected, Vertical, Geocentric, Engineering, Compound
CoordinateSystem	Cartesian, Ellipsoidal, GravityRelated, Spherical
Datum	Vertical, Geodetic, Engineering
CoordinateAxis	CoordinateAxis
PrimeMeridian	PrimeMeridian
Ellipsoid	Ellipsoid
CoordinateOperation	Transformation, ConcatenatedOperation, Conversion
CoordinateOperationMethod	CoordinateOperationMethod
CoordinateOperationParameter	CoordinateOperationParameter
UnitsOfMeasure	UnitsOfMeasure

and Vertical. Particular CRS instances are distinguished by EPSG codes, which are assigned internally by EPSG. The EPSG database also defines all necessary support components including datums, ellipsoids, prime meridians and operations.

Table 9 lists the types of components used in the EPSG database, almost all of which are already defined in GML, except for Geographic2D and Geographic3D. Although GeographicCRS type is defined in GML, no distinction is made between two- and three-dimensional types, the dimension is determined by the number of usesAxis properties in the corresponding EllipsoidalCS. Both of these components can be defined in an application schema. A type definition of Geographic2DCRS is provided earlier in this chapter in Section 15.9. Once you have created the required extensions for the base EPSG CRS types, you can create a dictionary as described in Section 15.3.

15.13   How do I handle satellite coordinate reference systems?

The following information is required to conduct a proper analysis of a satellite's trajectory:

1. The orbital element set that contains the information about the satellite's position and movement.
2. The CRS definition that the orbital element set is measured against.
3. The coordinate transformation from the satellite CRS to an Earth-fixed CRS, such as the Conventional Terrestrial System (CTS).

Complete `Satellite.xml`, `SatelliteCRS.xml` and `Satellite.xsd` examples are provided on the *Worked Examples CD*.

15.14 What about arbitrary affine coordinate reference systems?

There are many circumstances in engineering or Computer Aided Design (CAD) in which arbitrary affine CRSs are used. Arbitrary affine CRSs are usually not related to a geospatial CRS; that is, they are not geo-referenced. It is possible to make two-dimensional CRSs that are geo-referenced indirectly if a transformation to a projected CRS is defined, however, in engineering and CAD, non-georeferenced CRSs are typically required. GML 3 provides two CRS types that do not have to be geo-referenced: `EngineeringCRS` and `ImageCRS`. Both types can be easily used for creating CRS definitions for engineering or CAD.

An example of an `EngineeringCRS` is provided in Section 15.10. The following example shows a GML encoding of an `ImageCRS` definition:

```
<gml:ImageCRS gml:id="ImageCRS">
   <gml:srsName>Image CRS</gml:srsName>
   <gml:srsID>
      <code>00021</code>
      <codeSpace>OGC</codeSpace>
      <version>0.1</version>
   </gml:srsID>
   <gml:scope>CAD</gml:scope>
   <gml:usesCartesianCS>
      <gml:CartesianCS gml:id="CartesianforImage">
      <gml:csName>2D orthogonal Cartesian CS</gml:csName>
      <gml:csID>
         <code>00022</code>
         <codeSpace>OGC</codeSpace>
         <version>0.1</version>
      </gml:csID>
      <gml:usesAxis>
         <gml:CoordinateSystemAxis gml:id="A001"
            ⌐gml:uom="http://www.measures.org/
            ⌐uom.xml#cm">
            <gml:axisName>Axis 1</gml:axisName>
            <gml:axisAbbrev>X</gml:axisAbbrev>
            <gml:axisDirection>East</gml:axisDirection>
         </gml:CoordinateSystemAxis>
      </gml:usesAxis>
      <gml:usesAxis>
         <gml:CoordinateSystemAxis gml:id="A002"
            ⌐gml:uom="http://www.measures.org/
            ⌐uom.xml#cm">
            <gml:axisName>Axis 2</gml:axisName>
            <gml:axisAbbrev>Y</gml:axisAbbrev>
            <gml:axisDirection>North</gml:axisDirection>
         </gml:CoordinateSystemAxis>
      </gml:usesAxis>
      </gml:CartesianCS>
   </gml:usesCartesianCS>
   <gml:usesImageDatum>
      <gml:ImageDatum gml:id="ID1">
         <gml:datumName>Image Origin</gml:datumName>
```

```
          <gml:datumID>
              <code>00022</code>
              <codeSpace>OGC</codeSpace>
              <version>0.1</version>
          </gml:datumID>
          <gml:pixelInCell
              ⌣codeSpace="pixelInCellCodeList.xml">
              ⌣cellCorner</gml:pixelInCell>
      </gml:ImageDatum>
    </gml:usesImageDatum>
  </gml:ImageCRS>
```

15.15   Chapter summary

Coordinate Reference Systems provide the real-world context for the interpretation of the coordinate values of geographic objects. CRS dictionaries are documents that contain a list of CRS definitions. In GML, these CRS dictionary definitions are typically referenced from geometry objects using the srsName attribute. According to *GML Best Practices*, the value of the srsName attribute should be a URN because it is a location-independent resource identifier. Note that GML 3.1 will provide local visibility attributes for data users who cannot access a dictionary.

Most data collectors who encode GML instances only need to know two things about CRS definitions: how to use the srsName attribute to reference a CRS and the list of CRS dictionaries and definitions that can be referenced. Data administrators who specialize in CRS also need to know how to encode CRS and support component dictionaries. The dictionary.xsd schema provides the Dictionary object, which can be encoded as the root element of a CRS dictionary. The Dictionary object can have an unbounded list of dictionaryEntry properties, each of which contains a CRS or support component definition.

CRS dictionary entries consist of support component definitions, in particular CS and Datum definitions, and in some circumstances, Coordinate Operation definitions. These definitions also consist of additional support components. For example, a CS definition consists of a number of Coordinate System Axis definitions and a GeodeticDatum definition consists of Prime Meridian and Ellipsoid definitions.

The CRS definition elements and types are defined in the coordinateReferenceSystems.xsd schema. CS and Coordinate System Axis definitions are defined in coordinateSystems.xsd. Datum, Prime Meridian and Ellipsoid definitions are defined in datums.xsd, and Coordinate Operations, Transformations and Conversion definitions in coordinateOperations.xsd.

Additional CRS and support component definitions can be defined in GML application schemas. For example, GML does not provide specific two- or three-dimensional GeographicCRS definition types, but it is possible to create user-defined Geographic2dCRS and Geographic3dCRS definitions. User-defined CRS definition types must derive, directly or indirectly, from AbstractCoordinateReferenceSystemType, while user-defined support components must derive from the appropriate abstract base type. For example, a user-defined CS definition's content model must derive indirectly from AbstractCoordinateSystemBaseType.

If a CRS dictionary contains application-specific types, these must be defined in a GML application schema that imports the CRS types from the appropriate GML schemas. The `coordinateReferenceSystems.xsd` schema is usually the schema that should be imported, because it includes all of the other support component schemas, plus the basic geometry, temporal and units schemas. If additional GML types are required, the `gml.xsd` wrapper schema can be imported instead.

GML 3.0 provides most of the CRS elements required for encoding dictionaries (such as engineering and geographic CRSs) for various industries that require CRS definitions. For example, all but two of the EPSG CRS types are defined in GML. The missing types are `Geographic2DCRS` and `Geographic3DCRS`, both of which can be defined in GML application schemas. To encode a satellite's trajectory, you can create user-defined `SatelliteCRS` definition elements. GML also provides CRS definition elements and types for arbitrary affine CRSs, such as those used in engineering or CAD. These definition elements are `EngineeringCRS` and `ImageCRS`. Examples of all of these dictionaries are provided on the *Worked Examples CD*.

References

http://crs.opengis.org/crsportal (October 14, 2003).

http://member.opengis.org/tc/archive/arch01/01-014r5.pdf (October 17, 2003) OGC members-only access.

http://www.opengis.org/docs/02-102.pdf (October 17, 2003).

International Organization for Standardization. (2000) ISO Technical Committee 211 Geographic Information/Geomatics. ISO DIS 19107, *Geographic Information – Spatial Schema*.

http://www.opengis.org/docs/03-010r7.pdf (October 17, 2003).

http://www.geoconnections.org/developersCorner/devCorner_devNetwork/components/GML_bpv1.3_E.pdf (October 22, 2003).

http://www.faqs.org/rfcs/rfc2141.html (October 17, 2003).

http://schemas.opengis.net/gml/3.0.1/base/ (October 15, 2003).

Additional references

http://www.ietf.org/rfc/rfc3406.txt?number =3406 (October 12, 2003).

http://schemas.opengis.net/gml/3.0.1/base/coordinate-ReferenceSystems.xsd (October 22, 2003).

http://schemas.opengis.net/gml/3.0.1/base/coordinate-Systems.xsd (October 22, 2003).

http://schemas.opengis.net/gml/3.0.1/base/datums.xsd (October 22, 2003).

http://schemas.opengis.net/gml/3.0.1/base/coordinate-Operations.xsd (October 22, 2003).

http://schemas.opengis.net/gml/3.0.1/base/dataQuality.xsd (October 22, 2003).

http://schemas.opengis.net/gml/3.0.1/base/reference-Systems.xsd (October 22, 2003).

Chapter 16

Units of measure, values and observations

In many geospatial applications, feature properties are required to have designated units of measure and values that reflect conventional measurement scales (that is, nominal, ordinal, interval and ratio scales). For example, a building's height is expressed in metres or feet (a ratio scale), while the value of a species in a species count is a name from a list of possible species names (a nominal scale).

Observations often yield measured quantities; in fact, quantities without units are often meaningless. Examples of observations include measurements of water quality, rock samples or soil types or even a tourist's photograph of a point of interest. Not all observations produce measurements, since measurement requires the use of some kind of measuring instrument (for example, a pH sensor or a radiometer), whereas some acts of observation – such as classifying the soil type – do not.

To support these different geospatial applications, GML 3 introduces a set of schema components for representing measured quantities, units of measure and observations. These schema components are found in the schemas `units.xsd`, `measures.xsd`, `valueObjects.xsd` and `observation.xsd`. All of these schemas are available at http://schemas.opengis.net/gml/3.0.1/base/.

The creation of units of measure dictionaries is covered in the first section of this chapter. The second section covers values – including non-numeric values, aggregate values and value lists – and the third section covers observations. Note that Coverages provide the distribution of values of various quantities, with their units of measure, over a geographic surface. Coverages are discussed in Chapter 10.

Note: The generic term 'value' is used throughout this book to represent the content of GML properties. For example, the `span` property of a `Bridge` feature can have a 'value' of `200`. In this chapter, the term 'value' also refers to value objects that are used to represent specific kinds of measured quantities, such as a `Category` or a `Quantity`. Value objects are discussed later in this chapter.

Geography Mark-up Language (GML).   R. Lake, D. S. Burggraf, M. Trninić, L. Rae © 2004 Galdos Systems Inc.
Published by John Wiley & Sons, Ltd   ISBNs: 0-470-87153-9 (HB); 0-470-87154-7 (PB)

16.1 Units of measure

In many of the examples covered in this book, GML properties do not have values that are explicitly associated with a specific unit of measure. For example, consider the following simple instance of a `Bridge` feature with two properties, `span` and `heightAtCenterOf`:

```
<app:Bridge gml:id="Georgia">
   <app:span>200</app:span>
   <app:heightAtCenterOf>50</app:heightAtCenterOf>
</app:Bridge>
```

On the basis of this example, it is unclear what the units of measurement are for the values of the `span` and `heightAtCenterOf` properties. Although it is possible to simply assume that all distance measurements are in metres by default – it is after all the standard unit of length – such an assumption is much too restrictive for general interoperability.

In order to designate the unit of measurement for any property value, GML provides the uom attribute, which specifies (by reference) the unit of measure for a property, as shown in the following example:

```
<app:heightAtCenter uom="#m">50</app:heightAtCenter>
```

Note that to associate a user-defined property value with a unit of measure, the uom attribute must be included in the property's type definition in a GML application schema.

In the above example, the uom attribute indicates that the units of measure for the `heightAtCenter` property are represented by the shorthand XPointer #m. There are three possible interpretations:

1. The value #m is a pointer to an element within the current document (the one in which the `heightAtCenterOfSpan` property appears) to a definition with a `gml:id` whose value is m.

2. The fragment #m is relative to a base URI specified with an `xml:base` attribute, as described in the W3C *XML Base Recommendation* (http://www.w3.org/TR/xmlbase/).

3. The #m is a string that denotes a unit of measure whose meaning is assumed to be well known; for example, m indicates length in metres.

The first option requires the document to contain a definition of the unit of measure, such as the following:

```
<gml:UnitDefinition gml:id="m">
   <gml:name>meter</gml:name>
   <gml:quantityType>length</gml:quantityType>
   <gml:catalogSymbol>m</gml:catalogSymbol>
</gml:UnitDefinition>
```

As shown in this definition, the unit represented by m has the following properties: it is a unit of measure called `meter`, the quantity is a length and the preferred symbol is m. The above definition, however, does not state who created this definition, nor does it indicate the catalogue in which the symbol m is located.

To provide more explicit information about a unit of measure, it is necessary to introduce a units of measure dictionary that lives in the domain of a particular authority. The authority might be a national standards body – such as the Canadian Standards Board or the National Institute of Standards and Technology (NIST) in the United States – an international body, such as ISO, or simply a single corporation or government department.

Suppose that the above `UnitDefinition` for metre is contained in an XML document named `Units.xml` that is available at this URL: http://www.ukusa.org/UnitsDictionary.xml. The instance fragment for the `heightAtCenterOf-Span` property then becomes

```
<app:heightAtCenterOfSpan
    ⌣uom="http://www.ukusa.org/UnitsDictionary.xml#m">
    ⌣50</app:heightAtCenterOfSpan">
```

In this example, the `uom` reference clearly indicates that the unit definition identified by m, is located in the `UnitsDictionary.xml` document, a GML units of measure dictionary. A units dictionary can be managed as a simple file or through a Catalog Service that supports XPointer references.

As discussed above, you can also use the `xml:base` attribute to relate the unit to a base URI, as shown in the following example:

```
<app:Bridge xml:base="http://www.ukusa.org/">
    <app:heightAtCenterOfSpan uom="UnitsDictionary.xml#m">
        ⌣50</app:heightAtCenterOfSpan">
</app:Bridge>
```

Note that it is also possible to use a location-independent URN for the `uom` reference. As shown in the following example, the syntax is similar to that for `srsName` (as discussed in Chapter 15):

```
"urn:x-ukusa:uom:m"
```

In the above example, `uom` represents the resource category (that is, the resource is a unit of measure), and m represents the unique identifier of the unit definition. For additional information about URNs, please see http://www.faqs.org/rfcs/rfc2141.html.

16.1.1   What are units of measure dictionaries?

In GML 3.0, the `dictionary.xsd` schema provides a model for creating various kinds of dictionaries. `Dictionary`, `Definition`, `DefinitionCol-lection`, `DefinitionProxy`, `dictionaryEntry` and `indirectEntry` are the schema components provided for representing dictionaries in GML.

You only need be familiar with the contents of this section if

1. you are creating or maintaining a units of measure dictionary, or
2. you are a data provider and must be familiar with the contents of a units of measure dictionary.

The data provider is responsible for checking that all `uom` attribute references are correct. This is part of the basic contract between the data provider and data

Table 1   SI base units and fundamental quantities

Base Unit (symbol)	Quantity
metre (m)	Length
kilogram (kg)	Mass
second (s)	Time interval
ampere (A)	Electric current
Kelvin (K)	Thermodynamic temperature
mole (mol)	Amount of substance
candela (cd)	Luminous intensity

consumers. As units of measure dictionaries are deployed in online web services, it will become easier for the data provider to ensure that the uom references are correct.

The units.xsd schema provides additional constructs for defining units of measure entries in dictionaries, including UnitDefinition, a generic element whose type (UnitDefinitionType) derives from DefinitionType and has two additional properties: quantityType and catalogSymbol. The units.xsd schema also provides the following three specific unit-definition elements whose types derive from UnitDefinitionType: BaseUnit, DerivedUnit and ConventionalUnit. These three unit-definition elements are used to encode dictionary entries in units of measure dictionaries. Note that UnitDefinition is not typically used to create units of measure dictionary entries. These definition elements and properties are also discussed in Clause 7.10 of the *GML Version 3.00 OpenGIS® Implementation Specification* (http://www.opengis.org/docs/02-023r4.pdf).

16.1.1.1   Base unit dictionary entries

Base units are independent units of measurement used for fundamental quantities, such as length, mass and time. The list of base units is defined by the associated system of units. These base units form a kind of basis of generators from which all of the other units of measure in the system are derived. Table 1 lists the seven base units and their corresponding quantity types, as defined by the International System of Units (SI) (http://www.bipm.fr/pdf/si-brochure.pdf).

In GML, the BaseUnit element is used to encode dictionary entries for base units, such as those listed in the above table. The following example shows a BaseUnit dictionary entry for metre:

```
<gml:dictionaryEntry>
  <gml:BaseUnit gml:id="m">
    <gml:description>...</gml:description>
    <gml:name codeSpace="http://www.bipm.fr/
       ⌐en/3_SI/base_units.html">metre</gml:name>
    <gml:name>meter</gml:name>
    <gml:quantityType>length</gml:quantityType>
    <gml:catalogSymbol codeSpace="http://www.bipm.fr/
       ⌐en/3_SI/base_units.html">m</gml:catalogSymbol>
```

```
    <gml:unitsSystem xlink:href="http://www.bipm.fr/
        ↳en/3_SI"/>
  </gml:BaseUnit>
</gml:dictionaryEntry>
```

The `BaseUnit` definition in this example has an id of `m` and has a number of properties that are inherited from `UnitDefinitionType`, such as `quantityType` and `catalogSymbol`. In addition to these properties, the `BaseUnit` definition also has a `unitsSystem` property that provides a remote reference to the unit system that the unit belongs to; in the above example SI is the unit system.

Some of the properties have `codeSpace` attributes, which are optional attributes for identifying an authoritative source – such as a dictionary or some other resource – for the value of the property. For example, the `codeSpace` attribute for the `catalogSymbol` indicates that the SI is the source of the symbol m for metre. The above example also includes a second `name` property with the US spelling (`meter`).

If you encode a dictionary using the elements and types from `dictionary.xsd`, the root element of the dictionary is typically `Dictionary`. For a comprehensive example of a unit of measure dictionary with multiple definitions for base, derived and conventional units, please see `UnitsDictionary.xml` on the *Worked Examples CD*.

16.1.1.2   Derived unit dictionary entries

In contrast to base units, you can create derived units by combining other units. Derived units are used for quantities other than those that correspond to the base units. For example, mass density is a unit that is derived by combining metres and kilograms (kg/m^3). A dictionary entry for mass density can be encoded in GML as follows:

```
<gml:dictionaryEntry>
  <gml:DerivedUnit gml:id="massDensity">
    <gml:description> .. </gml:description>
    <gml:name codeSpace="http://www.bipm.fr/en/3_SI">
      ↳mass density </gml:name>
    <gml:quantityType>mass density</gml:quantityType>
    <gml:catalogSymbol codeSpace="http://www.bipm.fr/
      ↳en/3_SI">kg/m3</gml:catalogSymbol>
    <gml:unitDerivation>
      <gml:unitTerm uom="#kg" exponent="1"/>
      <gml:unitTerm uom="#m" exponent="-3"/>
    </gml:unitDerivation>
  </gml:DerivedUnit>
</gml:dictionaryEntry>
```

In this example, the derived units dictionary entry is part of a definition collection for derived units. The `DerivedUnit` element has a `gml:id` of `massDensity`. In addition to the properties that it inherits from `UnitDefinitionType`, the `DerivedUnit` element has the `unitDerivation` property, which contains a set of `unitTerm` elements. As shown in the above example, each `unitTerm` has `uom` and `exponent` attributes – for example, `kg` is the unit of measure and `1` is the exponent for the first `unitTerm` element.

Each referenced unit is raised to the power of its exponent and combined with the other unit to form the product – for example, kg/m^3.

Although the `Dictionary` element is not included in the above example, this is only to emphasize the content of the entry itself. As shown in the example for a base unit definition, the definition collection from the above example must also be contained within a `Dictionary` element, via a `dictionaryEntry` property.

16.1.1.3 Conventional unit dictionary entries

GML provides a third element, `ConventionalUnit`, for encoding dictionary entries of units that are neither base nor derived. Many application domains use conventional units; for example, the unit for length commonly used in the United States is `feet` instead of `meter`. Conventional units can be converted to a preferred unit, which is either a base or a derived unit. The following schema fragment shows the element declaration and type definition for `ConventionalUnit`:

```
<element name="ConventionalUnit"
    type="gml:ConventionalUnitType"
    substitutionGroup="gml:UnitDefinition"/>

<complexType name="ConventionalUnitType">
    <complexContent>
        <extension base="gml:UnitDefinitionType">
            <sequence>
                <choice>
                    <element ref=
                        "gml:conversionToPreferredUnit"/>
                    <element ref=
                        "gml:roughConversionToPreferredUnit"/>
                </choice>
                <element ref="gml:unitDerivation"
                    minOccurs="0"/>
            </sequence>
        </extension>
    </complexContent>
</complexType>
```

In addition to deriving from `UnitDefinitionType`, the `Conventional Unit` content model has a choice of two properties – `ConversionToPreferredUnit` and `roughConversionToPreferredUnit` – plus an optional `unitDerivation` property. The first two properties contain the elements for encoding the conversion to a preferred unit, and the `unitDerivation` property can be used to indicate how this unit may be derived from other units. The conversion is done with either a scaling factor or a formula.

Consider the following example of a `ConventionalUnit` dictionary entry:

```
<gml:dictionaryEntry>
    <gml:ConventionalUnit gml:id="ft">
        <gml:name codeSpace="http://www.bipm.fr/en/3_SI">
            foot</gml:name>
        <gml:quantityType>length</gml:quantityType>
        <gml:catalogSymbol codeSpace="http://www.bipm.fr/
            en/3_SI">ft</gml:catalogSymbol>
```

```
    <gml:conversionToPreferredUnit uom="#m">
        <gml:factor>0.3048</gml:factor>
    </gml:conversionToPreferredUnit>
  </gml:ConventionalUnit>
</gml:dictionaryEntry>
```

In GML, the `conversionToPreferredUnit` property is used to encode the conversion to the preferred unit, in this example, m, as indicated by the value of the uom attribute. In this case, the property has a `factor` element, which contains a scale factor of `0.3048`. The scale factor is multiplied by the value of the conventional unit of measure to obtain the corresponding value for the preferred unit of measure.

Consider again the simple example of the `heightAtCenterOf` property with a uom reference, except this time the reference is as follows:

```
<app:heightAtCenter
  ⌐uom="http://www.ukusa.org/UnitsDictionary.xml#ft">164
</app:heightAtCenter>
```

The `heightAtCenter` property references the conventional unit, ft, instead of m. The scale factor `0.3048` can be multiplied with the value `164` to obtain the corresponding value in metres, which is approximately `50`.

Some conventional units include conversion parameters instead of a scale factor, as shown in the following example of a dictionary entry for Celsius degrees:

```
<gml:dictionaryEntry>
  <gml:ConventionalUnit gml:id="degC">
    <gml:name codeSpace="http://www.bipm.fr/
      ⌐en/3_SI">degree Celsius</gml:name>
    <gml:quantityType>Celsius temperature
      ⌐</gml:quantityType>
    <gml:catalogSymbol codeSpace="http://www.
      ⌐bipm.fr/en/3_SI">°C</gml:catalogSymbol>
    <gml:conversionToPreferredUnit uom="#K">
      <gml:formula>
        <gml:a>273.15</gml:a>
        <gml:b>1</gml:b>
        <gml:c>1</gml:c>
      </gml:formula>
    </gml:conversionToPreferredUnit>
  </gml:ConventionalUnit>
</gml:dictionaryEntry>
```

The `formula` element can contain the four elements, a, b, c and d, whose values provide the parameters for converting the value of the conventional unit of measure to the corresponding value for the preferred unit of measure. The values of the elements a, b, c and d are used in the formula $y = (a + bx)/(c + dx)$, where x is a value using the current unit, and y is the corresponding value using the preferred unit. Note that other more general conversion expressions are not currently supported in GML.

Note that elements a and d are optional, and if values are not provided, those parameters are considered to be zero, as shown above where a value is not provided for d. In the above example of the conversion of Celsius to Kelvin, the formula becomes the following:

$$K = 273.15 + C$$

16.1.2 Can I reference a units of measure dictionary from all GML objects and properties?

Units of measure dictionaries can only be referenced from objects and properties that have the uom attribute. MeasureType is a base GML type that has the uom attribute, as shown in the following schema fragment:

```
<complexType name="MeasureType">
   <simpleContent>
      <extension base="double">
         <attribute name="uom" type="anyURI"
            ⌣use="required"/>
      </extension>
   </simpleContent>
</complexType>
```

In addition to including the uom attribute, the MeasureType content model extends the XML schema built-in double type. Many GML elements that can reference units of measure dictionaries are of MeasureType. For example, the speed property of MovingObjectStatus (discussed in Chapter 14) is of MeasureType, as shown in the following element declaration:

```
<element name="speed" type="gml:MeasureType"
   ⌣minOccurs="0"/>
```

16.1.3 How do I create user-defined types that reference units of measure dictionaries?

As discussed in Chapter 11, if you want a property to have a uom attribute, the attribute must be added to the property's type definition in an application schema. A simple approach is for the property to be of MeasureType, as shown in the following:

```
<element name="heightAtCenterLine" type="gml:MeasureType">
```

This approach is appropriate for properties that do not require additional attributes or other content. There are circumstances, however, in which the property requires a different content model. Consider the following example of a heightAtCenterLine property that includes a uom attribute:

```
<element name="heightAtCenterLine"
   ⌣type="app:HeightAtCenterLineOfType">

<complexType name="HeightAtCenterLineType">
   <simpleContent>
      <extension base="double">
         <attribute name="uom" type="anyURI"
            ⌣use="optional"/>
      </extension>
   </simpleContent>
</complexType>
```

In contrast to the MeasureType definition, the uom attribute is optional in this example. It is only included here to demonstrate one way in which the content model can differ from the MeasureType content model. The above approach is recommended if you need to create a user-defined property that is qualified with a uom attribute, but whose content is different from MeasureType.

It is also possible to create user-defined objects that have the uom attribute. The rules for creating these objects are essentially the same as for creating user-defined properties with uom attributes. Note, however, that not all GML objects should be able to have the uom attribute. For example, uom attributes should not be attached to user-defined GML features, but only to their measurable properties. Similarly, uom attributes cannot be attached to derived or user-defined geometry or topology objects. CoordinateSystemAxis (from the coordinateSystems.xsd schema) is one example of a GML object with a uom attribute.

16.2   Values

GML provides a set of objects that can be used to represent the results of an observation. These objects are called value objects. Note that the need for value objects arises, in particular, in the recording of measurements that may be geographically referenced. For an example of a GML application schema that uses GML value objects and measurements, see Observations and Measurements (http://www.opengis.org/docs/03-022r3.pdf).

The valueObjects.xsd schema provides various elements and types for recording values of measured quantities. This schema can be viewed as a framework for constructing schema components for recording measurement data. The types from basicTypes.xsd are used in valueObjects.xsd to define the value objects listed in Table 2. All of these value objects are also discussed in Clause 7.10 of the *GML Version 3.00 OpenGIS® Implementation Specification* (http://www.opengis.org/docs/02-023r4.pdf).

The valueObjects.xsd schema has three abstract elements – _Value, _ScalarValue and _ScalarValueList – that can be used as variables in application schemas. For example, the range set of a Coverage can contain any value that is substitutable for _Value. This is covered in Chapter 10. Note that all of the value objects listed in Table 2 are substitutable for _Value, because both _ScalarValue and _ScalarListValue are substitutable for _Value.

16.2.1   What is a category?

A GML category is a value object for expressing distinct kinds of non-numeric values that reflect a nominal or ordinal scale (if the categories are ordered). For example, soils may be assigned a soil type from a list of soil-type categories, or rocks may be classified according to the British Geological Survey (BGS) and other rock classification schemes. These kinds of values can be encoded as categories in GML using the Category element.

In GML, the Category element is of CodeType, which can be viewed as an alternative to the enumerated type from XML schema. CodeType is useful when the enumerated list is

- so long that it is impractical to represent it as an XML Schema enumerated type;
- supported through a web service.

Table 2   Value objects

Value Object	Type	Substitutable for...	Used for Recording...
Boolean	boolean	_ScalarValue	A value from two-valued logic
Category	CodeType	_ScalarValue	A term representing a classification
Count	integer	_ScalarValue	An integer representing a rate of occurrence
Quantity	MeasureType	_ScalarValue	A numeric value with a scale
BooleanList	BooleanOrNullList	_ScalarValueList	A list of values from two-valued logic
CategoryList	CodeOrNullListType	_ScalarValueList	Terms representing a classification
CountList	integerOrNullList	_ScalarValueList	Integers representing a rate of occurrence
QuantityList	MeasureOrNullListType	_ScalarValueList	Numeric values with a scale
CategoryExtent	CategoryExtentType	_Value	Value extent for categories. This value can only contain two items
CountExtent	CountExtentType	_Value	Value extent for counts. This value can only contain two integers
QuantityExtent	QuantityExtentType	_Value	Value extent for quantities. This value can only contain two items

To create a property whose value is a category, you need to use the `Category` type with a `codeSpace` attribute, as shown in the following example of a `soilType` property:

```
<app:LandParcel gml:id="..">
  <app:soilType>
    <gml:Category codeSpace="urn:x-ukusa:soils:234">
```

```
     ‿podzolic</gml:Category>
  </app:soilType>
</app:LandParcel>
```

The `codeSpace` attribute refers to a resource (for example, a dictionary, a list or a namespace) in which the `Category` values are either stored or defined.

16.2.2   How do I define a feature with value object content in a GML application schema?

For a feature to have value object content, it must have a property that can contain value objects. Consider the following element declaration and type definition for a user-defined `LandParcel` feature:

```
<element name="LandParcel" type="app:LandParcelType"
  ‿substitutionGroup="gml:_Feature"/>

<complexType name="LandParcelType">
  <complexContent>
    <extension base="gml:AbstractFeatureType">
      <sequence>
        . . .
        <element ref="app:soilType" minOccurs="0"/>
      </sequence>
    </extension>
  </complexContent>
</complexType>
```

The complex type definition for `LandParcelType` includes the `soilType` property, which is a global property that can have the following element declaration:

```
<element name="soilType" type="gml:ValuePropertyType"/>
```

Note that `ValuePropertyType` is a GML property type that contains a group called `Value`, which is a utility group that comprises all values that are substitutable for `_Value`, plus objects from the geometry, temporal and measures schemas. By including `ValuePropertyType` in the element declaration for a user-defined property, you do not need to define the property's content. This pattern can be used to define all value object properties in GML application schemas. Note that this allows `soilType` to have a very wide range of possible value types.

16.2.3   How do I restrict a feature property to contain only one kind of value object?

GML provides the following convenience property types that can be used to restrict a property to contain a specific value object:

- `BooleanPropertyType`
- `CategoryPropertyType`
- `CountPropertyType`
- `QuantityPropertyType`.

For example, the `soilType` property from the previous section can be of `CategoryPropertyType`, as shown in the following element declaration:

```
<element name="soilType" type="gml:CategoryPropertyType"/>
```

Note that the above-listed property types are only for objects that are substitutable for _ScalarValue, and cannot be used for _ScalarValueList objects.

16.2.4 What are aggregate values?

GML 3 provides several constructs for creating aggregate values, such as lists of rock types, arrays of sensor values and lists of measured property values. These constructs include the following:

- QuantityList
- CategoryList
- CountList
- BooleanList
- CompositeValue
- ValueArray.

16.2.4.1 Scalar value lists

As shown in Table 2, the first four elements in this list are substitutable for _ScalarValueList, and their respective encodings are similar to those shown in the following examples for CategoryList and QuantityList. A CategoryList is a list of values of category type that are drawn from the same codeSpace.

```
<app:rockType>
    <gml:CategoryList codeSpace="urn:x-ukusa:rocks:345">
        ⌣SyeniteGranite Tuff</gml:CategoryList>
</app:rockType>
```

A QuantityList is a list of quantities that share a common scale (unit of measure, ordinal or ratio scale reference), as shown in the following examples:

```
<app:rockHardness>
    <gml:QuantityList uom="urn:x-ukusa:uom:massDensity">
        ⌣4.5 3.2</gml:QuantityList>
</app:rockHardness>

<app:temperature>
    <gml:QuantityList uom="urn:x-ukusa:uom:degreesC">
        ⌣100.1 20.4 25.6</gml:QuantityList>
</app:temperature>
```

16.2.4.2 Composite values

CompositeValue is an aggregate value that can have a valueComponents property or an unbounded list of valueComponent properties. This element can be used for collections of arbitrary values, such as the one shown in the following example:

```
<gml:CompositeValue>
    <gml:valueComponents>
        <gml:Point srsName=
```

```
          ⌐"urn:epsg:v6.1:coordinateReferenceSystem:1234">
          <gml:pos>10.2 -45.6</gml:pos>
       </gml:Point>
       <gml:TimeInstant>
          <gml:timePosition>2003-03-06T00:21:22.3
             ⌐</gml:timePosition>
       </gml:TimeInstant>
       <gml:Quantity uom="urn:x-si:v1999:uom:ppm">12.3
          ⌐</gml:Quantity>
       <gml:Quantity uom="urn:x-si:v1999:uom:ppm">18.2
          ⌐</gml:Quantity>
    </gml:valueComponents>
 </gml:CompositeValue>
```

Note that the values in a `CompositeValue` can be heterogeneous – that is, of more than one type – and can even include geometry and temporal objects, such as `Point` and `TimeInstant`. In the above example, the `valueComponents` property is used to encapsulate the different values. A `CompositeValue` cannot contain more than one `valueComponents` property, however, it can contain an unlimited number of `valueComponent` properties, as shown below:

```
<gml:CompositeValue>
   <gml:valueComponent>
      <gml:QuantityList uom="urn:x-si:v1999:uom:kg">
         ⌐44.2 76. withheld 30.</gml:QuantityList>
   </gml:valueComponent>
   <gml:valueComponent>
      <gml:Category>poor</gml:Category>
   </gml:valueComponent>
   <gml:valueComponent>
      <gml:Null>template</gml:Null>
   </gml:valueComponent>
</gml:CompositeValue>
```

Note that `CompositeValue` structures are very weakly typed. This means that schema validators cannot determine, for example, which `QuantityType` objects might or might not belong in a given `CompositeValue`.

Writers of GML schema parsers can detect the `CompositeValue` and its value components, but cannot determine the identities of these components in general. For example, a GML schema parser cannot determine the identities of the two `Quantity` objects in the first `CompositeValue` example, because they both have units of `ppm` and might be turbidity values, salinities or any other such quantity. Applications must have other means to determine the meaning of the `Quantity` objects. One possibility is to create user-defined value objects and value components; this is covered in the following section.

16.2.4.3   Defining value objects in application schemas

To define a value object in an application schema, you can create elements that are substitutable for `_Value`, as shown in the following element declarations:

```
<element name="Temperature" type="gml:MeasureType"
   ⌐substitutionGroup="gml:_Value"/>
<element name="Salinity" type="gml:MeasureType"
   ⌐substitutionGroup="gml:_Value"/>
```

```
<element name="Turbidity" type="gml:MeasureType"
   ⌐substitutionGroup="gml:_Value"/>
<element name="DissolvedOxygen" type="gml:MeasureType"
   ⌐substitutionGroup="gml:_Value"/>
```

These user-defined value objects can be contained within the GML-defined `valueComponent` and `valueComponent` properties. If you want to restrict the properties to only contain one or more of the above-mentioned value objects, you need to create a new abstract element, such as `_WaterQualityValue`, which can be the head of the substitution group to which the values belong. Then you need to create user-defined value-component properties that are restricted to contain the new abstract element. You also need to create a user-defined composite value that can contain these new value-component properties.

The following schema fragment shows a `WaterQualityMeasurement` object that derives by restriction from `CompositeValueType` and has two user-defined properties: `waterQualityValueComponent` and `waterQuality ValueComponents`.

```
<element name="WaterQualityMeasurement"
   ⌐type="app:WaterQualityMeasurementType">

<complexType name="WaterQualityMeasurementType">
   <complexContent>
      <restriction
         ⌐base="gml:CompositeValueType">
         <sequence>
            <element ref="gml:metaDataProperty"
               ⌐minOccurs="0" maxOccurs="unbounded"/>
            <element ref="gml:description" minOccurs="0"/>
            <element ref="gml:name" minOccurs="0"
               ⌐maxOccurs="unbounded"/>
            <element ref="app:waterQualityValueComponent"
               ⌐minOccurs="0" maxOccurs="unbounded"/>
            <element ref="app:waterQualityValueComponents"
               ⌐minOccurs="0"/>
         </sequence>
      </restriction>
   </complexContent>
</complexType>
```

The `waterQualityValueComponent` and `waterQualityValue-Components` properties can be defined as follows:

```
<element name="waterQualityValueComponent"
   ⌐type="app:WaterQualityValuePropertyType"/>

<complexType name="WaterQualityValuePropertyType">
   <sequence minOccurs="0">
      <element ref="app:_WaterQualityValue"/>
   </sequence>
   <attributeGroup
      ⌐ref="gml:AssociationAttributeGroup"/>
</complexType>

<element name="waterQualityValueComponents"
   ⌐type="app:WaterQualityValueArrayPropertyType"/>
```

```
<complexType name="WaterQualityValueArrayPropertyType">
   <sequence>
      <element ref="app:_WaterQualityValue"
         ⌣maxOccurs="unbounded"/>
   </sequence>
</complexType>
```

The abstract element _WaterQualityValue can be defined as follows:

```
<element name="_WaterQualityValue" abstract="true"
   ⌣substitutionGroup="gml:_Value">
```

For a value object to be contained within either the waterQualityVal-
ueComponent or waterQualityValueComponents property, it must be
substitutable for _WaterQualityValue, as shown below:

```
<element name="Temperature" type="gml:MeasureType"
   ⌣substitutionGroup="app:_WaterQualityValue"/>
<element name="Salinity" type="gml:MeasureType"
   ⌣substitutionGroup="app:_WaterQualityValue"/>
```

With the above schema fragments, the following is true:

1. WaterQualityMeasurement can be detected as a kind of CompositeValue, because WaterQualityMeasurementType derives from
   CompositeValueType.
2. The only children of WaterQualityMeasurement are properties
   whose values are substitutable for _WaterQualityValue.

Note that a complete version of the Water Quality example is available in the
WaterQuality.xsd and WaterQuality.xml documents on the *Worked
Examples CD*.

16.2.4.4 Value arrays

A ValueArray can also contain multiple valueComponent properties or one
valueComponents property, but unlike CompositeValue, it can only contain homogenous values – that is, values of the same type. The following example
shows a ValueArray with a valueComponents property that contains a
series of Temperature values:

```
<gml:ValueArray>
   <gml:valueComponents>
      <app:Temperature uom="urn:x-si:v1999:uom:degreesC">
         ⌣3</app:Temperature>
      <app:Temperature uom="urn:x-si:v1999:uom:degreesC">
         ⌣5</app:Temperature>
      <app:Temperature uom="urn:x-si:v1999:uom:degreesC">
         ⌣7</app:Temperature>
      ...
   </gml:valueComponents>
</gml:ValueArray>
```

Note that Temperature is a user-defined value object that is used in some
of the examples covered in Chapter 17. Note also that if the value objects in an
array are scalar values, it is more efficient to encode them as a scalar value list.

16.3 Observations

A GML observation is used to model the act of observing, using instruments, such as sensors or cameras. For example, the acquisition of an aerial photo of a city is a kind of observation. The `observation.xsd` schema defines basic constructs for creating schemas for various kinds of observations. These constructs are also discussed in Clause 7.12 of the *GML Version 3.00 OpenGIS® Implementation Specification* (http://www.opengis.org/docs/02-023r4.pdf).

16.3.1 How do I encode an observation?

In GML, observations are modelled as GML features with a valid time and a result, as shown in the following example:

```
<gml:Observation gml:id="Vancouver1">
   <gml:location xlink:href=
      ⌴"http://www.ususa.org/weather/
      ⌴GrouseMountain"/>
   <gml:timeStamp>
      <gml:TimeInstant>
         <gml:timePosition>2003-06-23T08:51:232
            ⌴</gml:timePosition>
      </gml:TimeInstant>
   </gml:timeStamp>
   <gml:using xlink:href=
      ⌴"http://www.ukusa.org/thermometers/
      ⌴thermometer2"/>
   <gml:target xlink:href=
      ⌴"http://www.ususa.org/weather/
      ⌴GrouseMountain"/>
   <gml:resultOf>
      <app:Temperature uom="urn:x-si:v1999:uom:degreesC">
         ⌴21</app:Temperature>
   </gml:resultOf>
</gml:Observation>
```

In the above example, the GML `Observation` element is used to encode an observation with `location`, `timeStamp`, `using`, `target` and `resultOf` properties. The `timeStamp` property contains the time at which the observation was made. For more information about this property, refer to Chapter 14. It is likely to be renamed `validTime` in GML 3.1. The `using` property references the instrument used in the observation, the `target` property references the target of the observation, and the `resultOf` property contains or references a value object. Note that the `using`, `target` and `resultOf` properties can either contain in-line or remote values.

> *Note:* A user-defined observation – for example, `CityTemperature` – could also be used instead of the `Observation` element to encode the above example. The rules for creating user-defined observations are covered in Section 16.3.4.

As shown in the following element declaration and type definition for `Observation`, the `timeStamp` and `resultOf` properties are required:

```
<element name="Observation" type="gml:ObservationType"
    substitutionGroup="gml:_Feature"/>

<complexType name="ObservationType">
    <complexContent>
        <extension base="gml:AbstractFeatureType">
            <sequence>
                <element ref="gml:timeStamp"/>
                <element ref="gml:using" minOccurs="0"/>
                <element ref="gml:target" minOccurs="0"/>
                <element ref="gml:resultOf"/>
            </sequence>
        </extension>
    </complexContent>
</complexType>
```

Note also that `Observation` is substitutable for `_Feature`, and its content model derives from `AbstractFeatureType`; in other words, an observation is a feature.

16.3.2   What is a directed observation?

A directed observation is essentially an observation with a `direction` property for indicating the direction (for example, NW) in which the observation was made. Directed observations should only be used for specific kinds of observations that have a directional orientation, such as visual observations made by individuals.

The `DirectedObservation` element can be used to encode a directed observation, however, user-defined directed observations can also be used, as shown in the following example of a `DirectedCityPhoto` observation:

```
<app:DirectedCityPhoto gml:id="Toronto1">
    <gml:location
        xlink:href="http://www.ususa.org/tourism/CNTower"/>
    <gml:timeStamp>
        <gml:TimeInstant>
            <gml:timePosition>2002-11-12T09:12:00
                </gml:timePosition>
        </gml:TimeInstant>
    </gml:timeStamp>
    <gml:using
        xlink:href="http://www.ususa.org/cameras/
        pentaxMZS"/>
    <gml:target
        xlink:href="http://www.ususa.org/tourism/CNTower"/>
    <gml:resultOf
        xlink:href="http://www.ususa.org/photos/
        CNTower23.jpg"/>
    <gml:direction>
        <gml:CompassPoint>NW</gml:CompassPoint>
    </gml:direction>
</app:DirectedCityPhoto>
```

In the above example, the directed observation is a picture that was taken in a direction facing northwest, as indicated by the value of the `direction` property.

Note that the `direction` property is defined in the `directions.xsd` schema and can contain or reference one of the following objects: `DirectionVector`, `CompassPoint`, `DirectionKeyword` or `DirectionString`. In the above example, the directions property contains a `CompassPoint` property, which is a simple enumeration string type.

16.3.3 What is an observation collection?

Observations can be encoded within a feature collection or an observation collection, which is a set of observations. The following example shows a user-defined observation collection `AerialPhotos` containing a list of `CityPhoto` and `StatePhoto` observations:

```
<app:AerialPhotos gml:id="Canada1">
   <app:observationMembers>
      <app:CityPhoto gml:id="Vancouver1">
      ...
      </app:CityPhoto>
      <app:DirectedCityPhoto gml:id="Toronto1">
      ...
      </app:DirectedCityPhoto>
      <app:ProvincePhoto gml:id="..">
      ...
      </app:ProvincePhoto>
      ...
   </app:observationMembers>
</app:AerialPhotos>
```

The rules for creating user-defined observations and observation collections are discussed in the following section.

16.3.4 What are the rules for creating user-defined observations?

In addition to the basic rules discussed in Chapter 11 for defining GML application schemas, the following rules apply specifically to creating application schemas for observations:

- A user-defined observation must derive, directly or indirectly, from `ObservationType`.
- A user-defined observation collection should derive from or follow the pattern of `AbstractFeatureCollectionType`.

16.3.4.1 Observations are features

All user-defined observations must adhere to the GML object-property rules covered in Chapters 9 and 11. For example, observation properties must be encoded as child elements, not attributes, and these properties must describe the observation.

16.3.4.2 User-defined observations must derive, directly or indirectly, from `ObservationType`

The content model of all user-defined observations must derive either directly or indirectly from `ObservationType`. The following schema fragment shows an element declaration for a user-defined `CityPhoto` observation:

```
<element name="CityPhoto" type="gml:ObservationType"/>
```

This element is of `ObservationType`, and therefore, its content is already defined in GML. If you want the user-defined observation to have different content, however, the type definition must be supplied in the application schema and must derive from `ObservationType`, as shown in the following:

```
<element name="CityPhoto" type="app:CityPhotoType"
   ↳substitutionGroup="gml:Observation"/>

<complexType name="CityPhotoType">
   <complexContent>
      <extension base="gml:ObservationType">
         <sequence>
            <element name="resolution" type"string"/>
         </sequence>
      </extension>
   </complexContent>
</complexType>
```

The above content model derives directly from `ObservationType`. In this example, the `CityPhotoType` content model includes a `resolution` property.

A similar pattern can be followed for encoding directed observations, except that the type should derive from `DirectedObservationType`. Note that only directed observations can derive from this type, and therefore all other observations should derive from `ObservationType`.

16.3.4.3 Observation collections should derive from `AbstractFeatureCollectionType`

An observation collection is essentially a set of observations, but it is not necessarily an observation itself. It is, however, a type of feature collection since observations are features. Because GML 3.0 does not define a specific observation collection, a user-defined element is required. The recommended approach is to derive by restriction from `AbstractFeatureCollectionType`.

The following example shows an `ObservationCollection` element that can be used as a model for other observation collections:

```
<element name="ObservationCollection"
   ↳type=app:ObservationCollectionType"
   ↳substitutionGroup="gml:_FeatureCollection"/>

<complexType name="ObservationCollectionType">
   <complexContent>
      <restriction
         ↳base="gml:AbstractFeatureCollectionType">
         <sequence>
            <element ref="app:observationMember"
               ↳minOccurs="0" maxOccurs="unbounded"/>
            <element ref="app:observationMembers"
               ↳minOccurs="0"/>
         </sequence>
```

```
          </restriction>
        </complexContent>
      </complexType>
```

In contrast to `AbstractFeatureCollectionType`, the above content model has specific `observationMember` and `observationMembers` properties, both of which can contain only observations and directed observations. The element declaration and type definition for `observationMember` can be written as follows:

```
<element name="observationMember"
    ⌐type="app:ObservationMemberType"
    ⌐substitutionGroup="gml:featureMember"/>

<complexType name="ObservationMemberType">
   <complexContent>
      <restriction base="gml:FeaturePropertyType">
         <sequence>
            <element ref="gml:Observation" minOccurs="0"/>
         </sequence>
         <attributeGroup
             ⌐ref="gml:AssociationAttributeGroup"/>
      </restriction>
   </complexContent>
</complexType>
```

The following schema fragment shows how `observationMembers` can be defined:

```
<element name="observationMembers"
    ⌐type="app:ObservationMembersType"
    ⌐substitutionGroup="gml:featureMembers"/>

<complexType name="ObservationMembersType">
   <complexContent>
      <restriction base="gml:FeatureArrayPropertyType">
         <sequence>
            <element ref="gml:Observation"
                ⌐maxOccurs="unbounded"/>
         </sequence>
      </restriction>
   </complexContent>
</complexType>
```

Note that directed observations can also be included in the `observationMember` and `observationMembers` properties, because `DirectedObservation` substitutes for `Observation`.

The following example of an `AerialPhotos` element declaration shows how additional user-defined observation collections can be declared:

```
<element name="AerialPhotos"
    ⌐type=app:ObservationCollectionType"
    ⌐substitutionGroup="app:ObservationCollection"/>
```

The `AerialPhotos` observation collection is of the type, `ObservationCollectionType`, and therefore it has the same content model.

16.4 Chapter summary

GML 3.0 provides a number of schemas that are concerned with the observation and measurement of geographic objects. These schemas include `units.xsd`, `measures.xsd`, `directions.xsd`, `valueObjects.xsd` and `observations.xsd`. These schemas are all available online at http://schemas.opengis.net/gml/3.0.1/base/.

The `units.xsd` extends the dictionary model defined in `dictionary.xsd` to provide constructs for encoding units of measure dictionaries that can be referenced from different properties and objects in GML instances. For example, a `Bridge` feature can have a `heightAtCenterOf` property that references the definition of `metre` in a units of measure dictionary.

The following three kinds of unit definitions can be encoded in units of measure dictionaries: base unit, derived unit and conventional unit. All three elements are defined in the `units.xsd` schema. The `BaseUnit` element is used to encode base unit dictionary entries, which represent the seven base units from the SI, including metre and kilogram.

`DerivedUnit` is used for derived unit dictionary entries. Derived units are units that are typically derived by combining other units; for example, mass density (kg/m^3). The `unitDerivation` property contains the different units from which the `DerivedUnit` derives. If a unit is not a base or derived unit, the `ConventionalUnit` element should be used. This element has a `conversionToPreferredUnit` property that contains the formula used for converting the unit to a preferred unit of measure.

To reference a units of measure dictionary, an object or property must use the uom attribute, can be inherited from `MeasureType`. If a GML object or property does not have the uom attribute, it cannot reference a units of measure dictionary. The same rule applies to user-defined objects and properties. Note that uom attributes have no meaning in GML when attached to user-defined features, geometries and topologies.

Value objects are used to encode the different kinds of values that can be used in GML. These value objects include `Boolean`, `Category`, `Count` and `Quantity`, plus various aggregate and extent values. A category is a non-numeric value object that can be used to provide a distinct classification for a geographic object. For example, a `LandParcel` feature can have a `soilType` property that contains a `Category` value object called `podzolic`.

Geography objects can only have value object content if they have a property that can contain value objects. For a property to contain any kind of value object, it can be of `ValuePropertyType` type. It is also possible to restrict a property to only contain one kind of value object. For example, to restrict the `soilType` property to only contain a `Category` object, the property can be of `CategoryPropertyType` type.

Aggregate value objects are lists or arrays of multiple values. Most aggregate values can only contain values of the same type – such as a `CategoryList`, which can only contain a list of categories. `CategoryList` and other similar value objects are from the `_ScalarValueList` substitution group, which means that the values are all listed inside the opening and closing tags of the value object. A `CompositeValue` object can contain different kinds of value objects as well as temporal and geometry objects, and a `ValueArray` object can contain multiple value objects of the same type.

The constructs for modelling observations in GML are provided in the `observations.xsd` schema. An observation is the act of observing something, usually with an instrument, such as a camera. There are two different GML elements for encoding observations, `Observation` and `DirectedObservation`, both of which have the same properties, except that the latter also has a `direction` property. These elements and their type definitions can also be extended to create user-defined observations and directed observations. All user-defined observations must derive, directly or indirectly, from `AbstractObservationType`.

Observation collections are sets of observations and are not observations themselves. GML 3 does not define base types for encoding observation collections, but user-defined observation collections derive from `AbstractFeatureCollectionType` and contain specific properties for observations, such as `observationMember` and `observationMembers`.

References

http://schemas.opengis.net/gml/3.0.1/base/ (October 25, 2003).

http://www.w3.org/TR/xmlbase/ (October 10, 2003).

http://www.faqs.org/rfcs/rfc2141.html (October 20, 2003).

http://www.opengis.org/docs/02-023r4.pdf (October 15, 2003).

http://www.bipm.fr/pdf/si-brochure.pdf (October 16, 2003).

http://www.opengis.org/docs/03-022r3.pdf (February 13, 2004).

Additional references

http://schemas.opengis.net/gml/3.0.1/base/units.xsd (October 21, 2003).

http://schemas.opengis.net/gml/3.0.1/base/valueObjects.xsd (October 21, 2003).

http://schemas.opengis.net/gml/3.0.1/base/measures.xsd (October 21, 2003).

http://schemas.opengis.net/gml/3.0.1/base/observation.xsd (October 21, 2003).

Chapter 17

GML Coverages

A Coverage, as defined by the OGC and ISO/TC 211, is a function describing the distribution of some set of properties over a spatial-temporal region. GML 3 support for Coverages is based on a subset of *ISO/TC 211 19123* (ISO, 2001). A GML Coverage is implemented as a GML feature that has domain and range sets. The domain set can either contain geometry or temporal objects, and the range set contains arbitrary value objects. This versatility makes Coverages very useful. Although GML provides support for both spatial and temporal Coverages, this chapter focuses primarily on spatial Coverages.

17.1 What is a Coverage?

A Coverage is a mathematical function from a spatial-temporal set (called the domain of the function) to some value set (called the range of the function). This idea is illustrated in Figure 1.

It is also possible to regard the Coverage as the graph of the function $f(x)$, that is, as the following set:

$$\{(x, f(x))|x \; in \; A\}$$

This set corresponds to the usual notion of a graph for a simple function, such as $y = f(x) = x^2$. With this interpretation, it is also possible to regard a Coverage as a set of (geometry, value) pairs, since in many cases the domain A is either a finite set of points or a set of geometry elements that cover the region of interest. You might also want to think of A as a collection of sets that cover a spatial-temporal region.

Note that in Topic 6 of the *OGC Abstract Specification* (http://www.opengis.org/docs/00-106.pdf) and the *ISO/TC 211 19123*, a Coverage is a subtype of feature. This makes sense because a Coverage has a spatial extent (for example, the extent

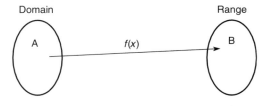

Figure 1 Coverage as a mathematical function, $f : A \rightarrow B$.

Geography Mark-up Language (GML).   R. Lake, D. S. Burggraf, M. Trninić, L. Rae © 2004 Galdos Systems Inc.
Published by John Wiley & Sons, Ltd   ISBNs: 0-470-87153-9 (HB); 0-470-87154-7 (PB)

of the set *A*). Note further that an OGC/ISO Coverage is not a Coverage in the sense of ESRI's ArcInfo.

17.2   Are Coverages always gridded data?

Coverages are not always gridded data. Although a Coverage might be based on a grid (for example, the set *A* in Figure 1 might be a uniform grid of points) or rectified grid, it might also be based on a collection of curves, triangles or other geometries. Common examples of Coverages include remotely sensed images and aerial photographs, and soil, rock type, temperature and elevation distributions.

17.3   Is a Coverage a feature collection?

Conceptually, a feature collection is a Coverage, if its members have at least one property in common and a homogeneous geometry set – that is, the geometry properties are all of the same type. Consider, for example, a feature collection with members that represent the provinces of Canada and with a birth rate property common to all of the feature members. This feature collection represents a birth rate distribution function across the provinces of Canada – that is, a birth rate Coverage. In this case, the domain of the distribution function consists of the provinces of Canada, and the range is the birth rate.

In general, however, the domain of a spatial Coverage can consist of a large set of geometry locations. In this case, a feature collection encoding is inefficient. A Coverage encoding is much more efficient because all of the geometry elements are listed as part of a domain, and the associated values for each geometry element are listed as part of a range.

Note also that a feature collection is a collection of features. The geometries in a Coverage may correspond to features (that is, as values of geometry-valued properties), but they may also just be arbitrary geometry elements without any feature interpretation. The lines on a contour map or the polygons on a soil distribution are examples of arbitrary geometric elements.

17.4   How is a GML Coverage encoding structured?

A GML Coverage consists of three parts described by the properties: `domainSet`, `rangeSet` and `coverageFunction`, as shown in the following schema fragment:

```
<complexType name="AbstractCoverageType" abstract="true">
   <complexContent>
      <extension base="gml:AbstractFeatureType">
         <sequence>
            <element ref="gml:domainSet"/>
            <element ref="gml:rangeSet"/>
            <element ref="gml:coverageFunction"
               ⌣minOccurs="0"/>
         </sequence>
         <attribute name="dimension"
            ⌣type="positiveInteger" use="optional"/>
      </extension>
   </complexContent>
</complexType>
```

All Coverages must derive, directly or indirectly, from `AbstractCoverageType`.

17.4.1   About the `domainSet` property

In GML instances, the `domainSet` property contains a set of geometry or temporal objects. The following encoding shows a `domainSet` property with `Multipoint` geometry:

```
<gml:domainSet>
    <gml:MultiPoint srsName="..">
        <gml:pointMembers>
            <gml:Point gml:id="p1">
                <gml:coordinates>1,1</gml:coordinates>
            </gml:Point>
            <gml:Point gml:id="p6">
                <gml:coordinates>2,2</gml:coordinates>
            </gml:Point>
            <gml:Point gml:id="p11">
                <gml:coordinates>3,3</gml:coordinates>
            </gml:Point>
            <gml:Point gml:id="p16">
                <gml:coordinates>4,4</gml:coordinates>
            </gml:Point>
        </gml:pointMembers>
    </gml:MultiPoint>
</gml:domainSet>
```

Note that the `MultiPoint` object has a `pointMembers` property that contains a list of `Point` instances with `coordinate` properties. The `pointMembers` property is typically used when all of the `Point`s are encoded as in-line values. If a `Point` is referenced remotely, then the `pointMember` property must be used.

`MultiPoint` is only one kind of Coverage geometry that can be contained within a `domainSet`. The property can also contain a `MultiSurface`, `Grid` or `RectifiedGrid`. A `domainSet` for a simple `Grid` can be encoded as follows:

```
<gml:domainSet>
    <gml:Grid dimension="2">
        <gml:limits>
            <gml:GridEnvelope>
                <gml:low>1 1</gml:low>
                <gml:high>2 2</gml:high>
            </gml:GridEnvelope>
        </gml:limits>
        <gml:axisName>x</gml:axisName>
        <gml:axisName>y</gml:axisName>
    </gml:Grid>
</gml:domainSet>
```

Examples of `MultiSurface` and `RectifiedGrid` domains are provided later in this chapter and in the Average Temperature Coverage examples on the *Worked Examples CD*. GML 3 also provides specific Coverage elements and domain properties for all four kinds of spatial Coverage geometries. These elements and properties are covered later in this chapter.

Note that GML 3.0 does not include a `MultiCurve` domain set. This was an omission from the specification and is likely to be rectified in GML 3.1. Data modellers can create their own `MultiCurve` Coverage in a GML application schema.

17.4.2   About the `rangeSet` property

The `rangeSet` property contains the values to which the geometry elements are mapped in the `domainSet` property. There are three different methods for encoding the range set values. These methods are covered in Section 17.7.

17.4.3   About the `coverageFunction` property

The `coverageFunction` property can contain a `mappingRule` or a `grid-Function`, both of which are used to assign a corresponding value from the range to each geometry element in the domain.

17.4.3.1   Mapping rules

In GML instances, the value of the `MappingRule` can be a string or a remote reference to a specific mapping rule. If a mapping rule is not specified, the default mapping is linear. That is, the first geometry element (in document order) in the domain corresponds to the first value in the range and so on. This is the simplest mapping rule from the domain to the range, and it is one that will be used in most cases.

Figure 2 shows an example of a linear mapping rule for a `MultiPoint` Coverage. The rule maps each geometry element in the domain – represented, in this case, by `Point` objects – to a corresponding temperature value in the range. The mapping rule assigns the domain elements to the range values in the order that they appear in the document.

17.4.3.2   Grid functions

If you are working with grid or rectified grid Coverages, `GridFunction` is used instead of `MappingRule`. Figure 3 shows an example of a linear grid function with an increment order of '+x+y' for a grid Coverage, where the first geometry element is at the bottom left corner of the grid.

The `GridFunction` element has a `sequenceRule` property whose value can be one of the following: `Linear`, `Boustrophedonic`, `Cantor-diagonal`, `Spiral`, `Morton` and `Hilbert`. `Linear` is the default. Every

Figure 2 Linear mapping rule for a multipoint temperature coverage.

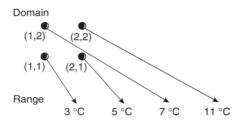

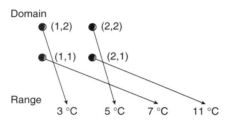

Figure 3 Linear grid function for a grid coverage with '+x+y' increment order.

Figure 4 Linear grid function for a grid coverage with '+x−y' increment order.

sequence rule can have an optional `order` attribute that specifies an increment order. For example, Figure 3 has an increment order of '+x+y', which means that the points are navigated from the lowest to the highest point on the x-axis and from the lowest to the highest on the y-axis – in this case, from (1,1) to (2,2), via (2,1) and (1,2).

The other increment orders are: '+x−y', '+y+x' and '−x−y'. With an increment order of '+x−y', the points are navigated from the lowest to the highest on the x-axis and from the highest to the lowest on the y-axis. In the case of the grid shown in Figure 4, the points are navigated from (1,2) to (2,1), via (2,2) and (1,1).

If a `sequenceRule` is supposed to start at a point other than the default (the bottom left corner of the grid), the `startPoint` property is required, as shown in the following example:

```
<gml:domainSet>
   <gml:Grid dimension="2">
      <gml:limits>
         <gml:GridEnvelope>
            <gml:low>1 1</gml:low>
            <gml:high>2 2</gml:high>
         </gml:GridEnvelope>
      </gml:limits>
      <gml:axisName>x</gml:axisName>
      <gml:axisName>y</gml:axisName>
   </gml:Grid>
</gml:domainSet>
<gml:rangeSet>...</gml:rangeSet>
<gml:coverageFunction>
   <gml:GridFunction>
      <gml:sequenceRule
         ↵order="+x-y">Linear</gml:sequenceRule>
      <gml:startPoint>1 2</gml:startPoint>
   </gml:GridFunction>
<gml:coverageFunction>
```

In the above example, the `startPoint` property's value is (1,2), which is the top left corner of the grid in Figure 4. If the `startPoint` is not specified, the default starting point is the point defined in the `low` property value in the `domainSet`, for example (1,1). Note that most of the examples in this chapter use the default grid function, `Linear`, and the '+x+y' increment order.

17.5   How do I use a Coverage encoding in a data request?

To use a Coverage encoding in a data request, you can request a Coverage that meets certain criteria on the `domainSet` and/or the `rangeSet` of the Coverage. For example, the criteria for the domain set of a spatial Coverage could specify the geometric extent – that is, the area on which the values are associated, which might be a grid or a set of polygons. Criteria on the range could specify certain range parameters – such as temperature or pressure – that satisfy a mathematical constraint – such as, `temperature > 20 degrees Celsius`. Note that the grid and rectified grid elements do not currently support irregular boundaries. Note further that these data requests, to some form of Coverage Service, would be based on the service's Coverage schema.

17.6   Which Coverage geometries are supported by GML?

Two GML core schemas, `coverage.xsd` and `grids.xsd`, provide the elements and types that support GML Coverages. Table 1 lists the four specific discrete Coverage elements currently supported by GML. Note that all four of the Coverages listed in this table are discrete Coverages. It is possible to create user-defined Coverage elements by using the content model of one of these Coverages (for example, `MultiPointCoverageType`) in the element declaration for the new Coverage. This is covered in Section 17.10.3.

> *Note:* GML does not currently provide a specific Coverage element for Multi Curves, however, it is likely that a `MultiCurveCoverage` element will be provided in a future version.

Each of the Coverage elements listed in Table 1 specifies a different `domain-Set` property. The name of the domain property suggests the kind of Coverage geometries that the property can contain. For example, the `rectifiedGrid-Domain` property can only contain Rectified Grid Coverage geometries. In the

Table 1   GML 3.0 Coverages and `domainSet` properties

Coverage	domainSet Property
MultiPointCoverage	multiPointDomain
MultiSurfaceCoverage	multiSurfaceDomain
GridCoverage	gridDomain
RectifiedGridCoverage	rectifiedGridDomain

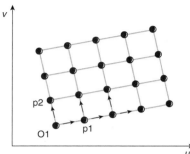

Figure 5 A rectified grid coverage domain.

following paragraphs, the `RectifiedGridCoverage` element is used to illustrate how these specific Coverages are encoded in GML. All of the constructs for encoding Coverages in GML are covered in Clause 7.13 of the *GML Version 3.00 OpenGIS® Implementation Specification* (http://www.opengis.org/docs/02-023r4.pdf), and the schemas are available online at http://schemas.opengis.net/gml/3.0.1/base/. Complete examples of the different Coverage elements and their domain properties are provided on the *Worked Examples CD*.

Coverage data often consists of measurements taken at discrete geometric locations, for example, at a discrete set of points on a grid or, more frequently, on a rectified grid. The domain in Figure 5 is a rectified grid and can be described using a `RectifiedGridCoverage` element.

In this figure, the origin is denoted as O, and the offset vectors are `p1` and `p2`. The `domainSet` for this Coverage contains a `RectifiedGrid`, which has the following properties: `limits`, `axisName`, `origin` and `offsetVector`. The following shows a `RectifiedGridDomain` encoding of a `RectifiedGrid-Coverage`:

```
<gml:rectifiedGridDomain>
   <gml:RectifiedGrid srsName=".." dimension="2">
      <gml:limits>
         <gml:GridEnvelope>
            <gml:low>0 0</gml:low>
            <gml:high>9 6</gml:high>
         </gml:GridEnvelope>
      </gml:limits>
      <gml:axisName>u</gml:axisName>
      <gml:axisName>v</gml:axisName>
      <gml:origin>
         <gml:Point gml:id="01" srsName="..">
            <gml:coordinates>2,1</gml:coordinates>
         </gml:Point>
      </gml:origin>
      <gml:offsetVector>1 0.2</gml:offsetVector>
      <gml:offsetVector>-0.2 1</gml:offsetVector>
   </gml:RectifiedGrid>
</gml:rectifiedGridDomain>
```

The `limits` property contains a `GridEnvelope` element that uses the `low` and `high` properties to describe the extent of the grid. Both of these properties contain an `integerList` that describes the lower-left and upper-right corner points of the grid.

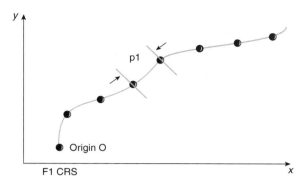

Figure 6   A one-dimensional RectifiedGrid with a linear reference system.

Each axisName property contains a string that relates to an axis of the grid; for example, u and v are the axis names for the rectified grid shown in Figure 5. The origin property contains a Point, which specifies the origin in some coordinate system. For each axis, the offsetVector property contains a vector that indicates the direction of the axis and the distance between grid points on that axis.

The coordinates of the grid points are given by the following simple vector equation:

$$\mathbf{P}_{n,m} = n\mathbf{p}_1 + m\mathbf{p}_2 + \mathbf{O}_1$$

Both $\mathbf{p}_1$ and $\mathbf{p}_2$ are the offset vectors shown in Figure 5, n and m are integers that range from low (low_x, low_y) to high (high_x, high_y) and $\mathbf{O}_1$ is the origin vector.

Note that although the example shows only two offset vectors, the number of offset vectors has to be the same as the dimension of the grid, and the offset vectors can have any number of entries. Similarly, the number of entries in the low and high elements of the GridEnvelope must be equal to the dimension of the grid.

The Coordinate Reference System (CRS) for the RectifiedGrid is determined by its srsName attribute value (please see Chapter 15 for a discussion of the use of srsName and CRSs). Note that the srsName for the Origin (the Point) can use a different CRS than that of the RectifiedGrid.

Note that it is also possible for the RectifiedGrid to be one dimensional (where only a single offset vector of dimension 1 is provided), and the CRS can be a one-dimensional linear reference system along a curve. In this case, the rectified 'grid' looks as shown in Figure 6.

Note that the Origin O is given relative to the CRS F1 (a two-dimensional affine frame) and not the linear reference system used for the offset vector. Note further that the origin of the linear reference system is not necessarily the same as the Origin O for the rectified grid. In the Figure 6, p1 is a one-dimensional vector or scalar representing the arc length between successive points on the curve. The entire curve is defined by the linear reference system.

17.7   Why does GML provide three models for range value encoding?

As mentioned above, the rangeSet property contains the values for the Coverage's range. The rangeSet property can contain one of the following: an unbounded list of ValueArray elements or of elements from the

_ScalarValueList value choice group, or a single DataBlock or File element, as shown in the following content model for RangeSetType:

```
<element name="rangeSet" type="gml:RangeSetType"/>

<complexType name="RangeSetType">
    <choice>
        <choice maxOccurs="unbounded">
            <element ref="gml:ValueArray"/>
            <element ref="gml:_ScalarValueList"/>
        </choice>
        <element ref="gml:DataBlock"/>
        <element ref="gml:File"/>
    </choice>
</complexType>
```

The three different encodings provide a trade-off between visibility in XML and efficiency. The ValueArray encoding – which is the most explicit from an XML perspective, because all values are available to XML parsers – is applicable to small and highly irregular Coverages.

The File encoding is the most opaque (i.e., none of the values are visible to an XML parser) but it is the most efficient of the three encodings and is the encoding of choice for remotely sensed images and photographs. Note that the file encoding can be used to capture and transport other popular image file formats such as Geo-Tiff and JPEG. The location of the file is specified in the value of the fileName property of the File element. This value is a URI reference to the file. Note that this can be a reference to a SOAP attachment where SOAP is being used to send the Coverage message content. Note further that although the File encoding is commonly used in conjunction with gridded geometries, it can be used with any of the possible geometries for the domainSet.

The DataBlock encoding strikes a middle ground in terms of efficiency and visibility. All of the values of the tupleList are visible in XML. Additional information is provided in the following section on the encoding of the Coverage range.

17.8   How do I encode the description of the range of the Coverage?

This section provides examples of three different rangeSet encodings for a Rectified Grid Coverage with temperature and pressure values. Note that the temperature and pressure values in these examples are defined in a GML application schema. This is covered in Section 17.10.5.

17.8.1   Encoding range data as an aggregate value

The following members of the _Value substitution group can be used to encode range data as an aggregate value:

- A ValueArray in which the members are homogeneously typed values.
- A member of the _ScalarValueList substitution group, such as CategoryList or CountList.

These aggregate values are defined in valueObjects.xsd, which is covered in Chapter 16. The following example shows how temperature and pressure values can be encoded as a value array:

```
<gml:rangeSet>
  <gml:ValueArray>
    <gml:valueComponents>
      <app:Temperature
        ⌣uom="urn:x-si:v1999:uom:degreesC">
        ⌣3</app:Temperature>
      <app:Temperature
        ⌣uom="urn:x-si:v1999:uom:degreesC">
        ⌣5</app:Temperature>
      <app:Temperature
        ⌣uom="urn:x-si:v1999:uom:degreesC">
        ⌣7</app:Temperature>
      ...
    </gml:valueComponents>
  </gml:ValueArray>
  <gml:ValueArray>
    <gml:valueComponents>
      <app:Pressure
        ⌣uom="urn:x-si:v1999:uom:kPa">
        ⌣101.2</app:Pressure>
      <app:Pressure
        ⌣uom="urn:x-si:v1999:uom:kPa">
        ⌣101.3</app:Pressure>
      <app:Pressure
        ⌣uom="urn:x-si:v1999:uom:kPa">
        ⌣101.4</app:Pressure>
      ...
    </gml:valueComponents>
  </gml:ValueArray>
</gml:rangeSet>
```

In the above example, the rangeSet has two ValueArray objects, one with temperature values and another with pressure values. Note that each ValueArray has a valueComponents property that contains a list of value objects. Each Temperature and Pressure object has a uom attribute that references SI unit of measure definitions, as discussed in Chapter 16.

17.8.2 Encoding range data as a DataBlock

The DataBlock element describes the range as a block of XML Schema double values and has two properties: rangeParameters and tupleList. The rangeParameters property contains a value object (that is substitutable for _Value), which describes the quantities in the tupleList property, including the units of measure used.

The following example shows how temperature and pressure values can be encoded in a DataBlock:

```
<gml:rangeSet>
  <gml:DataBlock>
    <gml:rangeParameters>
      <gml:CompositeValue>
        <gml:valueComponent>
          <app:Temperature
            ⌣uom="urn:x-si:v1999:uom:degreesC">
            ⌣template</app:Temperature>
        </gml:valueComponent>
```

```
            <gml:valueComponent>
                <app:Pressure uom=
                    ⌣"urn:x-si:v1999:uom:kPa">template
                    ⌣</app:Pressure>
            </gml:valueComponent>
        </gml:CompositeValue>
    </gml:rangeParameters>
    <gml:tupleList>3,101.2 5,101.3 7,101.4 11,101.5
        ⌣...</gml:tupleList>
    </gml:DataBlock>
    </gml:rangeSet>
```

Note that the `rangeParameters` property contains a `CompositeValue` object, which in turn has two `valueComponent` properties. The user-defined value objects, `Temperature` and `Pressure`, are contained within the `value-Component` properties. The `tupleList` property contains a list of coordinate tuples, where the entries of the coordinate tuples provide the numerical values of the range parameters. The tuples are interpreted according to the `valueCom-ponent` properties in document order. In the above example, the values for the first tuple are `Temperature`=3 degrees Celsius and `Pressure`=101.2 kPa.

17.8.3   Encoding the range set as a binary file

For maximum efficiency, the range set can also be encoded as a binary file using the `File` element, whose content model is defined in `FileType`, as shown in the following schema fragment:

```
<element name="File" type="gml:FileType"/>

<complexType name="FileType">
    <sequence>
        <element ref="gml:rangeParameters"/>
        <element name="fileName" type="anyURI"/>
        <element name="fileStructure"
            ⌣type="gml:FileValueModelType"/>
        <element name="mimeType" type="anyURI"
            ⌣minOccurs="0"/>
        <element name="compression" type="anyURI"
            ⌣minOccurs="0"/>
    </sequence>
</complexType>
```

The values of the range, in this case, are contained in a binary file that is referenced by the `fileName` property. The `fileStructure` property is defined by the `FileValueModelType`, which only supports a 'Record Interleaved' (also known as 'Parameter Interleaved') file structure in GML 3 (see Figure 7). Additional file structures may be supported in future releases of GML. Note also that all values must be enclosed in a single file, because multi-file structures for values are not supported in GML 3.0. The `mimetype` property points to the definition of a MIME type, while the `compression` property points to a compression algorithm. Both of these properties are optional.

The following binary file encoding can be used to record the temperature and pressure values:

```
<gml:rangeSet>
    <gml:File>
```

```
<gml:rangeParameters>
   <gml:CompositeValue>
      <gml:valueComponent>
         <app:Temperature
            ⌣uom="urn:x-si:v1999:uom:degreesC">
            ⌣template</app:Temperature>
      </gml:valueComponent>
      <gml:valueComponent>
         <app:Pressure
            ⌣uom="urn:x-si:v1999:uom:degreesC">
            ⌣template</app:Pressure>
      </gml:valueComponent>
   </gml:CompositeValue>
</gml:rangeParameters>
<gml:fileName>temperature.dat</gml:fileName>
<gml:fileStructure>Record Interleaved
      ⌣</gml:fileStructure>
   </gml:File>
</gml:rangeSet>
```

Note that the `rangeParameters` property is encoded in the same way as for the `DataBlock` example. The other two properties, `fileName` and `fileStructure`, contain the location and structure of the binary file in which the data is stored.

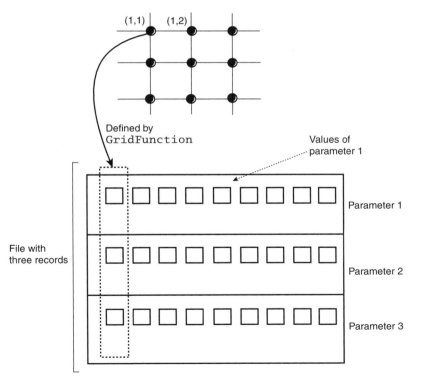

Figure 7 Record Interleaved file encoding.

17.8.4   Encoding the `AverageTempPressure` Coverage

The following example shows the entire temperature and pressure Coverage, `AverageTempPressure`, with range data encoded as a `DataBlock`:

```
<app:AverageTempPressure>
    <gml:rectifiedGridDomain>
        <gml:RectifiedGrid dimension="2">
            <gml:limits>
                <gml:GridEnvelope>
                    <gml:low>1 1</gml:low>
                    <gml:high>4 4</gml:high>
                </gml:GridEnvelope>
            </gml:limits>
            <gml:axisName>u</gml:axisName>
            <gml:axisName>v</gml:axisName>
                <gml:origin>
                    <gml:Point gml:id="P0001"
                        ⌐srsName="urn:epsg:v6.1:CRS:4327">
                        <gml:coordinates>1.2 3.3 2.1
                            ⌐</gml:coordinates>
                    </gml:Point>
                </gml:origin>
                <gml:offsetVector
                    ⌐srsName="urn:epsg:v6.1:CRS:4327">1 2 3
                    ⌐</gml:offsetVector>
                <gml:offsetVector
                    ⌐srsName="urn:epsg:v6.1:CRS:4327">4 5 6
                    ⌐</gml:offsetVector>
        </gml:RectifiedGrid>
    </gml:rectifiedGridDomain>
    <gml:rangeSet>
        <gml:DataBlock>
            <gml:rangeParameters>
                <gml:CompositeValue>
                    <gml:valueComponent>
                    <app:Temperature
                        ⌐uom="urn:x-si:v1999:uom:degreesC">
                        ⌐template</app:Temperature>
                    </gml:valueComponent>
                    <gml:valueComponent>
                    <app:Pressure
                        ⌐uom="urn:x-si:v1999:uom:kPa">
                        ⌐template</app:Pressure>
                    </gml:valueComponent>
                </gml:CompositeValue>
            </gml:rangeParameters>
            <gml:tupleList>3,101.2 5,101.3 7,101.4 11,101.5
                ⌐13,101.6  17,101.7  19,101.7  23,101.8  29,
                ⌐101.9 31,102.0  37,102.1  41,102.2  43,
                ⌐102.3  47,102.4 53,102.5  59,102.6
                ⌐</gml:tupleList>
        </gml:DataBlock>
    </gml:rangeSet>
</app:AverageTempPressure>
```

The `AverageTempPressure` Coverage is defined in a separate GML application schema, which is available on the *Worked Examples CD*. Note that this Coverage is a rectified grid Coverage with a `rectifiedGridDomain` property. The `coverageFunction` property is not included. This means that the default sequence rule, `Linear`, is used to map the geometry to the range values, and the increment order is the default, '+x+y'.

17.9 How do I process a GML Coverage in an application?

To understand how to process a GML Coverage in an application, this section examines two different examples: a Multi Point Coverage and a Rectified Grid Coverage.

EXAMPLE 1   Multi Point Coverage

First, consider the case of a simple Multi Point Coverage with a set of value arrays for the values in the range. This is an appropriate approach for small Coverage structures, but not for photographs or remotely sensed imagery.

To make things more concrete, consider an application that constructs a bar graph in which each point in the domain is represented by a bar cluster and each `value-Component` value (in the value array set) is represented by a bar. Figure 8 shows a possible result of such an application. Note that in this figure, the white bars represent the `valueComponent` values of one `ValueArray`, while the gray bars represent the `valueComponent` values of another `ValueArray`.

Assuming that the application is written in a common programming language like Java or C#, it will need to parse the GML schema and set up filters to look for things such as the `MultiPoint` geometry and the `RangeSet` with composite values in the Coverage data stream. Processing the schema should be done prior to receiving any data. The program can then build a data structure to hold the points in the domain (for example, an array of x, y coordinates) and the values in the range (a multidimensional array with one dimension for each `valueComponent`).

When the data is received, the 'filters' read the data into the `Point` and `ValueAr-ray` data structures. The program must also instantiate a mapping function (index), based on the mapping rule value obtained from the data stream. The program can create the plot using the `Point` array and the value of each `valueComponent`. Each point p1, p2, p3, p4 gives the location of each bar cluster, and each `valueComponent` value gives the height of each bar in the cluster.

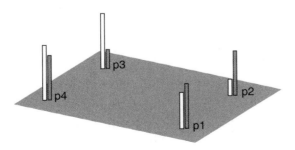

Figure 8 A simple bar graph of a Multi Point Coverage.

EXAMPLE 2   Rectified Grid Coverage

For the second example, consider the case of a Rectified Grid Coverage and a `File` encoding for the `Range` values, and assume that the file is in JPEG format with a single grayscale value. The objective of the application is to determine the average grayscale value over a polygon by computing the sum and dividing the area of the specified polygon within the extent of the Coverage.

In this example, it is assumed that this application only needs to deal with rectified grids. On parsing the schema for the Coverage, the application must find a `rectifiedGridDomain` or flag an error. On processing the actual data stream, the software looks for the `RectifiedGrid` and one of `File`, `DataBlock` or a set of GML values. Specific modules must be provided for processing data encoded within `File`, `DataBlock` and values.

To process a `File` element, do the following:

1. Allocate a multidimensional array by determining the following:
   - The array depth in the range by counting the number of `rangeParameter` values (e.g., `Temperature` and `Pressure`).
   - The other dimensions of the array from the size of the rectified grid using the parameters in the `RectifiedGrid` element.
2. Use the `File` information to locate, open and read the records in the file.
3. Read the data records from the file into the multidimensional array using the `coverageFunction` to map the elements in the desired order. Alternatively, use the `gridFunction` element to construct a function for reading the values from the array.
4. Process the multidimensional array to compute the average.

Note: The other range value encodings would be processed in an analogous fashion.

17.10   What are the rules for creating Coverage schemas?

The rules for creating application schemas for Coverages are similar to those for creating application schemas for features (covered in Chapter 11), with the following exceptions:

- The schema must import `coverages.xsd` from the GML namespace.
- New Coverage content models must derive, directly or indirectly, from `AbstractCoverageType`.
- If a Coverage's range value encoding has range parameters, they must be defined in an application schema.

17.10.1   Importing the Coverages schemas

Whenever you create application schemas with Coverages, the schemas must import `coverage.xsd`, either directly or indirectly. The following schema fragment shows how to import the schema directly:

```
<import namespace="http://www.opengis.net/gml"
  ⌐schemaLocation="http://schemas.opengis.net/gml/3.0.1/
  ⌐base/gml.xsd"/>
```

To import `coverage.xsd` indirectly, the application schema can import a GML schema document that includes `coverage.xsd`. For example, `gml.xsd` includes `coverage.xsd`.

The `schemaLocation` attribute must contain the path to `coverage.xsd`. This path can be to a local copy of the document, or it can be a URI reference to `coverage.xsd` in a remote repository, such as http://schemas.opengis.net on the OGC web site.

17.10.2   Declaring a Coverage element

Geographic Coverages are declared as global elements in the application schema, as shown in the following:

```
<element name="AverageTempPressure"
  ⌐type="gml:RectifiedGridCoverageType"
  ⌐substitutionGroup="gml:_Coverage/>
```

You can create a Coverage that has the same content model as one of the Coverage elements defined in GML – for example, `AverageTempPressure` has the `RectifiedGridCoverageType` content model. In other words, `AverageTempPressure` is of `RectifiedGridCoverage` type, and its domain property is `rectifiedGridDomain`.

The above element can be the root element of a document, or it can be part of a feature or feature collection. For example, the members of a `State` feature collection can contain the `AverageTempPressure` Coverage. Because all Coverages are substitutable for `_Feature` and derive indirectly from `AbstractFeatureType`, they can be contained within a `featureMember` property.

17.10.3   Creating user-defined Coverages based on existing GML Coverages

In many circumstances, you only need to create Coverages that are of one of the GML-defined concrete Coverage elements, as shown in the above element declaration for `AverageTempPressure`. The content model for new Coverage elements must derive, either directly or indirectly, from `gml:AbstractCoverage Type`, as shown in the type definition for `RectifiedGridCoverageType`:

```
<complexType name="RectifiedGridCoverageType">
  <complexContent>
    <restriction base="gml:AbstractCoverageType">
      <sequence>
        <element ref="gml:rectifiedGridDomain"/>
        <element ref="gml:rangeSet"/>
        <element ref="gml:coverageFunction"
          ⌐minOccurs="0"/>
      </sequence>
      <attribute name="dimension" type=
        ⌐"positiveInteger" use="required"/>
    </restriction>
  </complexContent>
</complexType>
```

17.10.4   Creating new Coverage content models

There are other circumstances, however, in which a Coverage with a completely new content model (type definition) needs to be defined. Consider the following instance of a feature Coverage for fish inventory:

```
<app:FishInventoryCoverage>
   <app:multiFeatureDomain>
      <app:FeatureBag>
         <gml:featureMember xlink:href="#Lake1"/>
         <gml:featureMember xlink:href="#River3"/>
         <gml:featureMember xlink:href="#Stream1"/>
      </app:FeatureBag>
   </app:multiFeatureDomain>
   <gml:rangeSet>
      <gml:DataBlock>
         <gml:rangeParameters>
            <gml:CompositeValue>
               <gml:valueComponent>
                  <app:FishCount>template</app:FishCount>
               </gml:valueComponent>
               <gml:valueComponent>
                  <app:FishSpecies>template
                     ↵</app:FishSpecies>
               </gml:valueComponent>
            </gml:CompositeValue>
         </gml:rangeParameters>
         <gml:tupleList>350,SteelHeadTrout 23,PinkSalmon
            ↵45,SockeyeSalmon</gml:tupleList>
      </gml:DataBlock>
   </gml:rangeSet>
</app:FishInventoryCoverage>
```

Note that unlike the other Coverages discussed in this chapter, the domain-Set property does not contain a geometry or temporal object, but a bag of features. Each feature in the FeatureBag might be described by a different geometry type, for example Polygon, LineString or Point. In a GML application schema, the element declaration and content model of FishInventoryCoverage can appear as the following:

```
<element name="FishInventoryCoverage"
   ↵type="app:MultiFeatureCoverageType"
   ↵substitutionGroup="gml:_Coverage"/>

<complexType name="MultiFeatureCoverageType">
   <annotation>
      <documentation>A discrete coverage type whose domain
         ↵is defined by a bag of features.</documentation>
   </annotation>
   <complexContent>
      <restriction base="gml:AbstractCoverageType">
         <sequence>
            <element ref="app:multiFeatureDomain"/>
            <element ref="gml:rangeSet"/>
            <element ref="gml:coverageFunction"
               ↵minOccurs="0"/>
         </sequence>
```

```
        </restriction>
      </complexContent>
    </complexType>
```

Note that the content model for `MultiFeatureCoverageType` follows a pattern similar to the content models of the other Coverage elements discussed in this chapter, in that it derives from `AbstractCoverageType` and has domain, range and Coverage function properties. The domain property, in this case, is the user-defined `multiFeatureDomain` property, which can be defined as follows:

```
<element name="multiFeatureDomain"
    ⌣type="app:MultiFeatureDomainType"
    ⌣substitutionGroup="gml:domainSet"/>

<complexType name="MultiFeatureDomainType">
    <complexContent>
      <extension base="app:FeatureDomainSetType">
        <sequence minOccurs="0">
           <element ref="app:FeatureBag"/>
        </sequence>
      </extension>
    </complexContent>
</complexType>
```

The `multiFeatureDomainType` derives by restriction from the user-defined `FeatureDomainSetType`, which can be defined as follows:

```
<complexType name="FeatureDomainSetType">
    <complexContent>
      <restriction base="gml:DomainSetType">
         <sequence/>
      </restriction>
    </complexContent>
</complexType>
```

The `multiFeatureDomainType` also contains a `FeatureBag` element, which can be defined as follows:

```
<element name="FeatureBag"
    ⌣type="app:FeatureBagType"
    ⌣substitutionGroup="gml:_GML"/>

<complexType name="FeatureBagType">
    <annotation>
      <documentation>An aggregate of Features, which is
         ⌣not itself a Feature.</documentation>
    </annotation>
    <complexContent>
      <extension base="gml:AbstractGMLType">
         <sequence>
            <element ref="gml:featureMember" minOccurs="0"
               ⌣maxOccurs="unbounded"/>
            <element ref="gml:featureMembers"
               ⌣minOccurs="0"/>
         </sequence>
      </extension>
    </complexContent>
</complexType>
```

A `FeatureBag` is different from a feature collection, because it is not a feature itself; it derives from `AbstractGMLType`, not `AbstractFeature-Type`. Also, a feature collection has a mandatory `boundedBy` property while the `FeatureBag` does not. The *Worked Examples CD* contains a complete example of the `FishCoverage.xsd` schema, including element declarations and type definitions for the range parameters.

17.10.5   Defining range parameters in GML application schemas

All of the range parameters that are used in a Coverage instance must be defined in or imported from an application schema. The range parameters must be substitutable for `gml:_Value`. The `Temperature` object can be defined as follows:

```
<element name="Temperature"
  ⌣type="gml:MeasureOrNullListType"
  ⌣substitutionGroup="gml:_Value"/>

<complexType name="TemperatureType">
  <simpleContent>
    <restriction base="gml:MeasureType">
      <attribute name="observable" type="anyURI"
        ⌣use="optional"/>
    </restriction>
  </simpleContent>
</complexType>
```

The `Pressure` object definition follows a similar pattern, as shown below.

```
<element name="Pressure" type="gml:MeasureOrNullListType"
  ⌣substitutionGroup="gml:_Value"/>

<complexType name="PressureType">
  <simpleContent>
    <restriction base="gml:MeasureType">
      <attribute name="observable" type="anyURI"
        ⌣use="optional"/>
    </restriction>
  </simpleContent>
</complexType>
```

These value objects can be used in all three range set value-encoding models – that is value sets, data blocks and binary files. Note that these value objects can be defined in the same application schema as the Coverage elements or in a separate schema. For more information about the different value objects that can be defined for Coverages, refer to Chapter 16 and to Clause 7.10.4 of the *GML Version 3.00 OpenGIS® Implementation Specification* (http://www.opengis.org/docs/02-023r4.pdf).

17.11   How do I define and encode temporal Coverages in GML?

GML does not define elements for temporal Coverages. The following schema fragment shows how a temporal Coverage, called `MultiTimeCoverage`, can be defined in an application schema:

```
<element name="MultiTimeCoverage"
  ⌣type="app:MultiTimeCoverageType"
```

```
 ⌐substitutionGroup="gml:_Coverage"/>

<complexType name="MultiTimeCoverageType">
   <annotation>
      <documentation>A discrete coverage type whose
         ⌐domain is defined by a collection of
         ⌐timePositions</documentation>
   </annotation>
   <complexContent>
      <restriction base="gml:AbstractCoverageType">
         <sequence>
            <element ref="app:multiTimeDomain"/>
            <element ref="gml:rangeSet"/>
            <element ref="gml:coverageFunction"
               ⌐minOccurs="0"/>
         </sequence>
      </restriction>
   </complexContent>
</complexType>
```

This `MultiTimeCoverage` contains a user-defined domain set called `multiTimeDomain` that can contain temporal objects. The element declaration and type definition can be specified as follows:

```
<element name="multiTimeDomain"
   ⌐type="app:MultiTimeDomainType"
   ⌐substitutionGroup="gml:domainSet"/>

<complexType name="MultiTimeDomainType">
   <complexContent>
      <restriction base="gml:DomainSetType">
         <sequence minOccurs="0">
            <element ref="app:MultiTimePosition"/>
         </sequence>
      </restriction>
   </complexContent>
</complexType>
```

The complex type specifies that domains of `MultiTimeDomainType` can contain a user-defined `MultiTimePosition` object that contains an array of temporal positions. Note that `MultiTimePosition` is not defined in GML, and, therefore, needs to be defined in an application schema, as shown below:

```
<element name="MultiTimePosition"
   ⌐type="app:MultiTimePositionType"
   ⌐substitutionGroup="gml:_TimeObject"/>

<complexType name="MultiTimePositionType">
   <complexContent>
      <extension base="gml:AbstractTimeType">
         <sequence>
            <element ref="gml:timePosition" minOccurs="0"
               ⌐maxOccurs="unbounded"/>
         </sequence>
      </extension>
   </complexContent>
</complexType>
```

This user-defined `MultiTimePositionType` can contain an unlimited list of `timePosition` properties, each of which has a temporal value. For more information about creating user-defined temporal objects, see Chapter 14. Future versions of GML will likely contain specific temporal Coverage elements and types, but this example provides a guide for defining and encoding user-defined temporal Coverages.

A temporal Coverage instance that imports the above-mentioned temporal Coverage elements can be encoded as follows:

```
<app:MultiTimeCoverage
    ⌴xmlns="http://www.MyCoverages.net/Coverages"
    ⌴xmlns:gml="http://www.opengis.net/gml"
    ⌴xmlns:xsi="http://www.w3.org/2001/XMLSchema-instance"
    ⌴xsi:schemaLocation="http://www.MyCoverages.net/
        ⌴Coverages TemporalCoverageExamples.xsd">
    <app:multiTimeDomain>
        <app:MultiTimePosition>
            <gml:timePosition>1999-12-02T11:01:00
                ⌴</gml:timePosition>
            <gml:timePosition>1999-12-02T11:11:00
                ⌴</gml:timePosition>
            <gml:timePosition>1999-12-02T11:21:00
                ⌴</gml:timePosition>
        </app:MultiTimePosition>
    </app:multiTimeDomain>
    <gml:rangeSet>
        <gml:ValueArray>
            <gml:valueComponents>
                <app:Temperature
                    ⌴uom="urn:x-si:v1999:uom:degreesC">
                    ⌴3</Temperature>
                <app:Temperature
                    ⌴uom="urn:x-si:v1999:uom:degreesC">
                    ⌴5</Temperature>
                <app:Temperature
                    ⌴uom="urn:x-si:v1999:uom:degreesC">
                    ⌴7</Temperature>
            </gml:valueComponents>
        </gml:ValueArray>
    </gml:rangeSet>
</app:MultiTimeCoverage>
```

17.12   Chapter summary

GML 3.0 supports certain kinds of Coverages, which are distribution functions defined on sets of geometry elements. In GML, Coverages are features that contain domains, ranges and mapping rules. GML 3 Coverages are based on a subset of the Coverage types defined in ISO 19123.

The domain of a Coverage can contain a set of geometry or temporal elements, though in the current version of GML, the elements are typically geometric. A Coverage's domain is encoded in GML with the `domainSet` property, which can contain geometry or temporal elements. The `rangeSet` property is used to encode the range of a Coverage.

Domain elements are mapped to range data with the `coverageFunction` property, which contains either a `MappingRule` or `GridFunction` element.

`MappingRule` applies to non-grid Coverages and is either a string or a remote reference. The default mapping rule is linear. `GridFunction` is used for grid Coverages, and it has a `sequenceRule` property, whose value can be one of six different sequence rules. The `sequenceRule` property can have an optional order attribute that is used to express the increment order of a sequence rule. `Linear` is the default sequence rule, and '+x+y' is the default increment order.

GML 3 provides four discrete spatial Coverage elements – `Multi PointCoverage`, `MultiSurfaceCoverage`, `GridCoverage` and `RectifiedGridCoverage` – each of which has its own domain set property that specifies the kind of geometry that the Coverage can contain. The content models of these Coverages can be used to create user-defined Coverages in GML application schemas. Examples are provided in the *Worked Examples CD*.

GML provides the following three models for encoding the values of a Coverage's range set: as a set of value arrays, a data block or a binary file. The value array model provides the most verbose encoding, while the binary file model is by far the most efficient. Data collectors can use any of these three models when they encode range set data in GML Coverage instances.

In addition to the rules for creating application schemas covered in Chapter 11, a few other rules apply to application schemas with Coverages. The schema must import `coverages.xsd` from the GML namespace. The content models of all new Coverage elements must derive, directly or indirectly, from `AbstractCoveragesType`. If a Coverage has range parameters, they must be defined in an application schema.

It is possible to create completely new Coverage elements and types, such as temporal Coverages. Currently, GML 3.0 only provides the most basic support for temporal Coverages. Although future versions will probably provide more specific models for encoding temporal Coverage data, you can define temporal Coverage elements in application schemas. The *Worked Examples* CD contains application schema and instance examples with temporal Coverage elements and types. The rules for modelling temporal elements are covered in Chapter 14.

References

International Organization for Standardization. (2001) ISO Technical Committee 211 Geographic Information/Geomatics. *ISO DIS 19123, Geographic Information – Schema for Coverage Geometry and Functions*.

http://www.opengis.org/docs/00-106.pdf (October 17, 2003).

http://www.opengis.org/docs/02-023r4.pdf (October 15, 2003).

Additional references

http://schemas.opengis.net/gml/3.0.1/base/coverage.xsd (October 15, 2003).

http://schemas.opengis.net/gml/3.0.1/base/grids.xsd (October 15, 2003).

Chapter 18

GML default styling

As elsewhere in XML, GML follows the principle of separation of content and presentation. Most of GML deals with the modelling and representation of geographic content, and only the `defaultStyle.xsd` schema – the subject of this chapter – is concerned with presentation. This chapter should be of particular interest to cartographers and anyone else concerned with the visual presentation of geographic information. The contents of this chapter are based on Clause 7.14 of the *GML Version 3.00 OpenGIS® Implementation Specification* (http://www.opengis.org/docs/02-023r4.pdf). Note that you can access the schema at http://schemas.opengis.net/gml/3.0.1/base/defaultStyle.xsd.

The separation of data and presentation provides many benefits. For example, it allows the same data to be presented in many different ways, and for different users and devices, simply by applying a different style. The content and presentation information can also be managed at different locations and by different authorities.

18.1 What is styling?

The process that applies a set of rules to geographic data to generate a presentation in the form of text, graphics or even voice is called styling. A style is just a set of rules that relates GML data to the presentation of that data. The current version of GML default styling is concerned only with styling for graphical presentation, and thus describes mappings from geographic objects to graphic objects.

The term 'default style' indicates that the style is a default option and may be overridden by other styling information or simply ignored. When attached to a feature collection, the default style applies to all features in the collection that meet the constraints of the style.

Default styling can also be used in situations that require persistent styling information. For example, it is very convenient to be able to draw a map in the same manner, regardless of where or when the drawing is made. To address this requirement, styling-information sets (that is, GML style instances) can be saved and attached to specific features or feature collections.

GML data styling is feature based; the styling information always applies to a feature, multiple features or a feature collection. It is not possible to apply styling

Geography Mark-up Language (GML). R. Lake, D. S. Burggraf, M. Trninić, L. Rae © 2004 Galdos Systems Inc.
Published by John Wiley & Sons, Ltd ISBNs: 0-470-87153-9 (HB); 0-470-87154-7 (PB)

information directly to other GML objects, such as geometry or topology objects. For example, if you want to apply a style to a geometry object, you must apply the style to the feature that contains the geometry object. The style can contain geometry style descriptions that apply to the feature's geometry, but a separate geometry style cannot exist independently of the feature-style description. This is discussed in Section 18.6.

18.2   What is the default style model?

To ensure that presentation and GML data are separate, the default style model has been designed so that there is only a minimal binding between the data and the style. To refer to a default style description from a feature or feature collection, simply attach the `defaultStyle` property to the feature or feature collection in question. This associates the feature to a `Style` object, which can be embedded within the feature – that is, inside the `defaultStyle` property – or remotely referenced using the `xlink:href` attribute. Figure 1 shows how the `defaultStyle` property can be used to associate a `Style` object with a `Feature`.

The `Style` object is a GML object that is the top-level styling entity, and it is a container for the style descriptors that apply to different components of the GML data. These style descriptors are discussed in detail further below.

The abstract element, `_Style` – which is the head element of the substitution group to which the `Style` object belongs – can be used to create user-defined style elements and types. A user-defined style type must derive, directly or indirectly, from the `AbstractStyleType`, and must be substitutable for `_Style`.

18.3   How do I encode the `defaultStyle` property in GML?

There are three important aspects of the `defaultStyle` property.

- First, default styling in GML is feature related and always applies to features.
- Second, the `defaultStyle` property, like many other properties in GML, has an `AssociationAttributeGroup` set of attributes that allows for remote referencing of the `Style` object.
- Third, two methods are supported for applying the `Style` to the GML data – implicitly, based on the position of the `defaultStyle` property

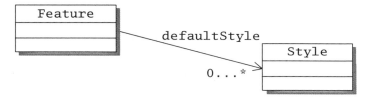

Figure 1 Binding of a `Style` to a `Feature` via the `defaultStyle` property.

within GML data, or explicitly, by using the property's about attribute. The about attribute's type is an anyURI and its value can point to any feature or features within the feature collection.

The following example shows a City feature collection that includes the Style object as an in-line value of the defaultStyle property.

```
<app:City gml:id="..">
   <gml:boundedBy>...</gml:boundedBy>
   <gml:featureMember>...</gml:featureMember>
   <gml:defaultStyle>
      <gml:Style>...</gml:Style>
   </gml:defaultStyle>
</app:City>
```

In the next example, the City feature collection uses the xlink:href attribute to reference a style that is located at a remote URI.

```
<app:City gml:id="..">
   <gml:boundedBy>...</gml:boundedBy>
   <gml:featureMember>...</gml:featureMember>
   <gml:defaultStyle
      ⌣xlink:href=
      ⌣"http://www.ukusa.org/hydroStyle.xml#Style001"/>
</app:City>
```

The following Bridge feature also references a remote style with the xlink:href attribute:

```
<app:Bridge gml:id="Bridge001">
   <gml:centerLineOf>...</gml:centerLineOf>
   <gml:defaultStyle
      ⌣xlink:href=
      ⌣"http://www.ukusa.com/transportationStyle.xml
      ⌣#Style002"/>
</app:Bridge>
```

The following example uses an about attribute to explicitly refer to a particular Bridge feature within the City feature collection:

```
<app:City gml:id="City001">
   <gml:boundedBy>...</gml:boundedBy>
   <gml:featureMember>...
      <app:Bridge gml:id="Bridge001">
         <gml:centerLineOf>...</gml:centerLineOf>
      </app:Bridge>
   </gml:featureMember>
   ...
   <gml:defaultStyle gml:about="#Bridge001"
      ⌣xlink:href=
      ⌣"http://www.ukusa.org/transportationStyle.xml
      ⌣#Style002"/>
   <gml:defaultStyle
      ⌣xlink:href="http://www.ukusa.org/hydroStyle.xml
      ⌣#Style001"/>
</app:City>
```

In the above example, the `defaultStyle` is only applied to the `Bridge` with the unique identifier of `#Bridge001` and not to any other `Bridge` features in the `City` feature collection. This approach should be used when you want specific feature collection members of a given type to have a different style than all other members of the collection.

18.4 Is there a relationship between default styling and SVG?

GML data can be graphically portrayed using a wide variety of graphics software and graphical representation standards, including proprietary drawing software and open standards, such as W3C Scalable Vector Graphic (SVG). GML default styling makes use of SVG and other standards to describe the graphical portrayal of GML features.

SVG provides GML with a grammar for the description of two-dimensional graphic objects such as points, lines, text and polygons. Note that this does not mean that GML default styling software is required to use SVG. Developers simply need to know how to interpret SVG with their own drawing primitives. For more information on SVG, please refer to http://www.w3.org/TR/SVG11/.

18.5 How do I include default styles in a GML application schema?

To associate GML data with styling definitions, application schema developers must include the `defaultStyle` property in their feature-type definitions. The following schema fragment shows how to define a `Bridge` with a `default-Style` property by including the default style in the content model:

```
<element name="Bridge" type="app:BridgeType"
   ⌐substitutionGroup="gml:_Feature"/>

<complexType name="BridgeType">
   <complexContent>
      <extension base="gml:AbstractFeatureType">
         <sequence>
            <element ref="gml:centerLineOf"/>
            <element ref="gml:defaultStyle"
               ⌐minOccurs="0"/>
         </sequence>
      </extension>
   </complexContent>
</complexType>
```

In the above example, the `maxOccurs` value is the default, which is 1. It is possible for the multiplicity of the element to be unbounded (that is, `maxOccurs="unbounded"`), however, this depends on how styles are managed within your application domain. There are no specific rules in GML for managing the number of default styles that can be applied to a GML data set.

18.6 What are the different style descriptors?

The `Style` object consists of two main style descriptors: `FeatureStyle` and `GraphStyle`. The `FeatureStyle` describes the style for a feature or a group of features and is composed of the style descriptors for feature geometry, topology and label. The `GraphStyle` describes the style for topology graphs, which

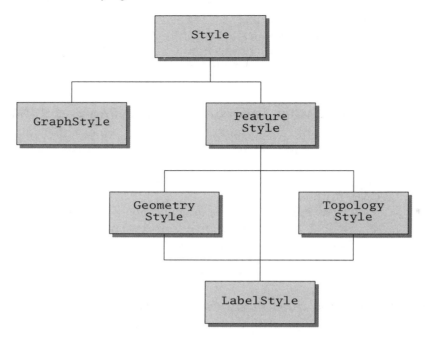

Figure 2 Detailed default style model diagram.

comprise a number of features. Figure 2 shows a diagram of the default style model, including the hierarchy of the different style descriptors.

In GML instances, the `FeatureStyle` and `GraphStyle` elements must be enclosed within a `Style` object. The `FeatureStyle` can have `GeometryStyle`, `TopologyStyle` and `LabelStyle` style descriptors. The `GeometryStyle` and `TopologyStyle` elements can also include a `LabelStyle` element.

The `GraphStyle` describes the overall appearance of graphs that represent multiple features, each of which can have an associated `FeatureStyle` element that determines the appearance of each individual feature type. Note that although a `Style` element can contain only a single `GraphStyle` element, it can contain any number of `FeatureStyle` elements.

In default styling, the GML object-property pattern applies only to the `defaultStyle-Style` pair. The style descriptors themselves do not follow the object-property model, and therefore, they are referred to as elements instead of objects or properties. All of the style-descriptor components are regarded as 'arbitrary complex types', and consequently, no extensibility mechanism is currently provided for them. In other words, at this time, you cannot create user-defined style-descriptor types. This is likely to change in GML 3.1.

18.7   How is the feature style encoded in GML?

To apply a feature style to a specific set of feature instances, use the `FeatureStyle` element with either of the following attributes: `featureType` or `baseType`.

- The `featureType` attribute is used to specify the element name of the feature, such as Road for all Road features or `Bridge` for all `Bridge` features within a given data set. This is the most common way to constrain a style to a set of feature instances. A feature style is typically created by manually or automatically examining the application schema and deciding which styles apply to which features.

- The `baseType` attribute can be used to specify a semantically different constraint. Instead of associating a style with a specific feature element name, the `baseType` attribute applies a style to all features whose content models derive from the same complex type. This is discussed in greater detail below.

The `FeatureStyle` element can contain a `featureConstraint` element, which provides additional filtering of the feature set to which the feature style is applied. The value of the `featureConstraint` is an XPath expression that results in a set of features. The following example shows a style that applies a feature-style descriptor to any Road instance that has more than two lanes.

```
<app:City gml:id="..">
    <gml:boundedBy>...</gml:boundedBy>
    <gml:featureMember>...</gml:featureMember>
    <gml:defaultStyle>
        <gml:Style>
            <gml:FeatureStyle featureType="app:Road">
                <gml:FeatureConstraint>
                    ⌐//app:Road[app:laneCount>2]
                    ⌐</gml:FeatureConstraint>
                ...
            </gml:FeatureStyle>
        <gml:Style>
    </gml:defaultStyle>
</app:City>
```

Note that the `featureType` attribute is `app:Road`, which applies this `FeatureStyle` to all `app:Road` instances within a given data set (feature collection). The `FeatureConstraint` element contains an XPath expression that limits the `FeatureStyle` to applying only to Road features with a `laneCount` property whose value is greater than 2. For more information about XPath, see http://www.w3.org/TR/xpath.

Styles are typically applied to GML data sets by associating graphic attributes with the properties of named concrete features, such as Road or Building. Although this is an accepted practice, it lacks flexibility and can be somewhat tedious if you need to create similar styles for many features. For this reason, GML default styling also provides the `baseType` attribute. This allows a single style to be specified for all features whose content models derive from a specified 'base' type.

To understand how this mechanism works, consider the following example. Imagine that you have an application schema that defines two features, each representing a specific type of road: Highway and Interstate. Each of these features is defined using complex types, for example, HighwayType and InterstateType. Both of these complex types derive from the common

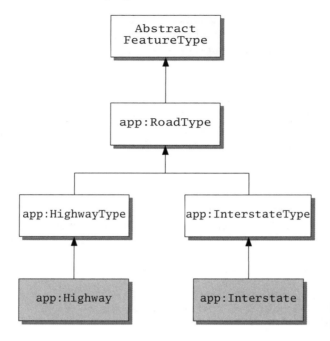

Figure 3 Road
complex-type hierarchy.

parent complex type, RoadType, which derives from AbstractFeature-
Type. Figure 3 shows the hierarchy of these different types. The white rectangles
represent the complex types, while the shaded rectangles represent element
names. For example, the Highway element is of HighwayType type.

Both the HighwayType and InterstateType are derived from the
same base complex type, RoadType, which can be encoded as the value of the
baseType attribute in a FeatureStyle element, as shown in the following
example:

```
<gml:defaultStyle>
   <gml:Style>
      <gml:FeatureStyle baseType="app:RoadType">
         <gml:styleVariation styleProperty="stroke">
            ⌣blue</gml:styleVariation>
         <gml:symbol symbolType="svg"
            ⌣xlink:href="http://www.symbols.com/
            ⌣Transportation.xml#Road"/>
      </gml:FeatureStyle>
   <gml:Style>
</gml:defaultStyle>
```

In the above example, the content of the FeatureStyle description is
associated with any feature instance that is ultimately based on the RoadType,
and therefore all Highway and Interstate features will have the same
style. Note that if you create new features whose content models derive from
RoadType, they will also have the same style.

18.7.1   Geometry and topology style

The `FeatureStyle` element can also include the following three kinds of style descriptors: geometry, topology and label styles. Geometry and topology styles are described in a similar manner. The geometry style is enclosed within the `GeometryStyle` element, and the topology style within the `Topolo-gyStyle` element. In both cases, `symbol`, `style` and `styleVariation` elements are used in the same manner to describe the graphic attributes of a feature's geometry or topology. These elements are discussed in greater detail below.

A feature can have multiple geometry and topology properties, and you can create distinct style descriptions for each of them by using a `GeometryStyle` or `TopologyStyle` element. Both of these style elements have attributes that specify the particular geometry or topology that the style is applied to. These attributes are shown in Table 1.

The `geometryType` and `topologyType` attributes are required because many geometry- and topology-valued properties can contain different kinds of objects. Note that many of the geometry and topology properties and objects are covered in Chapters 13 and 14.

The following example shows a `Lake` feature that has a geometry style element with a `geometryProperty` attribute whose value is `extentOf` and a `geometryType`, whose value is `Polygon`:

```
<gml:FeatureStyle featureType="app:Lake">
    <gml:GeometryStyle geometryProperty="gml:extentOf"
        ⌐geometryType="gml:Polygon">
    ...
    </gml:GeometryStyle>
</gml:FeatureStyle>
```

Table 1   Attributes for `GeometryStyle` and `TopologyStyle`

Element	Attributes	Value
GeometryStyle	geometryProperty	The name of the geometry-valued property to which the geometry style applies. For example, extentOf. Note that the geometry style is applied to the property, and not the geometry, since a given feature can have several properties with the same geometry.
	geometryType	The declared name of the geometry object contained within the geometry property. For example, Polygon.
TopologyStyle	topologyProperty	The name of the topology-valued property to which the topology style applies. For example, topoPointProperty.
	TopologyType	The declared name of the topology object contained within the topology property. For example, TopoPoint.

Now, consider an example of a `Station` feature that has a `TopologyStyle` element with a `topologyProperty` and a `topologyType` attribute.

```
<gml:FeatureStyle featureType="app:Station">
  <gml:TopologyStyle
    ⌐topologyProperty="gml:topoPointProperty"
    ⌐topologyType="gml:TopoPoint">
  ...
  </gml:TopologyStyle>
</gml:FeatureStyle>
```

> *Note:* The content of the `TopologyStyle` element specifies the styles that are applied to particular topology properties of selected features, while the `GraphStyle` element describes the styling information for an entire topology graph. This is discussed in greater in Section 18.8.

18.7.2   Label style

The label style is always applied to text content, and it controls text characteristics, such as size, font and colour. The text content is specified using the `label` element. The value of the `label` element can be a combination of text and element content, including, in particular, any number of `LabelExpression` elements. The content of the `LabelExpression` element is an XPath expression. Any valid XPath expression that evaluates to text content can be used to extract useful information from the data. Consider the following example of a `Road` feature with name and `direction` properties:

```
<app:Road gml:id="..">
  <gml:name>Highway 1</gml:name>
  <app:direction>East</app:direction>
  <gml:extentOf>
  ...
  </gml:extentOf>
</app:Road>
```

In the following example, the `name` and the `direction` are extracted from the `Road` feature and combined into a label.

```
<gml:LabelStyle>
  <gml:label>
    Road:
    <LabelExpression>//Road/name<LabelExpression>
    , Direction:
    <LabelExpression>//Road/direction<LabelExpression>
  </gml:label>
</gml:LabelStyle>
```

The resulting label then appears as:

Road: Highway 1, Direction: East

18.7.3 Symbolization, symbols and styles

In GML styling, the term 'symbol' refers to a generic graphic entity that has graphic properties and attributes, such as shape, colour and fill. Symbols can include traditional marker symbols, drawings, raster objects, line styles and area fills. Various formats, including binary or XML grammars such as SVG, can be used to encode these graphic entities. GML default styling relies on some of the existing style standards – in particular, SVG, CSS2 and Synchronized Multimedia Integration Language (SMIL) – and uses their grammars to represent specific graphical characteristics.

In GML default styling, symbols and features are not implicitly related. In general, any symbol can be used to represent any feature. Also, there can be more than one symbol associated with a feature in order to symbolize different aspects of the feature. For example, features with multiple geometric properties can have different symbols associated with each property.

Another common requirement is to have different representations of a feature for different map scales. For example, a Town feature can be drawn as a closed polygon on a large-scale map and as a famous bridge icon on a small-scale map. To relate a feature to a symbol, use the symbol element child of either GeometryStyle or TopologyStyle within the FeatureStyle for the associated feature.

The symbol element derives from AssociationType, which is used in GML to create properties that have in-line or remote values. As a result, the symbol element can have an in-line child that is the symbol description, or it can use the xlink:href attribute to reference the description remotely.

In default styling, it is likely that most symbols will be encoded as remote properties. Symbols are usually standardized and reused in various applications. Once reusable symbol sets are published on the Internet and made publicly available, it will be very easy to reference them using the xlink:href attribute. Note, however, that for rapid map drawing, early binding to these referenced symbols (for example, by de-referencing the link) will be necessary.

The symbol element has a symbolType attribute. In the default style schema, symbolType is an enumeration of two values, svg and other. Because SVG is XML-based, it is an ideal format for describing both symbols and individual graphic attributes. The following example shows how to relate a School feature to a symbol that is located at an external URI. Note that the symbolType is svg.

```
<gml:FeatureStyle featureType="app:School">
   <gml:GeometryStyle geometryProperty="gml:position"
      ⌙geometryType="gml:Point">
   <gml:symbol symbolType="svg" xlink:href=
      ⌙"http://www.symbols.com/
      ⌙citySymbols.xml#School"/>
   </gml:GeometryStyle>
</gml:FeatureStyle>
```

In this example, the FeatureStyle element uses the featureType attribute to apply the style to all School features in a data set. The FeatureStyle element also includes a GeometryStyle element, which contains the symbol element.

There are cases where the use of a symbol does not apply. For example, a set of graphic properties can be so simple and common that it does not make sense to create a distinct symbol. In such cases, the `style` element can be used to specify the style description directly. The `type` of the `style` element is `string`. The value of the string is based on CSS2 and must consist of semicolon-separated CSS2 property declarations of the form, `name:value`.

The following example shows how the `style` element can be used to specify the font family and size of the Highway 1 label:

```
<gml:FeatureStyle featureType="app:Road">
   <gml:LabelStyle>
      <gml:style>font-family:Arial;font-size:14;
         ⌣</gml:style>
      <gml:label>
         Road:
         <LabelExpression>//Road/name<LabelExpression>
         , Direction:
         <LabelExpression>//Road/direction
            ⌣<LabelExpression>
      </gml:label>
   </gml:LabelStyle>
</gml:FeatureStyle>
```

The resulting label resembles the following:

Road: Highway 1, Direction: East

Note: The `style` element is not the same as the upper-camel-case `Style` object discussed earlier in this chapter. The `Style` object is used to encapsulate the style description of a feature, while the `style` element can be contained within a `GeographyStyle`, `TopologyStyle` or `Label-Style` element.

18.7.4   Style variations and parameterized styles

In some applications, the same symbol is often reused with only slight changes to its style description. Because it is redundant to create and maintain a large number of symbols that are almost identical, GML default styling provides the `styleVariation` element. Different style descriptions can use this element to reference the same symbol and override some of its graphic properties.

The `styleVariation` element can also be used to create parameterized styles. These styles have graphic properties that depend on the values of feature properties. For example, earthquake features may be represented using circles of different radii, with the location specified by the earthquake epicentre, and the size depending on the earthquake's magnitude.

Note that `styleVariation` has two attributes, `featureProper-tyRange` and `styleProperty`. The `featurePropertyRange` is an XPath expression that expresses a constraint on the feature's properties. The `styleProperty` is a string that identifies a graphic property, such as 'r' (radius)

Figure 4 Style variations for earthquake features.

in SVG. Its value must be either a symbol property – as specified in the graphic language for the symbol – or a CSS2 property name.

For features whose properties satisfy the featurePropertyRange constraint, the value of styleProperty (for example, r for the circle radius) is given by the value of the styleVariation element. Note that this value is an XPath expression, and thus can be a constant or some expression involving feature properties (for example, Earthquake/magnitude div 2).

Figure 4 shows different style variations of the circle symbols that represent the epicentres for an Earthquake feature collection. There are three different radii. The style variations from Figure 4 can be encoded in GML as follows:

```
<gml:FeatureStyle featureType="app:Earthquake">
   <gml:GeometryStyle geometryProperty="gml:position"
     ⌐geometryType="gml:Point">

     <gml:styleVariation styleProperty="r"
       ⌐featurePropertyRange="//Earthquake/
       ⌐magnitude = 7">3</gml:styleVariation>

     <gml:styleVariation styleProperty="r"
       ⌐featurePropertyRange="//Earthquake/magnitude
       ⌐< 7 and //Earthquake/magnitude >= 3">
       ⌐//Earthquake/magnitude div 2
       ⌐</gml:styleVariation>

     <gml:styleVariation styleProperty="r"
       ⌐featurePropertyRange="//Earthquake/
       ⌐magnitude <3">1</gml:styleVariation>

     <gml:symbol
       ⌐xlink:href="http://www.Symbols.com/
       ⌐earthquake.xml#circle1"/>
   </gml:GeometryStyle>
</gml:FeatureStyle>
```

Note that `styleProperty="r"` references the r(radius) attribute of the circle symbol referenced through the `symbol` element (`<gml:symbol xlink:href="http://www.Symbols.com/earthquake.xml#circle1"/>`).

18.8 How is the graph style encoded in GML?

The graph style is not part of a feature style, because it is used to draw topology graphs, which always comprise a number of features; topology graphs cannot be constructed from only a single feature. The elements of the graph correspond to features, and each feature's style is described within a feature-style descriptor.

In GML styling, the style of a topology graph is described in two parts. The first part, `TopologyStyle` – which was discussed in the previous section – describes the style that applies to individual elements of the graph, such as nodes, edges and faces. The second part of the graph style describes the styling properties that apply to a graph as a whole. These properties correspond to the inputs of common graph-drawing algorithms, such as those covered in *Graph Drawing: Algorithms for the Visualization of Graphs* (Di Battista *et al.*, 1999). The graph-drawing properties include the type of the graph, the maximum angles, the existence of bends or crossings of edges and the distance between nodes or edges. Note that these properties determine the appearance of the graph independent of the feature's geometry.

The graph style consists of a number of elements, which are enumerations that are used for specifying style constraints. The graph-layout algorithm uses these constraints to construct the resulting graph. Since coordinates are not used to determine a feature's topology, many different valid results can be produced, as is the nature of topological graphs.

Consider the example of a road network, where roads are represented as edges (for example, a Road feature has a topology property that is edge-valued) and intersections as nodes of a topological graph. Depending on your preferences, you can generate various kinds of drawings. The following style instructs the layout algorithm to create a bi-connected, non-directional graph with edges that do not cross:

```
<gml:GraphStyle>
    <gml:planar>true</gml:planar>
    <gml:directed>false</gml:directed>
    <gml:minDistance>15</gml:minDistance>
    <gml:minAngle>30</gml:minAngle>
    <gml:graphType>BICONNECTED</gml:graphType>
    <gml:drawingType>POLYLINE</gml:drawingType>
    <gml:lineType>BENT</gml:lineType>
    <gml:aestheticCriteria>MAX_BENDS
       ↵</gml:aestheticCriteria>
</gml:GraphStyle>
```

The encoding also specifies that the minimal distance between edges and non-incident nodes is 15 (the number may be interpreted arbitrarily). The angular resolution is at least 30 degrees, and the edges may be bent. To create a cleaner

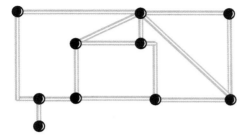

Figure 5 Topological graph example with bent edges.

image, a maximized number of edge bends is also specified. Figure 5 shows a possible visual result, based on the above style.

You may want the same graph to indicate the directions of the roads. In order to do this, the information about directions must exist in the GML data, and the layout manager must use this information. Note that, in general, the styling information is only a description of what the final result should look like. The actual result depends on the application that generates the layout and drawing.

In some types of topological analyses, such as traffic analysis, the length of the roads may be less important than the number of intersections. Consequently, you might want to portray the road network by making all roads the same length. To make the drawing even more appropriate for such an analysis, you might decide to draw all roads as straight lines. All these constraints are expressed in the following style:

```
<gml:GraphStyle>
    <gml:planar>true</gml:planar>
    <gml:directed>true</gml:directed>
    <gml:minAngle>15</gml:minAngle>
    <gml:graphType>BICONNECTED</gml:graphType>
    <gml:drawingType>POLYLINE</gml:drawingType>
    <gml:lineType>STRAIGHT</gml:lineType>
    <gml:aestheticCriteria>MIN_AREA</gml:aestheticCriteria>
</gml:GraphStyle>
```

The arrows on the lines indicate the direction of each line, thus reflecting that the value of the `directed` property is `true`, and the value of the `lineType` property is STRAIGHT, not BENT. The result can resemble the drawing shown in Figure 6.

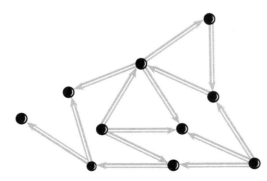

Figure 6 Topological graph example with straight directed edges.

Table 2   Animation elements

Element	Function
animate	Generic attribute animation
animateMotion	Moving an element along the path
animateColor	Animating colour attributes
set	Setting the value of an attribute for a specified duration

Section 18.10, provides an example of a map with a TopologyStyle element and a corresponding GraphStyle element.

18.9   What are some of the additional styling elements?

In addition to the items discussed in the previous sections, the default-Style.xsd also provides animation and spatial resolution elements. To describe animation, the default style uses a set of animation elements from the W3C SMIL, which specification that is available at http://www.w3.org./TR/smil20/. These elements are optional and can appear within the GraphStyle, GeometryStyle, TopologyStyle and LabelStyle elements. Table 2 shows the different animation elements and what they are used for. Styles with animation parameters may be applied to dynamic or time-varying data, which is discussed in Chapter 14.

SpatialResolution is a parameter that carries generic information about the map, including scale, size and density. In GML, this parameter is based on the definition of the equivalent spatial resolution attribute in the *ISO DIS 19115: Geographic information – Metadata* specification, where it is defined as a factor that provides a general understanding of the density of spatial data in the data set (ISO, 2003). Other than this definition, GML does not specify the exact use of this element.

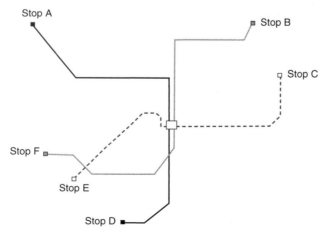

Figure 7 Simple subway map.

Application developers can use `SpatialResolution` in a number of different ways, one of which is to determine how to draw features at different scales. For example, a `City` feature collection can be represented as a symbol on a small-scale map, while it can be drawn in more detail for a large-scale map.

18.10   How can I use default styles to draw a map?

To understand how GML data elements can be styled to draw a map, consider the example of a simple subway map, as shown in Figure 7. Sample encodings are included for geometry and topology. Note that complete versions of these encodings are provided on the *Worked Examples CD*.

18.10.1   Drawing a map based on geometry

Imagine that the data is encoded using a number of `SubwayRoute` and `Sub-wayStation` features. A fragment of the GML data might appear as follows:

```
<app:Subway gml:id="..">
    <gml:featureMember>
        <app:SubwayRoute gml:id="RouteA">
            <gml:centerLineOf>
                <gml:LineString>
                    <gml:coordinates>...</gml:coordinates>
                </gml:LineString>
            </gml:centerLineOf>
        </app:SubwayRoute>
    </gml:featureMember>
    <gml:featureMember>
        <app:SubwayStation gml:id="StopA">
            <gml:position>
                <gml:Point>
                    <gml:coordinates>...</gml:coordinates>
                </gml:Point>
            </gml:position>
        </app:SubwayStation>
    </gml:featureMember>
    ...
</app:Subway>
```

To specify styles for all of the `SubwayRoute` and `SubwayStation` features, a number of `FeatureStyle` elements should be used, as shown below:

```
<gml:defaultStyle>
    <gml:Style>
        <gml:FeatureStyle featureType="app:SubwayStation">
            <gml:featureConstraint>@gml:id="MainStation"
                ↵</gml:featureConstraint>
            ...
        </gml:FeatureStyle>
        <gml:FeatureStyle featureType="app:SubwayStation">
            <gml:featureConstraint>@gml:id="StopA"
                ↵</gml:featureConstraint>
            ...
        </gml:FeatureStyle>
        ...
```

```
        <gml:FeatureStyle featureType="app:SubwayRoute">
          <gml:featureConstraint>@gml:id="RouteA"
            ↲</gml:featureConstraint>
            ...
        </gml:FeatureStyle>
        ...
      </gml:Style>
    </gml:defaultStyle>
```

Note that since all stations are encoded using the same feature element, Sub-wayStation, the featureConstraint element is used to provide feature identifiers to distinguish between the different stations. The same rule applies to the routes.

The main station is styled using a distinct symbol. Symbols are stored in the Transportation.xml symbol collection file at http://www.symbols.com, which is a fictional URI. The style for the main station simply references the desired symbol from the library, as shown below.

```
<gml:FeatureStyle featureType="app:SubwayStation">
    <gml:featureConstraint>@gml:id="MainStation"
      ↲</gml:featureConstraint>
    <gml:GeometryStyle
      ↲geometryProperty="gml:position"
      ↲geometryType="gml:Point">
      <gml:symbol symbolType="svg"
        ↲xlink:href="http://www.symbols.com/
        ↲transportation.xml#MainSubwayStation"/>
    </gml:GeometryStyle>
</gml:FeatureStyle>
```

The routes are styled using lines of the same width and different colors, and the stations are styled using the same rectangle symbol, which has a black outline and a shaded fill, depending on the route. Although the style for the other stations references a single symbol, the StyleVariation element is used to override the fill shade for each subway line. The following example shows a subway station whose symbol has a black fill.

```
<gml:FeatureStyle featureType="app:SubwayStation">
    <gml:featureConstraint>@gml:id="StopA"
      ↲</gml:featureConstraint>
    <gml:GeometryStyle geometryProperty="gml:position"
      ↲geometryType="gml:Point">
      <gml:styleVariation styleProperty="fill">black
        ↲</gml:styleVariation>
      <gml:symbol symbolType="svg"
        ↲xlink:href="http://www.symbols.com/
        ↲transportation.xml#SubwayStation"/>
    </gml:GeometryStyle>
</gml:FeatureStyle>
```

Finally, the style element is used to specify the line colour for each route, as shown in the following example:

```
<gml:FeatureStyle featureType="app:SubwayRoute">
    <gml:featureConstraint>@gml:id="RouteA"
      ↲</gml:featureConstraint>
```

```
        <gml:GeometryStyle geometryProperty="gml:centerLineOf"
          ⌞geometryType="gml:LineString">
          <gml:style>stroke:black;stroke-width:3</gml:style>
        </gml:GeometryStyle>
      </gml:FeatureStyle>
```

18.10.2   Drawing a map based on topology

A topology style is created in very similar manner as the geometry style. For
example, symbolization and style parameters are specified in the same way. The
only difference is that the styles apply to topology objects instead of geometry
objects, as shown in the following example:

```
<gml:defaultStyle>
  <gml:Style>
    <gml:FeatureStyle featureType="app:SubwayStation">
      <gml:featureConstraint>@gml:id="StopA"
        ⌞</gml:featureConstraint>
      <gml:TopologyStyle
        ⌞topologyProperty="gml:directedNode"
        ⌞topologyType="gml:Node">
        <gml:styleVariation styleProperty="fill">
          ⌞black</gml:styleVariation>
        <gml:symbol symbolType="svg"
          ⌞xlink:href="http://www.symbols.com/
          ⌞transportation.xml#SubwayStation"/>
      </gml:TopologyStyle>
    </gml:FeatureStyle>
    <gml:FeatureStyle featureType="app:SubwayRoute">
      <gml:featureConstraint>@gml:id="RouteA"
        ⌞</gml:featureConstraint>
      <gml:TopologyStyle
        ⌞topologyProperty="gml:directedEdge"
        ⌞topologyType="gml:Edge">
        <gml:style>stroke:black;stroke-width:3
          ⌞</gml:style>
      </gml:TopologyStyle>
    </gml:FeatureStyle>
    ...
  </gml:Style>
</gml:defaultStyle>
```

Given that the topology model does not carry any geometry information, the
actual layout of nodes and edges depends completely on the layout algorithm that
is used in drawing the topology network. That is, the layout is application specific.
You can, however, influence the layout by specifying a set of attributes in the
GraphStyle element, as shown in the following example:

```
<gml:defaultStyle>
  <gml:Style>
    <gml:GraphStyle>
      <gml:planar>true</gml:planar>
      <gml:directed>false</gml:directed>
      <gml:minAngle>30</gml:minAngle>
      <gml:drawingType>POLYLINE</gml:drawingType>
      <gml:lineType>BENT</gml:lineType>
```

```
            <gml:aestheticCriteria>MIN_CROSSINGS
            </gml:aestheticCriteria>
        </gml:GraphStyle>
        ...
    </gml:Style>
</gml:defaultStyle>
```

18.11 What about styling with XSLT?

The eXtensible Style Sheet Language (XSLT) is a common mechanism for styling XML data for presentation, most often as HTML. XSLT interpreters and compilers are widely available for most operating systems, web browsers and server-side web page generators – for example, Active Server Page (ASP) and Java Server Page (JSP). XSLT can also be applied to the styling of GML data for graphical presentation, especially into SVG, and also into other 'presentation' languages such as VoiceXML or HTML.

How does this fit in with the default styling discussed in this chapter? In many ways, XSLT can be regarded as a preferred implementation of GML default styling because XSLT is an open standard that can encode style transformation rules in a declarative manner.

Figure 8 shows the basic components of an XSLT-based web map architecture. The Map Style Editor component creates the styles (for example, GML default style) and stores them in a Map Style Library, which might be a collection of flat files, or an OGC Catalog Service. The styles are constructed using information in a symbol library – which can be managed in the same fashion as a style library – and GML schema information obtained either from a WFS or an OGC Catalog Service.

An XSLT Style Sheet can be generated by the style editor, the library or the map style engine. The XSLT script is executed within the Map Style Engine, and the resulting SVG is then rendered in a standard SVG web browser or desktop application.

Note that the execution of the XSLT can be handled either by client side (for example, Microsoft Internet Explorer containing an XSLT engine) or server side

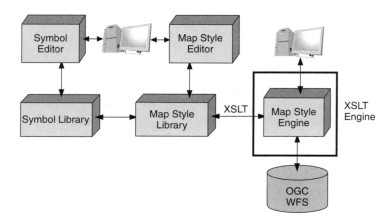

Figure 8 XSLT-based map styling architecture.

(for example, a Saxon XSLT engine in a Servlet). Note that XSLT can provide additional rule refinement of the default style and can be considered as a style extension, as well as an implementation. For more information about XSLT, please see http://www.w3.org/TR/xslt.

18.12 What is the OGC styled layer descriptor (SLD)?

The OpenGIS Consortium (OGC) has released another adopted specification called the Styled Layer Descriptor (SLD). In general, SLD describes similar kind of styling information as GML default styling. There are, however, a number of differences between SLD and GML default styling, including:

- SLD doesn't build on SVG, but on modified SVG grammar.
- SLD does not support graph styling.
- SLD is fundamentally based on the concept of map layers, whereas GML default styling is based on the more general concept of feature style.
- SLD does not support animation.
- SLD was developed with a focus on the Web Map Service (WMS) rather than on the graphical portrayal of arbitrary GML features.

It is anticipated that the SLD and GML default styling specifications will converge at some point in the future. For more information about SLD, please see http://www.opengis.org/docs/02-070.pdf.

18.13 Chapter summary

In addition to describing geographic objects, GML has a separate style description mechanism that can be used to specify a visual presentation style for a GML data set. These styles – which are encoded as `Style` objects – are typically applied to a feature or feature collection via the `defaultStyle` property.

The `defaultStyle.xsd` schema defines an abstract `_Style` object, from which the concrete `Style` object is derived. The `_Style` object can be used to create user-defined style objects; the rules for creating these objects are similar to those for creating user-defined features, except that style objects must derive from `AbstractStyleType` and be substitutable for `_Style`.

The `defaultStyle` property can contain a `Style` object or reference it remotely. Although the feature that has the `defaultStyle` property is usually the feature to which the style applies, the about attribute can also be used to apply a style to other objects. To attach the `defaultStyle` property to a feature (or feature collection), it must be included in the feature content model in a GML application schema.

The `Style` object can contain feature and graph style descriptions, the first of which can contain additional geometry, topology and label descriptions. In addition to these component-style descriptions, the `FeatureStyle` element can also contain the `featureConstraint` element – which applies an XPath constraint to a GML data set – and the `featureType` and `baseType` attributes. The `featureType` attribute constrains the feature style to apply to all features with a particular name, such as all Road features. The `baseType` attribute, on the other hand, applies the style to all features with a certain named content model, such as all features whose content model is of, or derives from, RoadType type.

The `GeometryStyle` also has the `geometryProperty` and `geometry-Type` attributes that indicate the geometry property and geometry object to which the style applies. `TopologyStyle` has similar properties, `topologyProperty` and `topologyType`. The geometry and topology style descriptors can also have `symbol` elements and label-style descriptions. The `symbol` element is used to relate a symbol to a feature; for example a bridge icon can represent a `Bridge` feature. This symbol can only be contained within, or referenced remotely, from the `GeometryStyle` or `TopologyStyle` elements. Note that SVG is the typical format for symbols in GML default styling. The `LabelStyle` element is used specifically for text content, and it can contain an unbounded sequence of `LabelExpression` elements, which are XPath expressions that extract information from the associated GML data set.

The geometry, topology and label styles can all have the `style` element for specifying simple graphic and text-layout properties. All style descriptors can also have the `styleVariation` element, which allows different style descriptions to use variations of the same symbol. The `styleVariation` element can also be used to create parameterized styles, which are styles in which graphic attributes depend on properties of the feature to which the styles apply.

The `GraphStyle` element is used for encoding topology graphs that comprise multiple features. Part of the style of a topology graph is specified in the `TopologyStyle` element, while the graph style contains the properties that apply to the whole graph. The elements in the `GraphStyle` element are `planar`, `directed`, `minDistance`, `minAngle`, `graphType`, `drawingType`, `lineType` and `aestheticCriteria`. All of these elements are of the enumeration type.

All of the style descriptors can also have elements that specify animation characteristics and the spatial resolution of the GML data. The animation elements – `animate`, `animateMotion`, `animateColor` and `set` – are from the W3C SMIL schemas. The `spatialResolution` element is defined in `defaultStyling.xsd`, and it can be used to specify generic information about a map, such as its scale or size.

GML default styling can be used to represent GML data in many visual formats. XSLT is a popular mechanism for implementing GML default styling. In addition, XSLT can also provide additional refinement of the default style.

The OGC Styled Layer Descriptor (SLD) can be also used instead of GML default styling to describe the graphical presentation of GML data. SLD styling is based on the 'map layer' concept from the WMS specification, and is useful for creating maps from GML data. SLD style descriptions can be referenced by the default style property. It is anticipated that GML default styling and SLD will converge at some point in the future.

References

http://www.opengis.org/docs/02-023r4.pdf (October 20, 2003).

http://schemas.opengis.net/gml/3.0.0/base/defaultStyle.xsd (October 20, 2003).

http://www.w3.org/TR/SVG (September 20, 2003).

http://www.w3.org/TR/xpath (October 21, 2003).

DI BATTISTA, G., EADES, P., TAMASSIA, R., and TOLLIS, I.G. (1999) *Graph Drawing: Algorithms for the Visualization of Graphs*. Prentice Hall, Upper Saddle River, NJ.

http://www.w3.org./TR/smil20/ (October 21, 2003).

International Organization for Standardization. (2003) ISO Technical Committee 211 Geographic Information/Geomatics. *ISO DIS 19115, Geographic Information – Metadata*.

http://www.w3.org/TR/xslt (October 17, 2003).

http://www.opengis.org/docs/02-070.pdf (October 23, 2003).

Additional reference

http://www.w3.org/TR/REC-CSS2/ (October 17, 2003).

Chapter 19

GML and geospatial web services revisited

Volume I: GML, An Introduction covers the basic concepts of geospatial web services, and the role of GML in relation to these services. This chapter contains a more technical discussion of web-service description and the OGC Web Feature Service (WFS). For more information about the WFS, please refer to the following OGC specifications: *OpenGIS® Web Feature Service Implementation Specification* (http://www.opengis.org/docs/02-058.pdf) and *OpenGIS® Filter Encoding Implementation Specification* (http://www.opengis.org/docs/02-059.pdf). Note that as of January 2004, these specifications do not explicitly support GML 3. The examples in this chapter have been adapted, however, to reflect how they might be encoded for GML 3. Note that the OGC has targeted WFS support for GML 3 for sometime in 2004.

A web service is a program that performs a task when directed by messages sent to the service over the Internet. It can also generate and send messages in response to changes in the web service's internal state or the world with which the service interacts. The term web service has a wide variety of definitions and web services are not necessarily restricted to particular protocols (such as SOAP) or payloads. Although most web services involve the transmission of messages encoded in XML, this is not a necessary condition for a web service. Geospatial web services deal with geospatial data, either as part of a request message (the message sent to the service), or the response message (the message sent from the service).

19.1 GML and web-service message payloads

GML can play several roles with respect to geospatial web services. GML-encoded data can form the payload of a request ('find all road features inside a test polygon') or the response ('here is the data for the requested road features'). GML can also play an important role in the description of a web service.

Consider the following example for a very simple routing service. The input message for the routing service contains a `RouteInput` feature with start and destination cities.

Geography Mark-up Language (GML). R. Lake, D. S. Burggraf, M. Trninić, L. Rae © 2004 Galdos Systems Inc. Published by John Wiley & Sons, Ltd ISBNs: 0-470-87153-9 (HB); 0-470-87154-7 (PB)

```
<app:RouteInput>
   <app:startCity>
      <app:City gml:id="Keremeos">
         <gml:position>
            <gml:Point srsName="#EPSG1234">
               <gml:pos>1432.3 2454.2</gml:pos>
            </gml:Point>
         </gml:position>
      </app:City>
   </app:startCity>
   <app:destination>
      <app:City gml:id="Winston">
         <gml:position>
            <gml:Point srsName="#EPSG1234">
               <gml:pos>5653.4 765.45</gml:pos>
            </gml:Point>
         </gml:position>
      </app:City>
   </app:destination>
</app:RouteInput>
```

The output message has a Route object that contains a list of Road features that make up the route, as shown in the following:

```
<app:Route gml:id="r1">
   <app:routeMember>
      <app:RoadSegment gml:id="s1">
         <gml:centerLineOf>
            <gml:LineString srsName="#EPSG1234">
               <gml:coordinates>...</gml:coordinates>
            </gml:LineString>
         </gml:centerLineOf>
         <app:classification>highway</app:classification>
         <app:length>32</length>
      </RoadSegment>
   </app:routeMember>
   <app:routeMember>
      <app:RoadSegment gml:id="s2">
         <gml:centerLineOf>
            <gml:LineString srsName="#EPSG1234">
               <gml:coordinates>...</gml:coordinates>
            </gml:LineString>
         </gml:centerLineOf>
         <app:classification>highway</app:classification>
         <app:length>23.5</app:length>
      </app:RoadSegment>
   </app:routeMember>
   <app:length>55.5</app:length>
</app:Route>
```

19.2   GML and web-service interface description

The description of a web service may encompass a variety of different viewpoints. This chapter discusses the description of a web service from a computational viewpoint, that is, in terms of the interfaces that the web service presents to the rest of the world. It is here that GML can be of particular importance because it

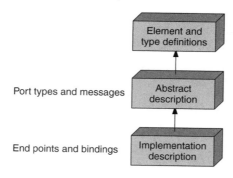

Figure 1 Structure of a WSDL document.

can describe the geospatial elements that form parts of the request and response messages. It is not, however, the role of GML to describe web services on its own.

The W3C Web Services Description Language (WSDL) is emerging as the standard way to describe web service interfaces. WSDL describes a web service on two levels: as an abstract description of the service in terms of its interfaces and as an implementation description in terms of the binding of these interfaces to web addresses and concrete transport protocols.

The abstract description is supported by a set of type and element definitions that are used to describe the input/output message arguments. Figure 1 summarizes the structure of a WSDL service description. The type and element definitions in Figure 1 can be drawn from GML application schemas. The structure shown in the figure has the following form in XML:

```
<definitions name="...">
<!-- this is the body of the WSDL document
   <!-- ... namespace declarations ...
<!-- Start of the Element and Type Definitions -->

   <types>  <!-- here we define supporting types -->
      <schema>
         .. XML schema type, element, and attribute
            declarations ..
      </schema>
   </types>

<!-- End of the Element and Type Definitions -->
<!-- Start of the Abstract Description -->

   <message>  <!-- message definitions -->
      ...
   </message>

   <message>  <!-- message definitions -->
      ...
   </message>

   <portType name=".."> <!-- port type description -->
      <operation name="..">
         <input message=".."/>
         <output message=".."/>
      </operation>
   </portType>
```

```
<!-- End of the Abstract Description -->
<!-- Start of the Implementation Description -->

   <binding name=".." type="..">
     <soap:binding style=".." transport="http://
       ⌐schemas.xmlsoap.org/soap/http"/>
     <operation name="..">
        <soap:operation soapAction=".."/>
        <input>
           <soap:body use=".."/>
        </input>
        <output>
           <soap:body use=".."/>
        </output>
     </operation>
   </binding>

   <service name="..">
      <documentation>My first service
        ⌐</documentation>
      <port name=".." binding="..">
         <soap:address location=".."/>
      </port>
   </service>

<!-- End of the Implementation Description -->

</definitions>
```

Examples of the different parts of a WSDL document are provided below. For additional information about the structure of a WSDL document, refer to the *Web Services Description Language (WSDL) 1.1* specification at http://www.w3.org/TR/wsdl.html. Note that WSDL Version 2.0 (previously called Version 1.2) became available as this book was going to press. The examples in this chapter are based on WSDL Version 1.1.

19.2.1 WSDL abstract description

WSDL describes a web service interface in terms of a set of port types or interfaces. A port type (interface) is the set of operations that the web service supports, with each operation consisting of one or more input, output or fault messages. Consider the following simple example of a port type (interface) description, in which the port type (interface) has a single operation consisting of a single input ('get shortest route') message and a single output message ('the shortest route').

```
<portType name="simpleroutePortType">
   <operation name="GetShortestRoute">
      <input message="app:GetShortestRouteInput"/>
      <output message="app:GetShortestRouteOutput"/>
   </operation>
</portType>
```

As shown in the following example, the WSDL `portType` description is supported by WSDL message definition elements that describe the messages for the above WSDL operation.

```
<message name="GetShortestRouteInput">
   <part name="body" element="app:RouteInput"/>
</message>

<message name="GetShortestRouteOutput">
   <part name="body" element="app:Route"/>
</message>
```

Note that GML can be used to describe all of the geospatial 'arguments' for the web service inputs and outputs. The two elements in the above example (RouteInput and Route) are geographic features that are defined in a GML application schema called SimpleRoute.xsd. This schema can be contained within the WSDL types element or referenced via the import element in the body of the WSDL document. The following example shows a WSDL document called SimpleRoute1.wsdl that references the SimpleRoute.xsd schema:

```
<definitions name="SimpleRouteService"
   ⌣targetNamespace="http://www.ukusa.org/transportation"
   ⌣xmlns="http://schemas.xmlsoap.org/wsdl/"
   ⌣xmlns:app="http://www.ukusa.org/ transportation"
   ⌣xmlns:soap="http://schemas.xmlsoap.org/wsdl/soap/"
   ⌣xmlns: xsi="http://www.w3.org/2001/
      ⌣XMLSchema-instance">
<import namespace=
   ⌣"http://www.ukusa.org/transportation"
   ⌣location="http://www.ukusa.org/transportation/
   ⌣SimpleRoute.xsd"/>

<message name="GetShortestRouteInput">
   <part name="body" element="app:RouteInput"/>
</message>

<message name="GetShortestRouteOutput">
   <part name="body" element="app:Route"/>
</message>

<portType name="simpleroutePortType">
   <operation name="GetShortestRoute">
      <input message="app:GetShortestRouteInput"/>
      <output message="app:GetShortestRouteOutput"/>
   </operation>
</portType>

</definitions>
```

For the above example, the following features and properties should be defined to support the above input and output messages: RouteInput, startCity, destination, City, Route, routeMember and RoadSegment. Note that the *Worked Examples CD* contains a complete version of the SimpleRoute.xsd schema and the SimpleRoute1.wsdl document, plus a SimpleRoute2.wsdl document with the types element.

While the OGC has focused on a small number of generic data-access, data-display and metadata services, many vertical geospatial services are anticipated in the future. These vertical services will provide specific business functions, such as forest-harvest planning, business-site location, underground cable location, urban planning and land-use planning. In addition to providing computational and analysis functions, these services will send and receive geographic data. The combination of WSDL and GML is critically important for this large body of future geospatial web services.

19.2.2 Implementation description

The implementation description part of the WSDL document contains the binding of the abstract interfaces to a concrete message transport, such as SOAP/HTTP POST or HTTP POST. Most OGC Web Services are currently implemented using HTTP POST or HTTP GET, although it is likely that SOAP will be gradually adopted. Note further that GML is not typically used in the Implementation Description component of the WSDL document.

19.3 Web feature services revisited

While it is expected that many geospatial web services will be deployed in the near future, the OGC WFS has a special relationship to GML and deserves additional treatment. The WFS provides a standardized means to access geospatial data that is both vendor- and storage-format neutral. Clients can interact with a WFS without knowledge of the underlying storage mechanisms. Furthermore, a given WFS can interact with any other WFS – enabling distributed querying and data replication across a network in which the server nodes are provided by many different vendors.

The WFS supports different kinds of requests, including the following:

- GetCapabilities
- DescribeFeatureType
- GetFeature
- Transaction.

Figure 2 shows a sample interaction sequence between a client application and a WFS. As shown in the figure, the client sends a GetCapabilities request

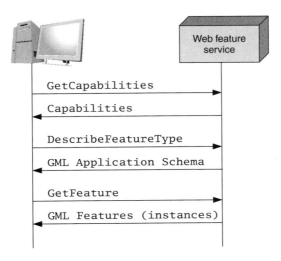

Figure 2 Sample client-WFS message exchange.

and receives a response with all of the WFS's capabilities. Note that the GetCapabilities response – which is an XML capabilities document – includes a list of the feature-type names known to the WFS.

The client then sends a DescribeFeatureType request, which can contain one or more feature type names (usually obtained from the GetCapabilities response), and the WFS sends a response message containing a GML application schema with the appropriate element and type definitions for the requested feature. Finally, the client sends a GetFeature request for one or more features and receives a GML feature collection with features that meet the constraints in the request.

19.3.1   GetCapabilities request

The GetCapabilities request is supported by all OGC web services. The following example shows a GetCapabilities request:

```
<GetCapabilities version="1.0.0" service="WFS"
   ⌣xmlns="http://www.opengis.net/wfs"/>
```

19.3.2   Capabilities response

The Capabilities response is a Capabilities XML document that includes a description of the service, plus information about the data content and other service features supported by the service endpoint. In addition to listing the available Distributed Computing Platform (DCP) and filter capabilities, the document includes a list of the GML feature names that are available from the WFS endpoint. These are listed in the following fragment from a Capabilities response:

```
<FeatureTypeList>
   <Operations>
      <Query/>
      <Update/>
      <Insert/>
      <Delete/>
      <Lock/>
   </Operations>
   <FeatureType
      ⌣xmlns:app="http://www.ukusa.org/transportation">
      <Name>app:Bridge</Name>
      <Title>Bridge</Title>
      <SRS>...</SRS>
      <LatLongBoundingBox maxx=".." maxy=".." minx=".."
         ⌣miny=".."/>
   </FeatureType>
   <FeatureType
      ⌣xmlns:app="http://www.ukusa.org/transportation">
      <Name>app:RoadSegment</Name>
      <Title>Road Segment</Title>
      <SRS>...</SRS>
      <LatLongBoundingBox maxx=".." maxy=".." minx=".."
         ⌣miny=".."/>
```

```
        </FeatureType>
      <FeatureType
        ⌣xmlns:app="http://www.ukusa.org/transportation">
        <Name>app:City</Name>
        <Title>City</Title>
        <SRS>...</SRS>
        <LatLongBoundingBox maxx=".." maxy=".." minx=".."
          ⌣miny=".."/>
      </FeatureType>
      <FeatureType
        ⌣xmlns:app="http://www.ukusa.org/transportation">
        <Name>app:County</Name>
        <Title>County</Title>
        <SRS>...</SRS>
        <LatLongBoundingBox maxx=".." maxy=".." minx=".."
          ⌣miny=".."/>
      </FeatureType>
    </FeatureTypeList>
```

Note that the grammar for this feature list is not GML. The content of the Name element must, however, be a valid XML element because it corresponds to an element defined in a GML application schema. On the basis of the above response, it is clear that the responding WFS can, for example, provide feature data for a `Bridge` feature from the http://www.ukusa.org/namespace.

A client application can then use the returned list to either request additional schema information from the WFS (for example, via `DescribeFeatureType`), or to initiate a query or update operation. The `DescribeFeatureType` request is discussed in the following section.

19.3.3 `DescribeFeatureType` request

The `DescribeFeatureType` request returns the content model of one or more features in a GML application schema. The following is a sample request:

```
<DescribeFeatureType outputFormat="XMLSCHEMA"
  ⌣xmlns:app="http://www.ukusa.org/transportation"
  ⌣xmlns="http://www.opengis.net/wfs" service="WFS"
  ⌣version="1.0.0">
  <TypeName>app:Bridge</TypeName>
</DescribeFeatureType>
```

This request returns GML schema information in XML Schema format for the `Bridge` feature in the http://www.ukusa.org/transportation namespace. Note that the `TypeName` element should contain the feature element name, not the name of the associated XML Schema type definition. That is, it should be `Bridge`, not `BridgeType`. The feature element name must be bound to its namespace, either via a prefix (`app` in this case), or without it, if the namespace is declared as the default namespace.

19.3.4 `DescribeFeatureType` response

The following example shows a WFS response to the above request:

```
<?xml version="1.0" encoding="UTF-8"?>
<schema attributeFormDefault="unqualified"
   ⌣elementFormDefault="qualified"
   ⌣targetNamespace="http://www.ukusa.org/transportation"
   ⌣xmlns="http://www.w3.org/2001/XMLSchema"
   ⌣xmlns:gml="http://www.opengis.net/gml"
   ⌣xmlns:app="http://www.ukusa.org/transportation"
   ⌣xmlns:xlink="http://www.w3.org/1999/xlink"
   ⌣version="1.0">
   <!-- =============================================== -->
   <include schemaLocation=".."/>
   <import namespace="http://www.opengis.net/gml"
      ⌣schemaLocation=".."/>
   <!-- ===============================================
   Global element Declarations
   =============================================== -->
   <element name="Bridge" type="app:BridgeType"
      ⌣substitutionGroup="gml:_Feature"/>
   <!-- ===============================================
   Type Declarations
   =============================================== -->
   <complexType name="BridgeType">
      <complexContent>
         <extension base="gml:AbstractFeatureType">
            <sequence>
               <element name="span"
                  ⌣type="gml:LengthType"/>
               <element name="height"
                  ⌣type="gml:LengthType"/>
               <element ref="gml:centerLineOf"/>
            </sequence>
         </extension>
      </complexContent>
   <complexType>
</schema>
```

The client can use schema information returned from the WFS and – possibly after receiving input from a human user – can construct a feature request or a transaction. A feature request is performed using GetFeature.

19.3.5   GetFeature request

The GetFeature request is used to retrieve features from the WFS that satisfy the specified attribute and geometric constraints described in the associated filter expression. The following example shows a sample GetFeature request:

```
<wfs:GetFeature>
   <wfs:Query typeName="app:Bridge"/>
</wfs:GetFeature>
```

Note that this request selects all bridge features from the particular WFS. It is also possible to request more than one feature type in a GetFeature request or to include filter expressions that specify constraints on the features returned in the response message. These filter expressions are covered in the *OpenGIS® Filter Encoding Implementation Specification* (http://www.opengis.org/docs/02-059.pdf).

19.3.6 `GetFeature` response

The following example shows a response message with a collection of bridge features:

```
<?xml version="1.0" encoding="UTF-8"?>
<wfs:FeatureCollection
    ⌐xmlns:gml="http://www.opengis.net/gml"
    ⌐xmlns:app="http://www.ukusa.org/transportation"
    ⌐xmlns:wfs="http://www.opengis.net/wfs"
    ⌐xmlns:xlink="http://www.w3.org/1999/xlink"
    ⌐xmlns:xsi="http://www.w3.org/2001/XMLSchema-instance"
    ⌐xsi:schemaLocation="...">
  <gml:boundedBy>
    <gml:Envelope srsName="#EPSG1234">
      <gml:pos>...</gml:pos>
      <gml:pos>...</gml:pos>
    </gml:Envelope>
  </gml:boundedBy>
  <gml:featureMember>
    <app:Bridge gml:id="Burrard">
      <gml:name>Burrard Street Bridge</gml:name>
      <app:span>...</app:span>
      <app:height>...</app:height>
      <gml:centerLineOf>...</gml:centerLineOf>
    </app:Bridge>
  </gml:featureMember>
  <gml:featureMember>
    <app:Bridge gml:id="GoldenGate">
      <gml:name>Golden Gate Bridge</gml:name>
      <app:span>...</app:span>
      <app:height>...</app:height>
      <gml:centerLineOf>...</gml:centerLineOf>
    </app:Bridge>
  </gml:featureMember>
  ...
</wfs:FeatureCollection>
```

Note that the `wfs:FeatureCollection` element is a container for the response to a `GetFeature` request. It is covered in the *OpenGIS® Web Feature Service Implementation Specification* (http://www.opengis.org/docs/02-058.pdf).

19.3.7 Transactions

The WFS protocol also supports transactions against the WFS, including `Insert`, `Update`, `Lock` and `Delete`. The following examples show how these transaction requests can be encoded.

19.3.7.1 Feature insert

The following example shows a transactional request for the insertion of an instance of a GML `Bridge` feature named `Hijiribashi`:

```
<wfs:Transaction service="WFS" version="1.0.0"
    ⌐xmlns="http://www.ukusa.org/transportation"
    ⌐xmlns:wfs="http://www.opengis.net/wfs"
```

```
⌒xmlns:ogc="http://www.opengis.net/ogc"
⌒xmlns:gml="http://www.opengis.net/gml"
⌒xmlns:app="http://www.ukusa.org/transportation">
<wfs:Insert handle="Insert Bridge features">
    <Bridge>
        <gml:name>Hijiribashi</gml:name>
        <span>60</span>
        <height>200</height>
        <gml:centerLineOf>
            <gml:CompositeCurve gml:id="CC1"
                ⌒srsName="#EPSG1234">
                <gml:curveMember>
                    <gml:LineString gml:id="C1">
                        <gml:coordinates>0,100 100,100
                            ⌒</gml:coordinates>
                    </gml:LineString>
                </gml:curveMember>
                <gml:curveMember>
                    <gml:LineString gml:id="C2">
                        <gml:coordinates>100,100 200,100
                            ⌒</gml:coordinates>
                    </gml:LineString>
                </gml:curveMember>
            </gml:CompositeCurve>
        </gml:centerLineOf>
    </Bridge>
</wfs:Insert>
</wfs:Transaction>
```

Note that the above sample encoding uses GML 3.0, which is not currently supported by the WFS specification.

19.3.7.2 Feature update

The Update transaction can be used to update the properties of features that are already stored in a WFS. In the following example, the span property of the Hijiribashi Bridge feature is updated from 60 to 150.

```
<wfs:Transaction service="WFS" version="1.0.0"
    ⌒xmlns="http://www.ukusa.org/transportation"
    ⌒xmlns:wfs="http://www.opengis.net/wfs"
    ⌒xmlns:ogc="http://www.opengis.net/ogc"
    ⌒xmlns:gml="http://www.opengis.net/gml"
    ⌒xmlns:app="http://www.ukusa.org/transportation">
<wfs:Update typeName="app:Bridge">
    <wfs:Property>
        <wfs:Name>app:Bridge/app:span</wfs:Name>
        <wfs:Value>150</wfs:Value>
    </wfs:Property>
    <ogc:Filter>
        <ogc:PropertyIsEqualTo>
            <ogc:PropertyName>app:Bridge/gml:name
                ⌒</ogc:PropertyName>
            <ogc:Literal>Hijiribashi</ogc:Literal>
        </ogc:PropertyIsEqualTo>
```

```
        </ogc:Filter>
      </wfs:Update>
  </wfs:Transaction>
```

19.3.7.3 Feature delete

The Delete transaction can be used to delete entire features based on certain constraints specified in a filter expression. In the following example, a Delete transaction shows how to delete any Bridge features named Hijiribashi:

```
<wfs:Transaction xmlns:wfs="http://www.opengis.net/wfs"
    ⌣xmlns:gml="http://www.opengis.net/gml"
    ⌣xmlns:ogc="http://www.opengis.net/ogc"
    ⌣xmlns:xlink="http://www.w3.org/1999/xlink"
    ⌣xmlns:app="http://www.ukusa.org/transportation"
        ⌣service="WFS" version="1.0.0">
    <wfs:Delete handle="delete the Bridge named
        ⌣Hijiribashi" typeName="app:Bridge">
      <ogc:Filter>
          <ogc:PropertyIsEqualTo>
            <ogc:PropertyName>app:Bridge/gml:name
                ⌣</ogc:PropertyName>
            <ogc:Literal>Hijiribashi</ogc:Literal>
          </ogc:PropertyIsEqualTo>
      </ogc:Filter>
    </wfs:Delete>
</wfs:Transaction>
```

19.4 Chapter summary

For clients to interact with a web service, they need to know the structure of the request and response messages that the service supports. WSDL provides an XML grammar for describing these messages in terms of input and output. This is independent of the implementation. WSDL also provides an implementation description that describes how the message is carried. A GML application schema can be contained inside the types element of a WSDL definition, and the input and output messages can also contain GML.

OGC web services, such as WFS, currently support GML 2, but GML 3 support should be available soon. WFS supports a number of requests, including GetCapabilities, DescribeFeatureType and GetFeature. The response to a GetCapabilities request includes a list of all features stored in a WFS. This response is not encoded in GML. A DescribeFeatureType request returns a GML application schema with element declarations and type definitions for the requested feature type(s). A GetFeature request returns a WFS feature collection containing GML features that meet the constraints specified in the request. A number of transactional operations – such as Insert, Update and Delete – are also supported.

WFS requests and responses are covered in the *OpenGIS® Web Feature Service Implementation Specification* (http://www.opengis.org/docs/02-058.pdf). The requests can include filter expressions containing specific constraints. The filter encoding rules are covered in the *OpenGIS® Filter Encoding Implementation Specification* (http://www.opengis.org/docs/02-059.pdf).

References

http://www.opengis.org/docs/02-058.pdf (October 17, 2003).

http://www.opengis.org/docs/02-059.pdf (October 17, 2003).

http://www.w3.org/TR/wsdl.html (October 17, 2003).

Additional references

http://schemas.opengis.net/gml/3.0.1/base/feature.xsd (October 17, 2003).

http://schemas.opengis.net/gml/3.0.1/base/geometry-Basic0d1d.xsd (October 17, 2003).

http://schemas.opengis.net/gml/3.0.1/base/geometry-Basic2d.xsd (October 17, 2003).

Chapter 20

GML, relational databases and legacy GIS

GML 3 is not an independent technology, but a mark-up language that provides an extensible and flexible means for encoding geographic information. For many individuals in the GIS industry, it is not enough to simply understand GML. There is also a need to understand how to map GML objects to existing data storage systems, such as relational databases and legacy GIS systems. This chapter provides an introduction to mapping GML features and geometries to some of these systems.

20.1 How do I store GML data in an RDBMS?

In this section we consider the storage of GML data in Relational Database Management Systems (RDBMS) without object or XML support. Object-relational databases are covered in a subsequent section. GML data, like all XML data, can be stored in ordinary relational databases, although there are certain restrictions and performance issues that developers should be aware of. It is currently not clear whether conventional RDBMS will play a long-term role in the persistent management of GML data.

It should be noted that more recent relational databases do provide specific support for XML data, including support for abstract data types for XML, query languages (such as, SQLX, SQL embedded in XML templates) and namespaces. There are many complex issues to consider regarding the storage of XML data in a relational database, and it is outside the scope of this book to go into them in detail.

20.1.1 GML feature data with simple properties and without geometry

Consider the following example of a `Vehicle` with `name`, `weight`, `manufacturer`, `model` and `horsepower` properties:

```
<app:Vehicle gml:id="v1">
    <gml:name>Fire Truck</gml:name>
    <app:weight>4300</app:weight>
    <app:manufacturer>Hale Equipment</app:manufacturer>
```

Geography Mark-up Language (GML). R. Lake, D. S. Burggraf, M. Trninić, L. Rae © 2004 Galdos Systems Inc. Published by John Wiley & Sons, Ltd   ISBNs: 0-470-87153-9 (HB); 0-470-87154-7 (PB)

```
        <app:model>Cobra-1</app:model>
        <app:horsepower>410</app:horsepower>
    </app:Vehicle>
```

This is a particularly simple GML feature, in that it does not include any geometry properties and all of the property values are XML Schema simple types. This type of GML feature can be readily stored in any type of RDBMS. The corresponding RDBMS schema might be written as

```
Vehicle(gml_id varchar(100), gml_name varchar,
    ⌐app_weight integer, app_manufacturer varchar,
    ⌐app_model string, app_horsepower integer)
```

Table 1 shows the corresponding database table for the Vehicle feature. This simple example illustrates a number of issues with respect to mapping GML to an RDBMS, including:

- The Vehicle feature type is mapped to a table in the RDBMS, and the feature's properties are mapped to columns in the table. This is easy to do because the properties are all XML Schema simple types.
- Many relational databases do not support XML namespaces. Table 1 illustrates a simple pseudo-solution by incorporating the namespace prefix (for example, app and gml) into the column name. Note that this is not a real solution, because it relies on an invariant one-to-one binding between the namespace prefix and the actual namespace.
- The mapping of XML Schema simple types into a RDBMS, while often straightforward, must be examined for each RDBMS. For example, although Unicode-character encoding is the norm for XML, it is not supported by all RDBMS.
- The identifier for Vehicle should be a primary key for the table, since it uniquely identifies each Vehicle in the XML document or dataset.

Table 1   Database table for app_Vehicle

app_Vehicle					
gml_id	gml_name	app_weight	app_manufacturer	app_model	app_horsepower
v1	Fire Truck	4300	Hale Equipment	Cobra-1	410
v2	Fire Truck	4500	Hale Equipment	Cobra-Vx	440

20.1.2   GML data with geometry and simple properties

Let's expand the previous example to include features with simple non-geometric and geometric properties. The rich variety of geometry types in GML 3 precludes any simple direct mapping from GML to relational tables. Consider the following example of a simple `Bridge` feature:

```
<app:Bridge gml:id="p1">
   <gml:name>Lions Gate</gml:name>
   <app:span>1000</app:span>
   <app:height>120</app:height>
   <gml:centerLineOf>
      <gml:LineString>
         <gml:coordinates>123.23,542.34 126.46,544.76
            ↳128.56,554</gml:coordinates>
      </gml:LineString>
   </gml:centerLineOf>
<app:Bridge>
```

As with Table 1, we can create a single relational schema, this time by using a Binary Large OBject (BLOB) type for the `gml:centerLineOf` column.

```
app_Bridge(gml_id varchar, gml_name varchar, app_span
   ↳integer, app_height integer, gml_centerLineOf blob)
```

Table 2 shows the corresponding table for the `Bridge` feature. Note that, in this example, the geometry data is stored as XML text in the BLOB field (`gml_centerLineOf`). Because there is no assumption that the database can represent XML, the application program must know how to interpret the contents of the `gml_centerLineOf` column.

Another more general solution – and one supported by many XML to RDBMS loading software applications – is to map each GML geometry type to its own table and use referential integrity constraints (keys) to link the tables to one another. Table 3 shows how this can be done with the Bridge feature. Note that this again assumes a BLOB type for the storage of the GML coordinates (`gml_coordinates`).

Note that the following join operation is required for the two tables:

```
App_Bridge.gml_centerLineOf=gml_LineString.geometry_key
```

This approach is a reasonable solution for simple databases with only `Point` and `LineString` elements, but it becomes more complex if polygons and more complex geometries are included, as illustrated by the following example:

```
<app:LandParcel gml:id="r1">
   <gml:name>Sick Childrens Hospital</gml:name>
   <app:area>500</app:area>
   <gml:extentOf>
      <gml:Polygon>
         <gml:exterior>
            <gml:LinearRing>
               <gml:coordinates>0.0,0.0 10.0,0.0 10.0,
                  ↳10.0 0.0,10.0 0.0,0.0
                  ↳</gml:coordinates>
            </gml:LinearRing>
         </gml:exterior>
      </gml:Polygon>
   </gml:extentOf>
</app:LandParcel>
```

Table 2   Database table for `app_Bridge`

app_Bridge				
gml_id	**gml_name**	**app_span**	**app_height**	**gml_centerLineOf**

p1	Lions Gate	1000	120	`<gml:LineString>`
				
				 `<gml:coordinates>123.23,` `542.34 126.46,` `544.76 128.56,` `554` `</gml:coordinates>`
				
				 `<gml:LineString>`

Table 3   Database tables for `app_Bridge` and `gml_LineString`

app_Bridge				
gml_id	**gml_name**	**app_span**	**app_height**	**gml_centerLineOf**
p1	Lions Gate	1000	120	14565

gml_LineString	
geometry_key	**gml_coordinates**
14565	123.23,542.34 126.46,544.76 128.56,554

The corresponding database tables are shown in Table 4. Note that joins must be made between the four tables just to retrieve the `Polygon`'s geometry. This is clearly not a solution that provides high performance.

The above approach can be extended to handle multiple inner boundaries, as well as the single outer boundary (`exterior`), by simply extending the boundary table, as shown in Table 5, or by adding an additional `interior` table.

Note that this approach assumes that the accessing application knows that the `exterior` and `interior` properties of the `Polygon` are `LinearRing` -valued and that `coordinates` is a property of a `LinearRing`. This approach does not handle the different coordinate approaches available in GML 3.0 (for example, `pos`). Nor does it handle the fact that `interior` and `exterior` properties can have `Ring` values that are composed of curve segments and composite curves.

Table 4 Database tables for app_LandParcel

app_LandParcel			
gml_id	gml_name	app_area	gml_extentOf
p1	Sick Childrens Hospital	500	14565

gml_Polygon	
geometry_key	boundary_key
14565	12010

app_exterior	
geometry_key	ring_key
12010	57676

gml_LinearRing	
geometry_key	gml_coordinates
57676	0.0,0.0 10.0, 0.0 10.0,10.0 0.0,10.0 0.0,0.0

Table 5 gml_Boundary table

gml_Boundary		
geometry_key	boundary_type	ring_key
12010	outer	57676

The following example shows a feature with complex geometry, including rings and arcs:

```
<app:LandParcel gml:id="p2">
  <gml:name>Sick Childrens Hospital</gml:name>
  <app:area uom="#m2">2135</app:area>
  <gml:extentOf>
  <gml:Polygon gml:id="g1" srsName="#myrefsys">
    <gml:exterior>
      <gml:Ring> <!--Ring #35-->
        <gml:curveMember>
          <gml:Curve gml:id=".."> <!--Curve 220-->
            <gml:segments>
              <gml:Arc> <!--Curve Segment 22-->
                <gml:pos>0.0 1.73</gml:pos>
                <gml:pos>6.79 0.0</gml:pos>
                <gml:pos>0.0 -1.73</gml:pos>
              </gml:Arc>
            </gml:segments>
          </gml:Curve>
        </gml:curveMember>
        <gml:curveMember>
          <gml:LineString> <!--Curve 14002-->
          <gml:coordinates>0.0,-1.73 0.0,
            ⌴1.73</gml:coordinates>
          </gml:LineString>
        </gml:curveMember>
      </gml:Ring>
    </gml:exterior>
    <gml:interior>
      <gml:Ring> <!--Ring #34-->
        <gml:curveMember>
          <gml:Curve gml:id=".."> <!--Curve 210-->
            <gml:segments>
              <gml:Arc>   <!--Curve Segment 21-->
                <gml:coordinates>0.0,0.87 3.31,
                  ⌴0.0 0.0,-0.87
                  ⌴</gml:coordinates>
              </gml:Arc>
            </gml:segments>
          </gml:Curve>
        </gml:curveMember>
        <gml:curveMember>
          <gml:LineString> <!--Curve 222-->
            <gml:coordinates>0.0,-0.87 0.0,
              ⌴0.87</gml:coordinates>
          </gml:LineString>
        </gml:curveMember>
      </gml:Ring>
    </gml:interior>
  </gml:Polygon>
  </gml:extentOf>
</app:LandParcel>
```

In the above example, each Ring object has two curveMember properties. In GML, the value of a curveMember property can be any geometry object from the _Curve substitution group, including LineString, Curve and

`OrientableCurve`. These geometries must also be present in the relational database model.

The first `LineString` can be represented as shown in Table 6. Table 7 shows how an `Arc` can be represented. Note that this example only works for two-dimensional coordinates. All of these columns need to be nullable, because in GML 3, the coordinates of the control points can be specified either by `pos` or `coordinates`. Note that the database attributes have been named `posX`, `posY`, even though these separate coordinates have no corresponding names in GML (for example, `<gml:pos>100.2 34.5</gml:pos>` is equivalent to `posX 100.2` and `posY 34.5` in the relational database).

Note that in GML a `Ring` is composed of curves, which are in turn composed of segments such as Arcs and B-splines. The segments of these curves are captured in the table `gml_Curve` (Table 8). Note that there should be multiple tables, one for each segment type. This means that an application needs to know the names of each of these tables. Note further that the join operation cannot be done in a single query. This implies a significant performance penalty.

Table 6   Database table for `LineString` with `srsName`

gml_LineString		
curve_key	srsName	gml_coordinates
222	#myrefsys	0.0,−0.87 0.0,0.87
14002	#myrefsys	0.0,−1.73 0.0,1.73

Table 7   Database table for `gml_Arc`

gml_Arc							
segment_key	gml_posX	gml_posY	gml_posX	gml_posY	gml_posX	gml_posY	gml_coordinates
21	null	null	null	null	null	null	0.0,0.87 3.31,0.0 0.0,−0.87
22	0.0	1.73	6.79	0.0	0.0	−1.73	null

Table 8   Database table for `gml_Curve`

gml_Curve		
curveID	Segment_type	Segment key
210	Arc	21
220	Arc	22

Table 9   Database table for gml_Ring

gml_Ring	
ringID	curve_key
34	210
35	220
35	140020
34	222

Table 10   Database table for gml_Polygon

gml_Polygon			
geometry_key	srsName	ringID	boundaryType
14565	#myrefsys	34	exterior
14565	#myrefsys	35	interior

Table 11   Database table for app_LandParcel

app_LandParcel			
gml_id	gml_name	app_area	gml_extentOf
p1	Sick Childrens Hospital	2135	14565

A Ring is then represented as shown in Table 9. A Polygon is then represented by Table 10. Finally, Table 11 shows how the LandParcel feature can be represented. Note that it is possible to put all of the curve types and segments (Arc, ArcByBulge, LineString) in a single table; however, this involves a large number of nullable columns and mutually exclusive columns (for example, if posX and posY are null, then coordinates must be null). With the table structure shown above, it is necessary for any accessing application to do a join on all of the curve and segment tables.

20.1.3   GML feature relationships

GML supports relationships between features, as illustrated in the Entity-Relationship (E-R) diagram shown in Figure 1. Entities in the E-R diagram map to features in GML, and the relationships map to GML properties.

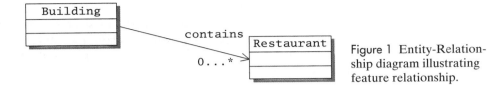

Figure 1 Entity-Relationship diagram illustrating feature relationship.

Figure 1 can be encoded in GML as

```
<app:Building gml:id="b1">
   <gml:name>Cerco</gml:name>
   <app:numFloors>10</app:numFloors>
   <app:contains>
      <app:Restaurant gml:id=""r2">
         <gml:name>Fluid</gml:name>
         <app:type>sea food</app:type>
         <app:menu
            ⌐xlink:href="http://www.placestoeat.
            ⌐com#Fluid.xml"/>
      </app:Restaurant>
   </app:contains>
</app:Building>
```

Note that both `Building` and `Restaurant` are GML features. The above encoding can be mapped into a relational model using the class and foreign key model for GML geometries, as shown in Table 12.

Table 12 Database tables for `app_Building` and `app_Restaurant`

app_Building		
gml_id	gml_name	app_numFloors
b1	Cerco	10

app_Restaurant			
gml_id	gml_name	app_type	app_menu
r2	Fluid	seafood	http://www.placestoeat.com#Fluid.xml

Containment	
building_id	restaurant_id
b1	r2

Note that GML allows the remote reference (for example, `app:contains xlink:href="#r2"`) to be external to the database and to point to anywhere on the Internet. This is not supported by most relational databases. Note further that the 'in-line relationship' in the above GML example would be typically used for a composition relationship in the UML sense.

20.1.4   GML feature collections

As discussed in Chapters 9 and 11, GML supports the construction of feature collections. In GML, a `FeatureCollection` is also a feature, and therefore it has its own properties in addition to a membership property. The following example shows a GML encoding of a `NationalPark` feature collection with `Island`, `Lake` and `Dock` features:

```
<app:NationalPark gml:id="BrokenIslands"/>
    <gml:name>Broken Islands National Park</gml:name>
    <gml:boundedBy>...</gml:boundedBy>
    <gml:featureMember>
        <app:Island gml:id="island01">
            <gml:name>Kings Island</gml:name>
            <app:area>5.1</app:area>
        </app:Island>
    </gml:featureMember>
    </gml:featureMember>
        <app:Dock gml:id="govdock">
            <gml:name>Government Dock</gml:name>
            <app:locatedOn xlink:href="#island03"/>
        </app:Dock>
    </gml:featureMember>
    <gml:featureMember>
        <app:Lake gml:id="lake10">
            <gml:name>unnamed</gml:name>
            <app:area>2.3</app:area>
        </app:Lake>
    </gml:featureMember>
    <app:created>11-June-1906</app:created>
</app:NationalPark>
```

This `NationalPark` feature collection can be represented in relational tables as shown in Table 13. Note that in GML, the members of the feature collection can be encoded as remote properties. This may not be supported in many relational database systems.

20.1.5   GML with geometry and complex properties

As discussed in Chapter 11, GML allows application schema developers to create their own arbitrary complex types. Instances of these types are allowed as values of GML properties. Consider the following simple example:

```
<app:Building gml:id="Chrysler">
    <gml:name>Chrysler</gml:name>
```

```
<app:numFloors>43</app:numFloors>
<gml:position>
   <gml:Point srsName="#myrefsys" gml:id="g1">
      <gml:pos>100.2 45.6</gml:pos>
   </gml:Point>
</gml:position>
<app:location>
   <app:Address>
      <app:streetNumber>311</app:streetNumber>
      <app:street>Walker</app:street>
      <app:city>St. London</app:city>
   </app:Address>
</app:location>
</app:Building
```

In this example, `location` is a property of the Building, and its value is an `Address` object that contains a number of properties. This can be represented in a relational database as shown in Table 14. Note that, as in the case of feature relationships, a separate table is used to represent the user-defined `Address` object.

20.1.6   Basic rules for mapping GML to RDBMS

The basic rules for mapping GML into the relational model can be summarized as follows. For each GML

1. feature, create a corresponding table in the RDBMS;
2. geometry object, create a corresponding table in the RDBMS. Canonical tables can be created for the core GML geometries;
3. feature property of simple type, there is a corresponding column in the associated RDBMS feature table (created in Step 1);
4. feature property of complex type, create
   a. a table in the RDBMS for the complex type;
   b. a relationship table corresponding to the property, with a foreign key pointer from the feature table to the table for the complex type.

Note that Step 4 applies, in particular, to GML feature relationships and GML geometry properties.

20.1.7   Spatial indexing

A conventional relational database has a restricted set of data types – including `integer`, `varchar` and `date` – and does not include any geometric types. In addition, it is not possible to make spatial queries, such as the following, against a relational database:

'Find all towns that lie inside this polygon.'

Table 13   Database tables for `app_NationalPark`

app_NationalPark		
gml_id	**gml_name**	**app_created**
BrokenIslands	Broken Islands National Park	11-June-1906

app_ParkMembers		
gml_id	**MemberTable**	**gml_featureMember**
BrokenIslands	**app_Island**	island01
BrokenIslands	**app_Dock**	govdock
BrokenIslands	**app_Lake**	lake10

app_Island		
gml_id	**gml_name**	**app_area**
island01	Kings Island	5.1

app_Dock		
gml_id	**gml_name**	**app_locatedOn**
govdock	Government Dock	island03

app_Lake		
gml_id	**gml_name**	**app_area**
lake10	unnamed	2.3

Even simple bounding box queries cannot be efficiently performed because of the lack of appropriate indexing structures. For example,

'Find all towns that lie inside this bounding box.'

Spatial databases typically employ some type of spatial index structure to make spatial searches more efficient. Such indexes include R-trees, R*-trees, k-d

Table 14   Database tables for `app_Building`

app_Building				
gml_id	gml_name	app_numFloors	gml_position	app_location
Chrysler	Chrysler	43	20012	20012

app_Address			
address_key	app_streetNumber	app_street	app_city
20012	311	Walker	St. London

gml_Point			
point_key	gml_posX	gml_posY	gml_coordinates
20012	100.2	45.6	null

trees and quadtrees, among many others. The spatial index must be stored in the database and managed by the search component of the spatial database.

20.2   How do I store GML data in object-relational databases?

Object-Relational Database Management Systems (ORDBMS) offer a more flexible means for storing GML data than conventional relational databases, in that they support SQL classes and permit the nesting of data type instances. This more closely conforms to the structure of GML itself. Note that the examples in this section are based on Oracle ORDBMS. Other object databases such as IBM DB2 and Informix (IBM) offer similar features.

20.2.1   GML feature data with simple properties and without geometry

To understand how to map feature data with simple properties to an ORDBMS, consider again the Vehicle example from Section 20.1.1,

```
<app:Vehicle gml:id="v1">
   <gml:name>Fire Truck</gml:name>
   <app:weight>4300</app:weight>
   <app:manufacturer>Hale Equipment</app:manufacturer>
   <app:model>Cobra-1</app:model>
   <app:horsepower>410</app:horsepower>
</app:Vehicle>
```

Oracle Objects can be used to create the following type for Vehicle. Note that the Oracle grammar is similar to other ORDBMSs.

```
CREATE TYPE app_Vehicle AS OBJECT
(gml_id VARCHAR2(20),
gml_name VARCHAR2(30),
app_weight NUMBER,
app_manufacturer VARCHAR2(50),
app_model VARCHAR2(20),
app_horsepower NUMBER);
```

We can then create an object table of vehicles by the statement:

```
CREATE TABLE Vehicles OF app_Vehicle;
```

To make a valid GML feature, it is necessary to introduce a type in the ORDBMS that corresponds to `AbstractFeatureType` and then inherit from this new type, as shown below. Note that the `boundedBy` property has been removed for simplicity.

```
CREATE TYPE gml_AbstractFeatureType AS OBJECT
(gml_id VARCHAR2(20),
gml_name VARCHAR2(256)};
```

We can then create `Vehicle` by inheriting from `gml_AbstractFeature-Type`.

```
CREATE TYPE app_Vehicle UNDER gml_AbstractFeatureType
{app_weight NUMBER,
app_manufacturer VARCHAR2(50),
app_model VARCHAR2(20),
app_horsepower NUMBER);
```

20.2.2 GML data with geometry and simple properties

As with relational databases, it is possible to map GML geometries and simple properties to an ORDBMS. Consider the `Bridge` example from Section 20.1.2.

```
<app:Bridge gml:id="p1">
    <gml:name>Lions Gate</gml:name>
    <app:span>1000</app:span>
    <app:height>120</app:height>
    <gml:centerLineOf>
        <gml:LineString>
            <gml:coordinates>123.23,542.34 126.46,544.76
              ⌄128.56,554</gml:coordinates>
        </gml:LineString>
    </gml:centerLineOf>
<app:Bridge>
```

Using Oracle Objects, the content model for the above encoding for `Bridge` becomes:

```
CREATE TYPE gml_LineString UNDER gml_AbstractCurveType
{gml_coordinates VARCHAR(512)};
CREATE TYPE app_Bridge UNDER gml_AbstractFeatureType
(app_span NUMBER,
app_height NUMBER,
gml_centerLineOf gml_LineString);

CREATE TABLE Bridges OF app_Bridge;
```

Note that this example has omitted the definition of `AbstractCurveType` and the rest of the GML geometry type hierarchy that would be set up in a real implementation. Note further that a BLOB or CLOB type might be used in place of the `VARCHAR(512)`.

Now consider the `LandParcel` feature example from Section 20.1.2.

```
<app:LandParcel gml:id="r1">
    <gml:name>Sick Childrens Hospital</gml:name>
    <app:area>500</app:area>
    <gml:extentOf>
        <gml:Polygon>
            <gml:exterior>
                <gml:LinearRing>
                    <gml:coordinates>0.0,0.0 10.0,0.0 10.0,
                    ↳10.0 0.0,10.0 0.0,0.0
                    ↳</gml:coordinates>
                </gml:LinearRing>
            </gml:exterior>
        </gml:Polygon>
    </gml:extentOf>
</app:LandParcel>
```

The content model for this encoding becomes the following:

```
CREATE TYPE gml_LinearRing UNDER gml_AbstractCurveType
{gml_coordinates VARCHAR(512)};

CREATE TYPE gml_Polygon UNDER gml_AbstractSurfaceType
(gml_exterior gml_LinearRing,
gml_interior gml_LinearRing);

CREATE TYPE app_LandParcel UNDER gml_AbstractFeatureType
(area NUMBER,
gml_extentOf gml_Polygon);

CREATE TABLE LandParcels OF app_LandParcel;
```

20.2.3 GML feature relationships

Feature relationships can also be supported by an ORDBMS, as shown in the Oracle Objects types that can be created for the following `Building` and `Restaurant` example from Section 20.1.3:

```
<app:Building gml:id="b1">
    <gml:name>Cerco</gml:name>
    <app:numFloors>10</app:numFloors>
    <app:contains>
        <app:Restaurant gml:id="r2">
            <gml:name>Fluid</gml:name>
            <app:type>seafood</app:type>
            <app:menu
                ↳xlink:href="http://www.placestoeat.
                ↳com#Fluid.xml"/>
        </app:Restaurant>
    </app:contains>
</app:Building>
```

The corresponding Oracle Objects types and tables are

```
CREATE TYPE app_RestaurantType UNDER
    ⌐gml_AbstractFeatureType
(app_type VARCHAR2(20),
app_menu VARCHAR(50));

CREATE TYPE app_RestaurantsType as table of
    ⌐app_restaurantType;

CREATE TYPE app_BuildingType UNDER
    ⌐gml_AbstractFeatureType
(app_numFloors NUMBER,
app_contains app_RestaurantsType);

CREATE TABLE Buildings OF app_BuildingType;
```

20.2.4   Basic rules for mapping GML to ORDBMS

The basic rules for mapping GML to an object-relational database are listed below. For each GML

1. feature type, create a corresponding type in the ORDBMS;
2. geometry type, create a corresponding type in the ORDBMS. Canonical types can be created for the core GML geometries;
3. feature type, create a corresponding table in the ORDBMS based on the ORDBMS feature type created in Step 1;
4. geometry type, create a corresponding table in the ORDBMS based on the ORDBMS feature type created in Step 2. Note that this step is not always required, given that in many cases (such as in the LandParcel example) the ORDBMS feature type table is automatically created;
5. property of simple type, there is a corresponding attribute in the associated ORDBMS feature type that was created in Step 1;
6. property of complex type, there is a corresponding attribute in the associated ORDBMS feature type that was created in Step 1.

20.3   GML and the geo-relational model

The geo-relational model was first popularized by ESRI and GeoVision. With this model, a relational or pseudo-relational database is used to store attributes and geometric connectivity. Separate flat files are used to store the actual geometry.

With this model, the term 'Coverage' refers to the partition of a geographic 'space' into a set of features – for example, roads, rivers and land parcels. Note that this term is a special case of what OGC calls a Coverage as used in GML. While ArcCoverages can be encoded as GML Coverages, in this chapter we only consider their encoding as GML features. Note further that in ESRI ArcInfo terminology, a feature class does not correspond to a kind of geographic entity but to a geometry, label or tic mark. Many GIS systems use this 'map oriented' terminology in which a feature is something that appears on a map.

Consider again the `Vehicle` feature instance discussed earlier in this chapter:

```
<app:Vehicle gml:id="v1">
   <gml:name>Fire Truck</gml:name>
   <app:weight>4300</app:weight>
   <app:manufacturer>Hale Equipment</app:manufacturer>
   <app:model>Cobra-1</app:model>
   <app:horsepower>410</app:horsepower>
</app:Vehicle>
```

In the case of an ArcInfo Coverage, the `Vehicle` is represented as a Coverage class. Although this particular example does not have a location, it is necessary to choose an ESRI feature class to represent the object. By choosing the `point(node)` feature class, the `Vehicle` can be represented as shown in Table 15.

Consider again the `LandParcel` example discussed earlier in this chapter:

```
<app:LandParcel gml:id="r1">
   <gml:name>Sick Childrens Hospital</gml:name>
   <app:area>500</app:area>
   <gml:extentOf>
      <gml:Polygon>
         <gml:exterior>
            <gml:LinearRing>
               <gml:coordinates>0.0,0.0 10.0,0.0 10.0,
                  ⌣10.0 0.0,10.0 0.0,0.0
                  ⌣</gml:coordinates>
            </gml:LinearRing>
         </gml:exterior>
      </gml:Polygon>
   </gml:extentOf>
</app:LandParcel>
```

An ESRI Coverage Class (`app_LandParcel`) can be created to represent this GML feature, and the `app_LandParcel` can be represented in the geo-relational model as a `Polygon` feature class. Table 16 shows an example of the `Polygon` table.

Note that in the geo-relational model, a Polygon must be bounded by arcs, and therefore the `LinearRing` must be decomposed from GML into one or more arcs (each bounded by nodes) in the geo-relational Coverage model. This is shown in Figure 2.

Since there are no other GML features in this example, the partition into arcs is completely arbitrary. Table 17 shows a possible set of node and arc tables.

Table 15   Geo-relational table for `Vehicle`

				Vehicle		
Node ID	gml_id	gml_name	app_weight	app_ manufacturer	app_model	app_ horsepower
1	v1	Fire Truck	4300	Hale Equipment	Cobra-1	410

Table 16   `Polygon` feature attribute table

Polygon feature attribute table			
Polygon ID	**gml_id**	**gml_name**	**app_area**
A	r1	Sick Childrens Hospital	500

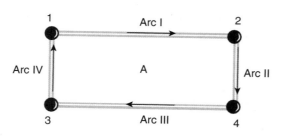

Figure 2 Geo-relational polygon bounded by four arcs.

Table 17   `Polygon, Arc and Node` tables

Polygon table	
Polygon ID	**Arc List**
A	I, II, III, IV

Arc table				
Arc ID	**From Node**	**To Node**	**Left Polygon**	**Right Polygon**
I	1	2		A
II	2	4		A
III	4	3		A
IV	3	1		A

Node table		
Node ID	**X-Coordinate**	**Y-Coordinate**
1	0.0	0.0
2	10.0	0.0
3	10.0	10.0
4	0.0	10.0

Note that the above tables contain information that is **not** in the GML encoding of the `LandParcel` feature, in particular, the coordinates of the nodes and the decomposition into directed edges. This information can be represented in GML using the GML topology model, which is covered in Chapter 13.

The node points in the above tables are the only points on the arcs, which is not typically the case. This means that in addition to the tables shown, the GML geometry needs to be mapped into the appropriate geometry representation for arcs and other geometries used in the implementation of the geo-relational model.

20.3.1 Basic rules for mapping GML to the geo-relational model

On the basis of these examples, the general process for mapping GML into the geo-relational model can be summarized as follows:

1. The GML geometry should be decomposed into polygons, arcs and nodes that correspond to the geo-relational model.
2. GML polygons can map directly to polygons in the geo-relational model.
3. The boundaries of GML polygons can map to arcs in the geo-relational model, but this mapping is not unique, since the geo-relational model adds an additional topology that may or may not be present in the GML encoding. The user must then make choices as to how the GML boundary should be represented in terms of geo-relational arcs.
4. Having decomposed the polygon boundaries into arcs, the directed nodes defining these arcs can be readily mapped to GML points.
5. Finally, the portions of GML curve elements that connect the directed nodes can be mapped according to the representation of curves in the particular implementation of the geo-relational model.

Although GML can represent any data in ArcInfo Coverages, the opposite is not true. GML is much more expressive than ArcInfo Coverages and it is possible to construct GML data – for example, features with multiple geometry properties of different types and time-dependent features – that cannot be easily mapped to an ArcInfo Coverage. This is also true of other spatial data representations.

Note that these restrictions do not prevent you from building a Web Feature Service (WFS) that serves GML data from an underlying ArcInfo Coverage. The WFS advertises the schemas that it supports, and this can include the underlying geometries supported by the WFS.

20.4 GML and ArcSDE

The most common spatial database, at the time of this writing, is most likely ESRI's ArcSDE, which can be used as a GML data store as long as you accept ArcSDE's limitations. Because it is more expressive than ArcSDE, GML can represent features and geometries that cannot be represented in ArcSDE. In particular,

- ArcSDE has no topology model. Only certain types of GML geometry properties and their values can be expressed in ArcSDE.

- The GML geometry model is compliant with ISO/TC 211 19107, and, therefore, can represent many types of geometries, including composite curves and surfaces, solids and various non-linear curve types that are not supported by ArcSDE.
- ArcSDE allows a feature to have only one type of geometry associated with it. This means that only GML features with a single geometry property can be mapped to ArcSDE. Note, however, that this can be circumvented to some degree by using multiple features in ArcSDE to represent a single feature in GML.
- ArcSDE/Geodatabase allows the representation of relationships between features. In principal, this representation can be used to construct features that are composed of other features. Essentially, this makes use of the underlying relational database's ability to express relationships. Note that a relationship is typically represented using a foreign key pointer between the tables that represent the related entity types. In the case of many-to-many relationships, the relationship is represented by a 'relationship table'. This table – or more correctly, half of this table – is represented by a GML property connecting the two GML objects that participate in the relationship.
- ArcSDE can also be used to represent certain types of time-dependent features. Note that in ArcSDE it may be challenging to model complex time types that can be represented in GML.

These restrictions do not prevent you from being able to create a WFS-based on an underlying ArcSDE data store. The WFS advertises the schemas that it supports, and these can include the restricted geometry types available from ArcSDE.

ArcSDE relies on an underlying relational database such as Oracle 8i or Microsoft SQL server for data storage. Geometry objects are restricted to the OGC Simple Feature Model and include `Point`, `LineString`, `Polygon`, `MultiPoint`, `MultiLineString` and `MultiPolygon`. These objects are stored as BLOBS in the underlying relational store and managed as geometric objects by the ArcSDE application. In this context, the primary function of ArcSDE is to support the spatial indexing of these geometry objects and to provide an associated set of spatial operators. These functions enable ArcSDE to support spatial queries and operations that are not supported by the underlying relational store.

ArcSDE non-geometric types rely on the attribute types of the underlying relational database, as shown in Table 18. ArcSDE organizes features into feature classes, which are collections of one or more features (of one geometric type) and are synonymous with the term 'layer' from SDE 3.x. In some cases, ArcSDE documentation refers to features as if they are geometries – for example, line feature and point feature – however, the reader should not be misled by this terminology. A feature in ArcSDE is a business object – for example, `Road` or `Well` – that can have only one geometric description.

ArcSDE implements a feature class as one or more tables, depending on the Database Management System (DBMS) and column type used for storing the geometry. ArcSDE does not change the existing DBMS or affect current database

Table 18 Mapping XML Schema simple types to ArcSDE simple types

Simple Types from XML Schema	Simple Types in Oracle 8i/9i
string	Varchar2 or char
double	number
integer	integer
positive integer	integer
datetime	date
ID	Varchar2 or char
enumeration	Lookup table

Table 19 Mapping GML data to ArcSDE

GML Entity	ArcSDE/Oracle Representation
Application feature type (for example, Road)	A business table with a 'spatial' column. The nature of the spatial column depends on how geometry is deployed in ArcSDE. There are three choices: • Compressed binary • Normalized OGC Geometry • Object-Relational
Simple properties of a feature (for example, numberOfLanes)	An attribute of the above business table.
Complex properties of a feature (for example, feature association or relationship)	A foreign key pointer from the feature to a table for the complex property value, or an intersection table for many-to-many relationships.

applications. It simply adds a spatial column to tables and provides tools for a client application, such as ArcMap, to manage and access the geometry data referenced by that column. In ArcSDE, a feature class or table can be mapped to a feature collection in GML. Table 19 shows how different GML entities can be mapped to ArcSDE.

As discussed above, ArcSDE does not support multiple geometry types for a single feature class. Note also that the geometry property (role name) in the GML data model may be lost in the translation to ArcSDE.

An additional restriction of ArcSDE is that it does not support XML Namespaces. There is no really satisfactory solution to this deficiency, however, it is possible to create a lexical convention for property and object names that enables the translation of Qualified Names (QNames) using their XML Namespace prefixes. For example, the XML element <app:Road> can be translated to app_Road. Since the XML Namespace prefix has no particular meaning, it is also necessary to maintain a mapping between the prefix and the actual namespace. This can be done by maintaining a prefix mapping table as shown in Table 20.

Table 20   Namespace mapping table

Prefix	Namespace
gml	http://www.opengis.net/gml
app	http://www.ukusa.org/
gde	http://www.galdosinc.com/

20.4.1   Representing GML features in ARCSDE relational tables

GML features can be modelled using relational tables. Table 21 demonstrates how simple user-defined Road features and their properties are represented in relational (for example, Oracle) tables.

Table 22 shows the separate table that contains the ArcSDE-defined geometry. Each Road feature from Table 21 has a gid key that points to a corresponding geometry record in the ArcSDE-defined geometry table. For example, the Road feature with the id a101 has a gid of 102, which points to the gid 102 record in the ArcSDE geometry table.

Table 21   User-defined road features and properties in relational tables

	Road		
id	numberoflanes	surface	gid
a100	2	gravel	101
a101	3	dirt	102
a103	2	gravel	103

Table 22   ArcSDE-defined geometry

gid	entity	...	points
101	2	...	\<compressed point binary coordinates\>
102	3	...	\<compressed multipoint binary coordinates\>
103	2	...	\<compressed binary coordinates\>

20.4.2   Modelling OGC simple geometries by tables

ArcSDE supports the OGC Simple Feature Model geometry types, for example Point, LineString, Polygon, MultiPoint, MultiLineString,

Table 23 Modelling points and multipoints

gid	entity	. . .	points
101	1	. . .	<compressed point binary coordinates>
102	2	. . .	<compressed multipoint binary coordinates>

Table 24 Modelling linestrings and multilinestrings

gid	entity	. . .	points
200	3	. . .	<compressed linestring binary coordinates>
201	9	. . .	<compressed multilinestring binary coordinates>

Table 25 Modelling polygons and multipolygons

gid	entity	. . .	points
300	5	. . .	<compressed polygon binary coordinates>
301	11	. . .	<compressed multipolygon binary coordinates>

`MultiPolygon` and `MultiGeometry`. The coordinate data is stored as a BLOB in a `points` column with the type of the geometry indicated by the value of the entity attribute (for example, `entity=1` => `points` contains the coordinates of a `Point`). This is illustrated for points (`entity="1"`) and multipoints (`entity="2"`) in Table 23. ArcSDE also supports linestrings (`entity="3"`) and multilinestrings (`entity="9"`), as shown in Table 24. ArcSDE supports polygons (`entity="5"`) and multipolygons (`entity="11"`), as shown in Table 25.

20.4.3 Representing the road feature in ArcSDE

Consider the following example of a simple GML Road feature:

```
<app:Road gml:id="p21">
   <app:surfaceType>gravel</abc:surfaceType>
   <app:numberOfLanes>2</app:numberOfLanes>
   <app:travelDirection>NE</app:travelDirection>
   <app:divided>False</app:divided>
   <gml:centerLineOf>
```

```
        <gml:LineString srsName="..">
          <gml:coordinates>100.1,20.3 120.2,22.4
             ⌐142.3,25.6</gml:coordinates>
        </gml:LineString>
      </gml:centerLineOf>
    </app:Road>
```

The Road feature can be modelled in ArcSDE as shown in Table 26.

Note that the `gid` in the first table contains keys that point to the corresponding geometry in a separate table. The `gid` key can be seen as equivalent to the GML geometry-valued property, `centerLineOf`. To model this correctly, you need to add an additional table that represents the correspondence between `gid` keys and property names.

Table 26   Road feature modelled in ArcSDE

	app_Road			
gml_id	app_surfaceType	app_noLanes	app_divided	gid
p21	paved	2	false	001

gid	entity	...	...	points
001	3	...	...	<binary coordinates>

20.5 Chapter summary

GML data can be mapped to many different existing data storage systems, including RDBMS, Object-Relational Database Management Systems (ORDBMS) and legacy GIS, such as ArcSDE. With RDBMS, it is relatively easy to map simple features whose properties are all of simple type; however, the mapping of simple types might differ, depending on the relational database. Because many relational databases do not support XML namespaces, the prefix must be incorporated into the column names (for example, `app_weight`). When mapping GML to relational databases, the unique identifier for a feature class should be a primary key.

GML geometry can be stored as XML text in a BLOB field, but the application program must know how to interpret the contents of the field. Another solution is to map each GML geometry to a separate table and link the tables together with keys and join operations. For example, a `Bridge` feature table can have a `gml_centerLineOf` column with a key pointing to a separate `gml_LineString` table. The join operation for this example is `gml_centerLineOf=geometry_key`. This approach becomes more complex for handling features with more complex geometry, because many tables are required.

GML feature relationships can also be modelled using the same class and foreign keys model that can be used for GML geometry. Feature collections can also be represented in relational tables. For example, a `NationalPark` feature collection can have foreign keys that link to separate tables for each member of the feature collection, such as `Island` and `Dock`. Separate relational tables can also be used to model properties of complex type.

ORDBMSs provide a more flexible option for storing GML data. Feature types require a corresponding type and object table in the ORDBMS, and geometry types should be modelled in a similar way. Properties of simple type and complex type have a corresponding attribute in the associated ORDBMS feature type.

To map GML data to the geo-relational model, GML geometry should be decomposed into the appropriate geometries (polygons, arcs and nodes) that correspond to the geo-relational model. GML polygons can map to polygons, and the boundaries of the GML polygons can map to arcs. The directed nodes that define the arcs can be mapped to GML points, and the different portions of GML curve elements that connect the directed nodes can be mapped according to the particular implementation of the geo-relational model. Note that more complex GML data cannot be easily mapped to the geo-relational model, however.

ESRI's ArcSDE is currently the most common spatial database, and it can be used as a GML data store, though with some limitations. There are certain aspects of GML data that cannot be represented in ArcSDE, for example, topology, some types of geometries and more complex time types. It is possible to represent feature relationships in ArcSDE by using relational tables with foreign key pointers.

In ArcSDE, feature classes are implemented as one or more tables. The feature tables contain a spatial column that points to a separate geometry table. For example, a `Road` feature can have a `gid` column with foreign keys that point to the corresponding `gid` in a separate geometry table. Complex properties can also be modelled with foreign keys that point to a separate table for the complex property.

Additional references

MacDonald, A. (1999) *Building a Geodatabase*. ESRI Press, USA.

Razavi, A.H. (2002) *ArcGIS Developer's Guide for Visual Basic Applications (VBA)*. Delmar Learning, Clifton Park, New York.

West, R. (2001) *Understanding ArcSDE*. ESRI Press, USA.

Appendix A

GML core schemas

This appendix provides you with a complete list of the core schemas in GML 3.0. The purpose and typical use of each schema are also included.

Schema	Purpose	Typical Use
`gmlBase.xsd`	Provides the base types and other required constructs for creating GML objects. With the exception of Xlinks and Basic Types, all other GML core schemas use this schema directly or indirectly.	Every application domain uses this schema, but often indirectly, given that most of the GML core schemas import the GML Base schema.
`feature.xsd`	Provides constructs for defining features and feature properties in GML. It defines the abstract feature type of GML, which is used to create new features in GML application schemas.	Most application domains use this schema to create GML application schemas.
`geometryBasic0d1d.xsd`	Defines the abstract geometry elements and types for GML and provides coordinate properties and simple geometries, such as points and lines.	This schema is required for all application domains that want to describe the geometry properties of features.
`geometryBasic2d.xsd`	Provides simple two-dimensional geometries, such as surfaces and polygons.	Used in conjunction with `geometryBasic0d1d.xsd` to describe features that have two-dimensional geometry properties.

Geography Mark-up Language (GML). R. Lake, D. S. Burggraf, M. Trninić, L. Rae © 2004 Galdos Systems Inc.
Published by John Wiley & Sons, Ltd ISBNs: 0-470-87153-9 (HB); 0-470-87154-7 (PB)

Schema	Purpose	Typical Use
geometryPrimitives.xsd	Provides additional one-, two- and three-dimensional primitives, including curves and solids.	Used in conjunction with geometryBasic0d1d.xsd to describe features that have one-, two- and three-dimensional geometry properties that are not provided by the basic geometry schemas.
geometryAggregates.xsd	Provides geometry aggregates, which are multiple geometries, such as multi-points and multi-polygons.	Used in conjunction with geometryBasic0d1d.xsd for features that have collections of geometry objects as property values.
geometryComplexes.xsd	Provides complex and composite geometries.	Used in conjunction with geometryBasic0d1d.xsd to describe features that have complex or composite geometries.
xlinks.xsd	Provides attributes for referencing data from other sources. This schema is an implementation of the W3C XLink specification. It is not in the GML namespace.	Used for remote properties and to reference other data sets, including those belonging to other authorities.
coverage.xsd	Defines elements and types for supporting Coverages and is based on ISO 19123.	Required for all application domains that want to create Coverages. For example, Coverages can be used for weather reports, remotely sensed images and soil maps.
valueObjects.xsd	Provides the foundation for creating user-defined quantities and values.	Used in association with the Coverages schema for defining coverage range values and for sensor data with observations.
units.xsd	Is used for defining units of measure, such as length and mass.	Used by all application domains that need to specify units of measure in their geographic data.
temporal.xsd	Is used for defining elements and types that deal with time. This schema is based on the ISO 19108 Temporal Schema. This schema is associated with the Dynamic Feature schema.	Used by all application domains that deal with objects that change over time. Use this schema if you want to provide an object with a time stamp, specify a creation time and ending time, or specify the duration of an event. The Temporal schema can be used in most application domains.

(continued overleaf)

Schema	Purpose	Typical Use
`dynamicFeature.xsd`	Provides a pattern for modelling features that change over time. This schema uses the Temporal schema.	Used for describing mobile objects, such as persons or vehicles. It can also describe the evolution of objects, such as floods or the construction of a building or roadway.
`grids.xsd`	Provides elements and types for simple grid structures. The schema is based on ISO 19123 and is typically used with the Coverages schema.	Used for defining grid coverages.
`topology.xsd`	Provides topology elements and types based on the topology section of the ISO 19107 Spatial schema. Used for expressing the topological properties of features.	Used by all application domains that want to describe topological relationships between features. For example, the transportation industry might use elements from the Topology schema to describe bus routes.
`measures.xsd`	Provides some standard measured quantities.	Can be used wherever measured quantities are used.
`basicTypes.xsd`	Defines a set of basic data types that can be used in GML and GML application schemas.	Used for all kinds of GML application schemas.
`direction.xsd`	Provides elements and types that can be used to express direction in various ways, for example, North.	Used for all schemas that need to express direction (for example, a directed observation).
`observation.xsd`	Provides the definition of the concept of an observation. The schema can be extended to create application-specific observations.	Used by all domains that need observations. For example, it can be used to describe observations at Points of Interest (for example, in Tourism) or to describe sensor observations from satellites.
`defaultStyle.xsd`	Provides the `defaultStyle` property that can be attached to a feature collection or a feature. This schema is different from the other GML schemas, because it is concerned with how data is presented, not with the data itself.	Used by all domains that need a mechanism for graphically presenting GML data.
`coordinateReference System.xsd`	Provides elements and types for creating Coordinate Reference Systems (CRS) dictionaries	Used for CRS descriptions for CRS data exchange. CRS dictionaries are available as CRS registry services.

Schema	Purpose	Typical Use
coordinateOperations.xsd	Provides elements and types for creating dictionaries of coordinate operations.	Used in association with the CRS schemas.
coordinateSystems.xsd	Provides elements and types for defining dictionaries of Coordinate Systems (CSs). It also provides types that are used in CS instances.	Used in association with the CRS schemas.
datums.xsd	Provides elements and types for creating datum dictionaries. It also provides definitions for ellipsoid and prime meridian objects.	Used in association with the CRS schemas.
referenceSystems.xsd	Provides elements and types for defining spatial reference systems, including the abstract base element and type for CRSs.	Used by all application domains that create reference system dictionaries.
dataQuality.xsd	Provides elements and types for expressing accuracy when defining CRS support components.	Used in association with the CRS schemas.
dictionary.xsd	Provides elements for creating dictionary instances. Also provides the base types for creating specialized dictionaries.	Used in association with units of measure (units.xsd) and CRS schemas.
gml.xsd	Provides a 'wrapper' for including all of the GML core schemas.	Used if your application needs to include more than one of the GML schema modules.
smil20-language.xsd smil20.xsd	Provided to fix a bug in the W3C SMIL schemas.	Used in conjunction with GML defaultStyle.xsd for animated graphical presentations of GML data.

Appendix B

Resources

There are many sources available if you need more information about GML. Given that GML is based on XML, you can also benefit from additional sources on XML. Since the Web is constantly changing, certain URLs might not be current.

Bibliography

Bureau International des Poids et Mesures (BIPM). *The International System of Units (SI)*. 1998. 16 October 2003. <http://www.bipm.fr/pdf/si-brochure.pdf>.

DI BATTISTA, G., EADES, P., TAMASSIA, R., and TOLLIS, I.G. (1999) *Graph Drawing: Algorithms for the Visualization of Graphs*, ISBN 0133016153. Prentice Hall, Upper Saddle River, NJ.

DICK, K. (2000) *XML: A Manager's Guide*, ISBN 0201433354. Addison-Wesley, Reading, MA.

FOWLER, M. and SCOTT, K. (2000) *UML Distilled: A Brief Guide to the Standard Object Modeling Language*, ISBN: 0-201-65783-X, 2nd Edition. Addison-Wesley Professional, Reading, MA.

Galdos Systems Inc. *Developing and Managing GML Application Schemas: Best Practices*. 15 May 2003. 22 May 2003. <http://www.geoconnections.org/developersCorner/devCorner_devNetwork/components/GML_bpv1.3_E.pdf>.

Internet Engineering Task Force (IETF). *RFC 3406: Uniform Resource Names (URN) Namespace Definition Mechanisms*. October 2002. 12 October 2003. <http://www.ietf.org/rfc/rfc3406.txt?number=3406>.

Internet Engineering Task Force (IETF). *RFC2141:URN Syntax*. May 1997. 17 October 2003. <http://www.faqs.org/rfcs/rfc2141.html>.

JELLIFFE, R. and Academia Sinica Computing Centre. *The Schematron Assertion Language 1.5*. October 2002. 16 October 2003. <http://www.ascc.net/xml/resource/schematron/Schematron2000.html>.

LANGRAN, G. (1992) *Time in Geographic Information Systems*, ISBN 0748400036. Taylor & Francis, London.

MARSH, D. (1999) *Applied Geometry for Computer Graphics and CAD*, ISBN 9624301123. Springer-Verlag, London.

MOLENAAR, M. (1998) *An Introduction to the Theory of Spatial Object Modelling for GIS*, ISBN 074840774X. Taylor & Francis, London.

Open GIS Consortium, Inc. *OpenGIS© Recommendation Paper: Recommended XML Encoding of Coordinate Reference System Definitions*. Wayland, MA, May 2003. 17 October 2003. <http://www.opengis.org/docs/03-010r7.pdf>.

Open GIS Consortium, Inc. *OpenGIS® Abstract Specification: Topic 5 – Features*. Wayland, MA, 24 March 1999. 17 October 2003. <http://www.opengis.org/docs/99-105r2.pdf>.

Open GIS Consortium, Inc. *OpenGIS® Abstract Specification: Topic 6 – The Coverage Type and its Subtypes*. Wayland, MA, 18 April 2000. 17 October 2003. <http://www.opengis.org/docs/00-106.pdf>.

Open GIS Consortium, Inc. *OpenGIS® Abstract Specification: Topic 2 – Spatial Referencing by Coordinates*. Wayland, MA, 8 March 2002. 17 October 2003. <http://www.opengis.org/docs/02-102.pdf>.

Open GIS Consortium, Inc. *OpenGIS® Catalog Services Specification*. Wayland, MA, 13 December 2002. 15 October 2003. <http://www.opengis.org/docs/02-087r3.pdf>.

Open GIS Consortium, Inc. *OpenGIS® Filter Encoding Implementation Specification (Version 1.0.0)*. Wayland, MA, September 2001. 17 October 2003. <http://www.opengis.org/docs/02-059.pdf>.

Geography Mark-up Language (GML).   R. Lake, D. S. Burggraf, M. Trninić, L. Rae © 2004 Galdos Systems Inc.
Published by John Wiley & Sons, Ltd   ISBNs: 0-470-87153-9 (HB); 0-470-87154-7 (PB)

Open GIS Consortium, Inc. *OpenGIS® Geography Markup Language (GML) 2.0*. Wayland, MA, 20 February 2001. 15 October 2003. <http://www.opengis.org/docs/01-029.pdf>.

Open GIS Consortium, Inc. *OpenGIS® Geography Markup Language (GML) Implementation Specification (Version 3.00)*. Wayland, MA, January 2003. 15 October 2003. <http://www.opengis.org/docs/02-023r4.pdf>.

Open GIS Consortium, Inc. *OpenGIS® Styled Layer Descriptor Implementation Specification (Version 1.0)*. Wayland, MA, September 2002. 23 October 2003. <http://www.opengis.org/docs/02-070.pdf>.

Open GIS Consortium, Inc. *OpenGIS® Web Coverage Service Implementation Specification (Version 1.0.0)*. Wayland, MA, 27 August 2003. 17 October 2003. <http://www.opengis.org/docs/03-065r6.pdf>.

Open GIS Consortium, Inc. *OpenGIS® Web Feature Service Implementation Specification (Version 1.0.0)*. Wayland, MA, September 2002. 17 October 2003. <http://www.opengis.org/docs/02-058.pdf>.

Open GIS Consortium, Inc. *Web Map Service Implementation Specification*. Wayland, MA, 27 November 2001. 15 October 2003. <http://www.opengis.org/docs/01-068r2.pdf>.

Open GIS Consortium, Inc. *Recommended Definition Data for Coordinate Reference Systems and Coordinate Transformations*. Wayland, MA, November 2001. 17 October 2003. <http://member.opengis.org/tc/archive/arch01/01-014r5.pdf>. OGC members-only access.

Open GIS Consortium, Inc. *OpenGIS Location Services (OpenLS™): Core Services*. Wayland, MA, 19 April 2003. 20 October 2003. <http://member.opengis.org/tc/archive/arch03/03-006r1.pdf>. OGC members-only access.

Open GIS Consortium, Inc. *OpenGIS® Location Services (OpenLS™): Part 6 – Navigation Service*. Wayland, MA, 19 April 2003. 20 October 2003. <http://member.opengis.org/tc/archive/arch03/03-007r1.pdf>. OGC members-only access.

SCHNEIDER, P.J. and EBERLY, D.H. (2002) *Geometric Tools for Computer Graphics*, ISBN 1558605940. Morgan Kaufmann, San Francisco, CA.

SHENE, Dr. C.-K. *Introduction to Computing with Geometry Notes*. December 2002. Michigan Technological University. 15 October 2003. <http://www.cs.mtu.edu/~shene/COURSES/cs3621/NOTES/notes.html>.

VALENTINE, C., DYKES, L., and TITTEL, E. (2002) *XML Schemas*, ISBN 0782140459. Sybex Inc., Alamed, CA.

World Wide Web Consortium. *Cascading Style Sheets, Level 2 CSS2 Specification*. May 1998. 17 October 2003. <http://www.w3.org/TR/REC-CSS2/>.

World Wide Web Consortium. *Extensible Markup Language (XML) 1.0 (Second Edition)*. October 2000. 20 September 2003. <http://www.w3.org/TR/REC-xml>.

World Wide Web Consortium. *Namespaces in XML*. January 1999. 20 October 2003. <http://www.w3.org/TR/1999/REC-xml-names-19990114/>.

World Wide Web Consortium. *Namespaces in XML 1.1*. 18 December 2002. 20 October 2003. <http://www.w3.org/TR/xml-names11/>.

World Wide Web Consortium. *RDF Vocabulary Description Language 1.0: RDF Schema*. January 2003. 20 September 2003. <http://www.w3.org/TR/rdf-schema/>.

World Wide Web Consortium. *Resource Description Framework (RDF) Model and Syntax Specification*. February 1999. 20 September 2003. <http://www.w3.org/TR/REC-rdf-syntax/>.

World Wide Web Consortium. *Scalable Vector Graphics (SVG) 1.1 Specification*. January 2003. 20 September 2003. <http://www.w3.org/TR/SVG11/>.

World Wide Web Consortium. *Simple Object Access Protocol (SOAP) 1.1*. 8 May 2000. 20 September 2003. <http://www.w3.org/TR/SOAP/>.

World Wide Web Consortium. *SMIL, Synchronized Multimedia Integration Language (SMIL 2.0)*. August 2001. 21 October 2003. <http://www.w3.org./TR/smil20/>.

World Wide Web Consortium. *W3C XML Pointer, XML Base and XML Linking*. April 2000. 20 September 2003. <http://www.w3.org/XML/Linking>.

World Wide Web Consortium. *Web Services Description Language (WSDL) 1.1*. March 2001. 20 September 2003. <http://www.w3.org/TR/wsdl>.

World Wide Web Consortium. *XML Base W3C Recommendation*. 27 June 2001. 20 September 2003. <http://www.w3.org/TR/xmlbase/>.

World Wide Web Consortium. *XML Schema Part 0: Primer*. May 2001. 20 September 2003. <http://www.w3.org/TR/xmlschema-0/>.

World Wide Web Consortium. *XML Schema Part 2: Datatypes*. May 2001. 20 September 2003. <http://www.w3.org/TR/xmlschema-2/>.

World Wide Web Consortium. *XPointer, XML Pointer Language (XPointer) Version 1.0*. August 2002. 20 September 2003. <http://www.w3.org./TR/xptr/>.

World Wide Web Consortium. *XML Path Language (XPath) Version 1.0*. 16 November 1999. 21 October 2003. <http://www.w3.org/TR/xpath>.

World Wide Web Consortium. *XSL Transformations (XSLT) Version 1.0*. 16 November 1999. 20 September 2003. <http://www.w3.org/TR/xslt>.

YUAN, May. *Temporal GIS and Spatio-Temporal Modeling*. 1996. 5 February 2003. <http://www.ncgia.ucsb.edu/conf/SANTA_FE_CD-ROM/sf_papers/yuan_may/may.html>.

Additional online resources

Geometry

> http://velab.cau.ac.kr/lecture/Bspline.ppt
> http://www.vrac.iastate.edu/~carolina/519/notes/13.bspline.pdf
> http://w3imagis.imag.fr/~Brian.Wyvill/course/notes/bspline.pdf

Topology

> http://www.math.toronto.edu/~drorbn/People/Eldar/thesis/index.html
> http://www.colorado.edu/geography/courses/geog_5003_s03/lecnotes/
> Modeling%20Our%20World.pdf
> http://www.cs.albany.edu/~amit/tutijcai.html

CRS examples and schemas

> http://crs.opengis.org/crsportal

Relevant ISO specifications

> The following documents can be purchased online at http://www.iso.org:
>
> *ISO 8601, Data Elements and Interchange Formats – Information Interchange – Representation of Dates and Times.* December 2001.
>
> *ISO 11404, Information Technology – Programming Languages, their Environments and System Software Interfaces – Language-Independent Datatypes.* December 1996. <http://www.iso.ch/iso/en/ittf/PubliclyAvailableStandards/s019346_ISO_IEC_TR_11404_1996(E).zip>
>
> *ISO PDTS 19103, Geographic Information – Conceptual Schema Language,* 2001.
>
> *ISO DIS 19107, Geographic Information – Spatial Schema.* December 2000.
>
> *ISO DIS 19108, Geographic Information – Temporal Schema.* September 2002.
>
> *ISO DIS 19111, Geographic Information – Spatial Referencing by Coordinates.* October 2000.
>
> *ISO DIS 19115, Geographic Information – Metadata.* May 2003.
>
> *ISO DIS 19123, Geographic Information – Schema for Coverage Geometry and Functions.* April 2001.

Organizations

Open GIS Consortium (OGC)

> http://www.opengis.org/

World Wide Web Consortium (W3C)

> http://www.w3.org/

Database Promotion Center, Japan (DPC)

> http://www.dpc.or.jp/

International Organization for Standardization (ISO)

> http://www.iso.org/

Bureau International des Poids et Mesures (BIPM)

> http://www.bipm.fr/enus/

OASIS

> http://www.oasis.org/

Mailing lists

gml-interest

> http://groups.yahoo.com/group/gml-interest/

Web sites

GML Central

> http://www.gmlcentral.com/

Topology and Geography

> http://mathworld.wolfram.com/

X Methods (web services)

> http://www.xmethods.com/

XML Schema

> http://www.xml.com
> http://www.xml.org

GXML

> http://gisclh.dpc.or.jp/gxml/contents-e/index.htm

Open source projects

GML for Java

> http://gml4j.sourceforge.net/

Extension Functions for JAVA, GML, SVG, XSLT

> http://gml-xslt-ext.sourceforge.net/

Geo-Tools Project (includes GeoServer)

> http://www.geotools.org/

Schema repositories

> http://schemas.opengis.net/

Tools

Geographic Data Servers (see http://www.opengis.org/resources)

> Galdos Cartalinea Geographic Data Server (http://www.galdosinc.com).
> Ionic WFS for Oracle 8i/9i Spatial (http://www.ionicsoft.com/).
> Intergraph's GeoMedia GML Data Server (http://www.intergraph.com/gis).
> Cadcorp SIS Version 6.0 – Spatial Information System (http://www.cadcorp.com/products.htm).

GML Schema Registries

> Galdos' INdicio Geographic Registry (http://www.galdosinc.com/).
> The ebXML Registry (http://www.ebxml.org/).
> CubeWerx Inc. (http://www.cubewerx.com/).
> Compusult Limited (http://www.compusult.nf.ca/).
> Polexis (http://www.polexis.com).
> Ionic Software (http://www.ionicsoft.com).

Style Engines and Editors

> Galdos' FreeStyler (http://www.galdosinc.com).
> Ionic Software (http://www.ionicsoft.com).
> CubeWerx Inc. (http://www.cubewerx.com/).
> Autodesk Inc. (http://usa.autodesk.com/).

Schema Design Tools

> Altova's XMLSpy (http://www.xmlspy.com/).
> Oxygen (http://www.oxygenxml.com).

Schema Parsers

> Galdos GML Schema Parser (http://www.galdosinc.com).
> GML4J (https://sourceforge.net/projects/gml4j/).
> GeoTools/GeoServer Project (https://sourceforge.net/projects/geotools/).

GML Data-Conversion Tools

Safe Software's Feature Manipulation Engine (FME) (http://www.safe.com/).
Snowflake's Go Loader (http://www.snowflakesoft.co.uk/).

The following companies have also developed GML data-conversion plug-ins for their applications:

- Ionic Software (http://www.ionicsoft.com).
- Intergraph Corporation (http://www.intergraph.com/).

Appendix C

Glossary of terms

abstract types

The types from the GML core schemas that serve as base types, from which new types can be derived in application schemas. For example, `AbstractFeature-Type` is an abstract type from the GML core Feature schema that is used to create new feature types. Note that abstract types cannot be instantiated as GML data.

aggregation relationship

The relationship between different components in which components are aggregates of other components that can also exist independently. For example, a CRS definition is an aggregation of support components, including datum and coordinate system definitions.

algebraic topology

A branch of pure mathematics that uses algebraic techniques to describe topological spaces.

API

Application Programming Interface. A set of interface definitions or program methods that exposes the functionality of a software system.

application schema

Schemas that extend abstract types from the GML core schemas. Before you deploy an application in GML, you need to create appropriate application schemas to provide a framework for defining the geographic data. GML application schemas are XML Schemas that follow GML rules. Can also be seen as an application vocabulary.

Geography Mark-up Language (GML). R. Lake, D. S. Burggraf, M. Trninić, L. Rae © 2004 Galdos Systems Inc.
Published by John Wiley & Sons, Ltd ISBNs: 0-470-87153-9 (HB); 0-470-87154-7 (PB)

array

In GML, an array is a collection of objects that are homogeneous in type and arranged in a particular order. In GML, properties that encapsulate the values of arrays are typically of `ArrayAssociationType`. For example, the `members` property is of `ArrayAssociationType` type.

association

A structural relationship describing the connections between instances of objects. GML properties are often used to express associations between different GML objects. For example, `centerLineOf` is a geometry-valued property that expresses an association between a feature and a geometric aspect of that feature.

attribute

An element modifier in XML that provides additional information about an element. For example, `gml:id` is an attribute that uniquely identifies an element instance. Note that in GML, attributes are not used to encode the values of properties. For example, you cannot express the span of a Bridge as `<app:Bridge gml:id="b1" span="50"/>`.

bag

In GML, a bag is an unordered collection of objects that can be repeated and have different content models.

base type

A generic term referring to the top-level type from which other XML Schema types (content models) derive. For example, `AbstractGMLType` is the base type from which all GML types derive.

base units

Independent units of measure that cannot be a combination of other units. Base units are the most commonly used units for fundamental quantities, such as length, mass and time. There are different systems of units for which base units are defined, one of which is the International System of units (SI). GML 3.0 provides the `BaseUnit` type for encoding base unit entries in a units of measure dictionary.

binding

An association between the abstract message components (for example, message arguments) and a specific transport protocol.

BLOB

Binary Large OBject. A special database attribute type that can contain large quantities of unstructured data. For example, a BLOB field can be used to store fragments of complex GML data in a RDBMS. Unlike other RDBMS types, the application type of a BLOB data element is unknown to the RDBMS.

boundary

The boundary of a region in space is the intersection of the closure of the region and the closure of its complement (or exterior).

catalog service

A geospatial web service that provides facilities for querying and managing various kinds of metadata, such as application schemas. See also *registry service*.

category

A kind of measurement scale defined in measurement theory. For example, soils may be assigned a soil type from a list of soil type categories. GML provides the `Category` element for encoding single-valued categories and `CategoryList` for lists of multiple category values.

client

In geospatial web services, a client is an application that sends messages to a web service. The client relies on the web service to perform specific operations, such as to perform calculations about geographic information.

closure

The union of a geometric or topological primitive with its boundary. The closure of a set A is the intersection of all the closed sets containing A.

coboundary

A coboundary of a topology or geometry primitive A consists of all primitives whose boundary contains the primitive A. See also *boundary*.

complex types

In XML Schema, complex types are a type of content model in which elements can contain elements and/or carry attributes. All GML objects have content models that are complex types in the sense of XML Schema.

complex-valued property

Any property whose content model is of complex type. It can have simple or complex content, and it does not need to derive from a GML object type. Note that if a content model is not derived from a GML object, it has no GML-specific meaning.

composite curve

A sequence of curves or a collection of curves that join at common end points. With the exception of the first curve, each curve in the sequence starts at the endpoint of the previous curve. The `CompositeCurve` type – which has an `curveMember` property of unbounded multiplicity – is used to encode composite curves in GML. (A `CompositeCurve` is a curve in GML.)

composite solid

A collection of solids that join on common boundary surfaces. The `CompositeSolid` type – which has a `solidMember` property with unbounded multiplicity – is used to encode composite solids in GML. Note that a `CompositeSolid` has all the geometric properties of a `Solid`.

composite surface

A collection of surfaces that join along common boundary curves. The `CompositeSurface` type – which has a `surfaceMember` property with unbounded multiplicity – is used to encode composite surfaces in GML. Note that a `CompositeSurface` has all the geometric properties of a `Surface`.

composition relationship

The relationship between different components in which components are composed of other components that cannot exist independently of the parent component. Note that in a composition, the parts of the composition cannot be shared.

concrete type

Any type from GML core or application schemas that can be instantiated in GML documents. Note that it is not an abstract type.

content model

A grammar that describes the rules governing the content of an element in an XML Schema definition.

conventional units

Units of measure that are neither base nor derived. Many application domains use conventional units; for example, the unit for length in the United States is 'feet' instead of 'metre'. Conventional units can be converted to a preferred unit, which is either a base or a derived unit. GML provides the `ConventionalUnit` element for denoting conventional unit dictionary entries.

coordinate reference system

A system for indicating the location of real world objects, such as features, using coordinates. GML 3.0 provides a number of schemas that contain the required components for creating Coordinate Reference Systems (CRS) dictionaries. The `srsName` attribute is used to relate GML geometry data to the CRS used to interpret its coordinate values.

coordinate system

A mathematical means of associating coordinates to points. A Coordinate System (CS) is a required component in a CRS. GML provides a number of types for encoding different kinds of CS dictionary entries.

core schema

The GML core schemas provide the abstract and concrete types that are used to create application schemas. In GML 3.0, there are 28 core schemas.

count

An integer that represents a rate of occurrence. In GML, `Count` can be used to encode a single count, and `CountList` can be used for multiple counts.

coverage

An OGC and ISO/TC 211 term that describes the distribution of some quantity or property over a portion of the surface of the earth. Digital Elevation Models (DEMs) are a kind of Coverage.

CRS

Coordinate Reference System

CS

Coordinate System

cubic spline

A cubic spline is a spline constructed of third-order polynomials, in a piecewise manner, that pass through a set of m control points.

curve

A one-dimensional geometric primitive. The first control point of a curve is the start point, and the last is the endpoint. In GML, an instantiated curve has a `segments` property that can contain or reference curve segment elements, such as `LineStringSegment`, `Arc` and `CubicSpline`. See also *curve segments*.

curve segments

A curve segment is a 'building block' from which a curve is constructed. There is a single interpolation method for each segment. For example, only circular interpolation is used within an `Arc` segment. The curve segments are connected to one another, and (with the exception of the last segment) the endpoint of each segment is the start point of the next segment in the list. GML 3.0 provides a number of curve segment elements, including `LineStringSegment`, `Arc` and `CubicSpline`. The `segments` property encapsulates the segments of the curve. A curve segment on its own is not a curve according to ISO/TC 211 19107.

data modeller

The individual, or group of individuals, responsible for modelling data on the basis of existing business practices and using these models to create GML application schemas. Systems analysts and database administrators fall into this category.

database administrator

A technical specialist responsible for developing and maintaining databases and database applications. With GML, a database administrator is typically responsible for creating GML application schemas.

datum

A datum is defined as any numerical or geometrical quantity or set of such quantities, which serve as a reference or base for other quantities. In GML, this is a support component for CRS definitions. GML supports different kinds of datum definitions, including geodetic and engineering datums.

DBMS

Database Management System. A system for managing database content, independent of its physical storage model.

definition

In GML, an object in a dictionary. A definition can be a value of a `dictionaryEntry` property or a property that derives from `dictionaryEntry`. For example, BaseUnit is a definition that can be in a units of measure dictionary, and CompoundCRS is a definition object for CRS dictionaries. Also called a *Dictionary Entry*.

DEM

Digital Elevation Model. A kind of Coverage that represents terrain relief.

derived units

Units that are derived from one or more base units within a particular system of units. For example, a Newton is derived from the SI base units with the following formula: $m \cdot kg \cdot s^{-2}$.

dictionaries

Dictionaries are used to store definitions, such as CRS and units of measure, which are referenced by other GML documents.

dictionary entry

An entry in a dictionary. See also *Definition*.

direction

In GML, direction is specified by the `direction.xsd` schema, which defines a `direction` property with a choice of the following values: `DirectionVector`, `CompassPoint`, `DirectionKeyword` or `DirectionString`.

domain

1. A particular field or area of interest that is concerned with a subset of information. For example, the transportation industry and local government are both application domains.
2. Coverages also have domains. In this context, a domain is a set on which a mathematical function is defined.

DPC

Database Promotion Center. A Japanese organization for promoting database standards in Japan. The DPC is responsible for the G-XML project. See also *G-XML*.

DTD

Document Type Definition. A non-XML language for describing XML documents. DTD is part of the W3C XML 1.0 specification. In GML 3.0, XML Schema is used instead of DTD to provide the framework for defining GML types. DTDs were used in GML 1.0.

edge

A one-dimensional topological primitive whose geometric realization is a curve.

element

In XML, an element is a unit of content. In GML, features and their properties are expressed as elements. An element has an element name, element content and, in some cases, attributes.

encoding

The expression of a data element or elements using a language written in XML (such as GML). In this guide, an encoding typically refers to GML data instances.

enumeration

A simple type in XML Schema with a specified set of values. For example, a Province simple type can be an enumeration that lists all of the Canadian provinces and territories. See also *simple type*.

face

A two-dimensional topological primitive whose geometric realization is a surface.

feature

A meaningful object in the world, usually with location or geographic extent in space and time. Examples of features include Roads, Rivers and Lakes. GML features are defined in GML application schemas and their content models must derive, directly or indirectly, from `AbstractFeatureType`.

feature collection

A collection of features that is also a feature. For example, a `City` element can be encoded as a feature collection with a number of feature members, such as Roads and Buildings. At the same time, the `City` is considered a feature. The content model of a GML feature collection must derive, directly or indirectly, from `AbstractFeatureCollectionType`.

feature relationship

A relationship between different features, expressed by a property. For example, a `Bridge` feature can span a `Gorge` feature. The `span` property expresses a relationship between these two features.

geodetic datum

In Geodesy, geodetic datums define the size and shape of the earth and the origin and orientation of the coordinate systems used to map the earth. Geodetic datums typically include an `Ellipsoid` definition.

geometry

In GML, geometry refers to the class of objects that are used to describe the geometric aspects of a geographic feature. The GML 3.0 geometry primitives are defined in the following five schemas: `geometryBasic0d1d.xsd`, `geometryBasic2d.xsd`, `geometryAggregates.xsd`, `geometryPrimitives.xsd` and `geometryComplex.xsd`. A geometry in GML is one of the following: a point, curve, surface or solid.

geometry properties

A property that is used to express the geometric qualities of a feature by referencing a Geometry object. The name of the property typically reflects the role of the Geometry object, for example, `centerLineOf`. Also called *geometry-valued properties*.

geometry schemas

A set of schemas in GML 3.0 that provides a number of different types, ranging from simple to complex, that can be used to express the geometry properties of a feature.

georelational model

The data model used in ESRI's ARCInfo, in which spatial and attribute information are handled separately and then linked together by unique identifiers. In this model, geographic features are represented as an interrelated set of spatial and descriptive data. Attribute information is stored in a relational or pseudo-relational database table.

geospatial web service

A web service that provides access to, or data processing on, geographic information. The OGC Web Feature Service (WFS) is an example of a geospatial web service.

Geo-Web

A globally integrated and accessible spatial infrastructure that comprises a number of interconnected geospatial datasets and web services. With the Geo-Web, a client can access standardized geographic information from anywhere in the world. GML and geospatial web services will play an important role in deploying the Geo-Web.

GIS

Geographic Information System. An organized collection of computer hardware, software, geographic data, and personnel designed to efficiently capture, store, update, manipulate, analyse and display all forms of geographically referenced information.

global property

In GML, a global property is a GML property, which is defined as an XML schema global element.

GML

Geography Mark-up Language, an OGC standard for encoding geographic information in XML.

GML gateway

A program on a network that provides access to other networks, including those that use different protocols. A GML gateway allows systems that do not recognize GML to exchange data with GML-enabled systems.

GML schema parser

A program that reads a GML application schema document and identifies the different parts, such as attributes, elements and content models.

G-XML

An XML language for describing geographic information, similar to GML. The G-XML Committee of the DPC developed G-XML. Note that G-XML Version 3.0 is a GML 3.0 application schema focused on location-based services. See also *DPC*.

heterogeneous

Composed of parts that are of different type. For example, in GML, a bag can contain elements of different type (content model).

history

A GML dynamic feature property that associates a feature instance with a sequence of time slices that encapsulate the evolution of the feature over time. Note that GML provides a built-in history property called `track`. A user-defined history property must have `history` as the head of its substitution group. See also *time slice*.

homogeneous

Composed of parts that are of one type. In GML, an array can only contain homogeneous content. For example, the `featureMember` property can only contain objects that are substitutable for `_Feature`.

HTML

Hypertext Mark-up Language. A mark-up language used to describe the content and appearance of web pages on the Internet.

HTTP

HyperText Transfer Protocol.

in-line property

In GML, a property whose value is enclosed between the opening and closing property tags. For example, `<app:span>30</app:span>` is an in-line property.

instance document

In GML, an instance document is an XML document that complies with a GML application schema (written in XML Schema).

interoperability

The capability for multiple information systems to exchange data with each other.

ISO

International Organization for Standardization.

ISO/TC 211

A committee of the ISO concerned with geomatics standards. The various GML schemas are based on standards established by ISO/TC 211.

LBS

Location-Based Services. Services that are based on information about the location of an individual, regardless of their mode of transport. These services are accessed by mobile devices, such as cellular phones or PDAs.

local property

A GML property defined by a local XML Schema element. See also *global property*.

metadata

Data that describes other data. GML provides a mechanism to denote metadata referenced from a GML feature or other GML object. Metadata may be specified in GML by a user-defined Metadata application schema. An example is provided in Chapter 11.

namespace

A concept from XML that is used to assign unique names to elements or attributes. XML namespaces are required in GML. Namespaces are denoted in XML using namespace prefixes bound to a namespace specified by a URI. Namespace prefixes are not namespaces. In GML, an application schema must declare a target namespace.

node

A zero-dimensional topological primitive whose geometric realization is a point.

object

In GML, an object is an element whose content model derives from `Abstract-GMLType`. GML objects include features, geometries, topologies, CRSs and styles.

observable

A value that is subject to observation.

ODBC

Open Database Connect. An industry standard for management of a session with a remote relational database.

OGC

Open GIS Consortium. International standards body headquartered in Washington, D.C. and Boston U.S.A. OGC is focused on defining standards to promote geographic systems interoperability. GML is an OGC standard.

ORDBMS

Object Relational DataBase Management System.

POI

Point of Interest. A term from location-based services.

point

A zero-dimensional geometric primitive, which can be specified in n-dimensional space by n ordinates. A point represents a position in space.

polygon

A two-dimensional geometric primitive described by a single outer boundary (exterior) and zero or more inner (interior) boundaries.

properties

In GML, properties are used to express the characteristics of an object. GML uses a naming convention for properties in which the first letter of a property is always lower case while all subsequent words are upper case; for example, `centerLineOf`. An `xlink:href` attribute attached to a property points to the value of the property.

range

In the definition of a mathematical function ($f : A \rightarrow B$), the set B is called the range of the function. In programming languages, a range is a kind of data type, sub-typed from some numeric type, such as integers or floating-point numbers, and defined by a pair of numbers from the numeric super type. In XML Schema, a range is one of the kinds of derived simple types.

range set

The part of a Coverage that specifies the range of the Coverage function. A range set is also referred to as the value set of the Coverage. For a DEM Coverage, the range set is a set of elevation values.

RDBMS

Relational DataBase Management System

RDF

Resource Description Framework. An XML language developed by the W3C for metadata description and semantic modeling.

RDFS

Resource Description Framework Schema (RDF Schema). A schema language for describing the content model of RDF documents. Also known as RDF/S.

realization

In *ISO/TC 211 19107 Spatial Schema*, a realization is a geometric object that represents the referencing topological object in terms of numerical coordinates. For example, a geometric `Point` is the realization of a topological `Node`.

rectified grid

A kind of grid geometry whose points are defined relative to a CRS. In GML, rectified grid also refers to a type of coverage whose geometry domain is a `RectifiedGrid`.

registry service

A registry is a type of catalog – typically operated by a designated registration authority – that complies with specific policies governing the use and management of metadata (for example, *ISO 11179-6* and *ISO 19136*).

remote property

In GML, a property that references a value located elsewhere, in the document, or at a remote location on the Internet. The `xlink:href` attribute attached to the property is used to point to the remote value of the property. See also *in-line property*.

root element

The top-level element in an XML (or GML) document that encapsulates all other elements. For example, a `Transit` feature collection might be the root element in a document.

schema component

The different 'building blocks' that are used to create schemas. For example, element declarations, complex types and simple types are all schema components.

semantic type

A classification of phenomena with common characteristics. Each semantic type represents some concept or object in a domain of discourse. In a GML application schema, the value of the name attribute of an element is the element's semantic type, for example, Road. See also *type*.

service offer

A description of a web service that describes the capabilities of the service; for example, the providers, the supported interfaces and any access constraints.

set

An unordered collection of non-repeated elements.

SI

International System of Units. A standardized system of units that has the following seven base units: metre, kilogram, second, ampere, kelvin, mole and candela.

simple type

An XML Schema structural type that only has text content. An element of simple type cannot contain child elements nor have attributes. See also *complex type*.

SLD

Styled Layer Descriptor. An OGC specification that provides an XML grammar describing the layout and appearance of a map. Also the name for the XML document describing the map style. Its functions partly overlap with GML default style.

SMTP

Simple Mail Transfer Protocol.

SOAP

Simple Object Access Protocol. An XML language for describing and implementing remote procedure calls and inter-process messaging. SOAP is a W3C specification.

spatial data warehouse

A database, or collection of databases, that stores geospatial information, often collected from multiple external data stores, in order to provide common access, distribution, integration or e-commerce functions.

spline

A function defined piecewise using control points and a specified interpolation schema. Splines are very useful for modelling and are used extensively in computer graphics.

SRS

Spatial Reference System. A system for referencing the location of objects, either by coordinates or by spatial identifiers, for example by building name. A CRS is a spatial reference system that is based on numeric coordinates.

styling

In GML, styling refers to the interpretation of geographic data for the purpose of producing a presentation that is typically visual, for example, a map. The `defaultStyling.xsd` schema provides the types and elements for encoding default styles for GML data.

substitution group

A collection of XML elements that are substitutable for a specific named element in the group. Substitution groups are used in GML to create template patterns especially for collection, bag or set structures. For example, if you want to say 'any feature can go here' when you define a property in a GML application schema, use a substitution group, such as `_Feature`.

surface

A two-dimensional geometry primitive representing a flat or curved connected region of space. See also *surface patch*.

surface patch

A geometric 'building block' from which surfaces are composed in GML. A surface object is composed of one or more surface patches that describe the structure of the surface. The `patches` property is used to associate a GML object with its surface patches. The GML 3.0 surface patches are `PolygonPatch`, `Triangle` and `Rectangle`. Note that there is a single interpolation method for each patch. Surface patches are not geometry objects (and not surfaces), and therefore, they cannot have unique identifiers.

SVG

Scalable Vector Graphics. This is a W3C standard for the specification of two-dimensional graphics in XML.

temporal

A way of denoting time. In GML, the `temporal.xsd` and `dynamicFeature.xsd` schemas are used to express the time-related properties of features.

time slice

In GML, a time stamped container for the time-varying properties of a dynamic feature. A history property contains a series of times slices that record how the state of a feature changes over time. `MovingObjectStatus` is a built-in GML time slice that is contained within the `track` property. See also *history*.

topology

A branch of pure mathematics that is concerned with the description of abstract surfaces or objects and their structural relationships to one another. Important branches include algebraic topology, point-set topology and differential topology. GML 3.0 has a schema for expressing the topology properties of geographic features.

track

In GML, a specific kind of history property that describes the state of a moving object over time. See also *history*.

tuple

An ordered list of numbers. In GML, the value of the `coordinates` property can be a tuple list.

type

In GML, the name of an element whose content model is derived from `AbstractGMLType` is referred to as the semantic type of the GML object. In XML Schema, each element name has an associated content model, which is indicated by the `type` attribute. In other words, the terms 'Feature Type' or 'Feature Type Name' refer to the semantic type or element name (for example, `Road`) that is used in GML instance documents. This is not to be confused with the XML Schema content model (structural type), as defined by the `type` attribute in the element declaration – for example `RoadType`. Throughout this book, 'element' is often used instead of 'type' (for example, the `Road` element).

UML

Unified Modeling Language. A graphical conceptual schema language standardized by the OMG.

union

A union type is a composite simple data type in XML Schema.

uom

An attribute for referencing units of measure dictionary entries. In GML, all properties or objects that are of MeasureType can have a uom attribute. The attribute can also be included in an element's type definition in a GML application schema.

URI

Uniform Resource Identifier.

URN

Uniform Resource Name. A location-independent resource identifier. Unlike resource locators, URNs are less likely to change in the short term. According to GML best practices, the use of URNs or some other kind of persistent identifier scheme is recommended for all external links, including links from the uom, srsName and attributes.

value

The content (or target) of a GML property that can be expressed in-line or remotely, via an XLink reference. In the GML object-property model, objects have properties, which in turn have values.

value objects

Objects from the valueObjects.xsd schema that are used to represent specific kinds of values, such as a Category or a Quantity. Value objects are used for sensor data and in the description of the range parameters of a Coverage.

varchar

A data type from SQL that represents variable length character data.

W3C

World Wide Web Consortium. International organization that is concerned with establishing standards for the World Wide Web.

WCS

Web Coverage Service. An OGC specification defining interfaces for requesting, inserting, or updating coverages.

web services

An application that accepts and processes requests from clients across the Internet. Web Services access data stored in web servers and return the data to the clients who send the requests. See also *Geospatial Web Services*.

WFS

Web Feature Service. An OGC Service specification that defines interfaces for requesting, inserting, deleting or updating GML features over the Internet. WFS 1.0 supports GML 2.0. Support for GML 3.0 is anticipated during 2004.

WMS

Web Map Service. An OGC specification that describes interfaces for requesting a map visualization of geographic data.

wrapper

A kind of schema that is used specifically to include other schemas. In GML, the gml.xsd schema is a wrapper schema that includes all of the other GML 3.0 schemas.

WSDL

Web Services Description Language. A W3C XML language standard for describing the input/output messages, protocol binding and address of a web service. WSDL 1.1 is a W3C Note. WSDL 2.0 (previously called 1.2) is currently a W3C Working Draft.

XLink

A W3C standard that specifies the syntax and behaviour for hyperlink traversal in a set of XML documents. Also known as *XLL*. In GML, the xlink:href attribute is used to reference remote values.

XML

eXtensible Mark-up Language. A W3C standard mark-up language for encoding a wide variety of data and documents. XML is also a language for creating other mark-up languages, such as GML.

XML Schema

A schema language defined by the W3C for defining the content and structure of XML documents. GML schemas are written in XML Schema. Also known as *XSD*.

XPath

A W3C language for navigating, searching and selecting from XML documents.

XPointer

A W3C language for pointing to or selecting elements from XML documents. XPointer extends XPath.

XSD

The file extension for XML Schema files.

XSLT

Extensible StyleSheet Language Transformations. XSLT is a W3C standard for manipulating and transforming XML documents. With GML, you can use XSLT to perform many useful tasks, such as transforming schema-based data or generating graphical maps in SVG.

Appendix D

XMLSpy tutorial

One of the advantages of GML over conventional geographic data formats is that it leverages XML tools and technology. This appendix demonstrates how to use Altova's XMLSpy to create a GML application schema and a GML instance, based on the schema. Note that it is also possible to use many other XML editors to perform these tasks. You can download a trial version XMLSpy at http://www.altova.com/download.html. Note that XMLSpy 5.0 was used to create the schema and instance documents covered in this tutorial. Note further that there are many other XML editors, such as Oxygen (see http://www.oxygenxml.com).

D.1 Creating GML application schemas with XMLSpy

In this section, XMLSpy is used to create a GML application schema with a `City` feature collection. The resulting schema (`City.xsd`) is available on the *Worked Examples CD*. Note that you should read Chapters 9, 10 and 11 before performing the procedures in this section.

D.1.1.1 Getting started

1. Start **XMLSpy**, select **File > New**, then select **W3C XML Schema** (`xsd`) as the file type and click **OK**.
   This creates a blank document with no schema content.
2. Select **Schema Design > Schema settings** to select some basic schema parameters, as shown in Figure 1.
3. Enter the target namespace (in this example, www.ukusa.org) in the **Target Namespace** field.
   When you type the namespace prefix in the **Target Namespace** field, it automatically appears beneath the **Namespace** column in the table at the bottom of the **Schema settings** dialog box.
4. Enter the target namespace prefix (for example, `app`) in the **Prefix** column, as shown in Figure 1, click the 🗐 icon to enter the `gml` prefix and namespace (http://www.opengis.net/gml), enter any other namespaces and namespace prefixes, and click **OK**.
   The **Schema Design** window appears. If it does not appear, select **View > Schema Design View**.

Geography Mark-up Language (GML).   R. Lake, D. S. Burggraf, M. Trninić, L. Rae © 2004 Galdos Systems Inc.
Published by John Wiley & Sons, Ltd   ISBNs: 0-470-87153-9 (HB); 0-470-87154-7 (PB)

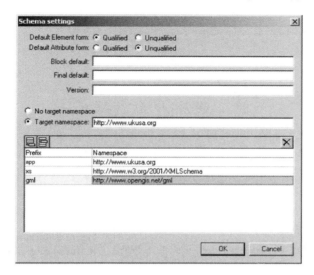

Figure 1 Setting namespaces and prefixes. Reproduced by permission of Altova.

5. To import the GML feature schema (`feature.xsd`) or all of the gml core schemas (`gml.xsd`), click the 曼 append icon (in the upper left corner of the schema window in XMLSpy) and select **Import** from the pop-up menu.

6. In the alert box, enter the location of the `feature.xsd` file and click **OK.**

   You can refer to schema locations that are local or anywhere on the Internet. You can get the GML core schemas at http://schemas.opengis.net/ gml. In this example, the schemas are from http://schemas.opengis.net/gml/ 3.0.1/base. For example, enter http://schemas.opengis.net/gml/3.0.1/base/ feature.xsd.

7. Enter the GML namespace (http://www.opengis.net/gml) in the field marked **ns:** to the right of the schema location.

D.1.1.2   Creating the root feature collection

Many GML application schemas have a root feature collection that is a container for a collection of feature instances. To create this root element, perform the following steps:

1. Rename the root element of the schema generated by XMLSpy, as shown in Figure 2, in which the root element is changed to `City`.

2. To set the type of this element, select the element (for example, `City`), select **type** in the **Details** window and then select `gml:Abstract-FeatureCollectionType` in the **type** drop-down list (see Figure 3). Click **derivedBy** in the **Details** window and select `extension`. Next, click **substGrp** in the **Details** window and select `gml:_FeatureCollection`.

   The **Details** window is typically located to the right of the main schema window (as shown in Figure 2). If the **Details** window is not displayed, select **Window>Entry Helpers**.

Figure 2 Creating the `City` root feature collection. Reproduced by permission of Altova.

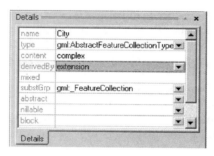

Figure 3 Setting the type of the root feature collection. Reproduced by permission of Altova.

Note: The root feature collection may derive from some other type (for example, `app:CityBaseType`), which in turn derives from `Abstract-FeatureCollectionType`. If this is the case, you should select this other type instead of `AbstractFeatureCollectionType`. Note that you can use the built-in convenience type `FeatureCollectionType` instead of `AbstractFeatureCollectionType`, in which case you do not need to select a value in the **deriveBy** field.

D.1.1.3   Creating features

To create a new feature, perform the following steps:

1. Click the 🗐 (Append) icon in the upper left corner of the Schema design window and select **Element** from the drop-down list.
2. Enter a name for the new element (for example, `Bridge`).
3. In the **Details** window on the right side of the **Schema Design** window, select the **type**, **derivedBy** and **substGrp** values.

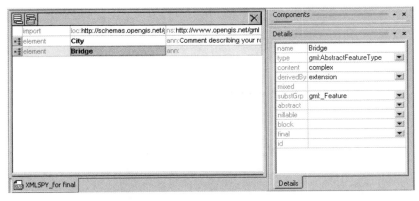

Figure 4 Creating a feature type. Reproduced by permission of Altova.

Figure 4 shows an example of the Bridge element, where the **type** is gml:AbstractFeatureType and the **substitutionGroup** is gml:_Feature. Initially, the **type** in the **Details** window must be set to gml:AbstractFeatureType. If you add additional properties, XMLSpy creates a locally defined anonymous complex type and derives the content model from gml:AbstractFeatureType. Note that if you want to define a feature with a named complex type, you need to create a separate complex type. This is discussed further below.

To create additional feature types, repeat steps 1 to 3. Once the feature name and **Details** have been set, add existing properties to the features, as discussed in the following section.

D.1.1.4   Adding feature properties

The simplest way to add properties to a feature is to select the feature and add a sequence of property elements. To do this, perform the following steps.

1. In the **Schema Design** window, select the feature by clicking on the schema detail icon to the left of the feature (for example, Bridge).
   This provides a graphical display of the schema of the selected feature, as shown in Figure 5.

   In Figure 5, the feature type is already derived from gml:AbstractFeatureType and inherits the boundedBy and location properties and the gml:id attribute. It also inherits name, description and metaDataProperty from gml:AbstractGMLType, which is higher up in the hierarchy.

2. Before adding the first property, insert a sequence into the feature type by right-clicking the feature type, for example Bridge. From the drop-down list, select **Add Child** and then **Sequence**.
   Note that the created sequence is automatically highlighted.

3. Add properties to the feature as follows:
   a. Right-click the sequence, select **Add Child** and then select **Element**.

   Figure 6 shows the Bridge feature with the following additional properties: span, height and centreLineOf.

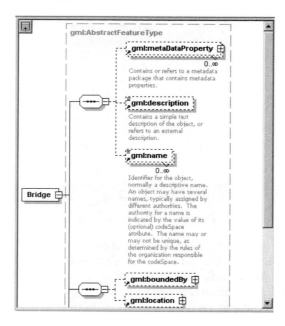

Figure 5 Graphical display of the Bridge feature. Reproduced by permission of Altova.

b. Do one of the following:
- To add a local property, enter the name of the property (for example, span) and then set the **type** of the property in the **Details** window on the right. For example, in this tutorial, the span **type** is xs:integer. Edit the **minOcc** and **maxOcc** fields, as required. In this example, the span and height property should have a **minOcc** value of 0, which means that they are optional properties, and the default **maxOcc** value of 1.

- To add a global property, enter or select the name of the property (for example, gml:centerLineOf) from the drop-down list at the right side of the element graphic, as shown in Figure 6. Edit the **minOcc** and **maxOcc** fields, as required. In this example, gml:centerLineOf is optional.

4. Add additional properties to your feature type by repeating step 3 for each new property.

The following section shows how to create a user-defined property (mobility) whose type extends a built-in XML Schema type.

D.1.1.5 Extending a simple type

You can create new simple properties by extending the XML Schema built-in types like xs:integer or xs:float. In this tutorial, the mobility property is of the simple type, app:MobilityType, which is created by the following steps.

To create a simple type, first create a new property element (in this case, mobility) by doing the following:

1. Click the 📇 (**Append**) icon in the upper left corner of the **Schema design** window and select **Element** from the drop-down list.
2. Enter a name for the new element (for example, `mobility`).

Now you need to define the type for the `mobility` property. To do this, perform the following steps:

1. Select the **Schema Design** view in XMLSpy and click on the 📇 (**Append**) icon in the upper left corner of the window.
2. Select the **SimpleType** menu item.
   A new item with the word '`simpleType`' is displayed in the main window, as shown in Figure 7.
3. Enter the name of the simple type (for example, `MobilityType`).
4. Select **restr** in the **Details** window (this is the default).
   Note that **restriction** is used here to illustrate how to create an enumerative simple type.
5. Select the data type to be restricted.
   In this example, select `xs:string` from the drop-down list to the right of **restriction** in the **Details** window.

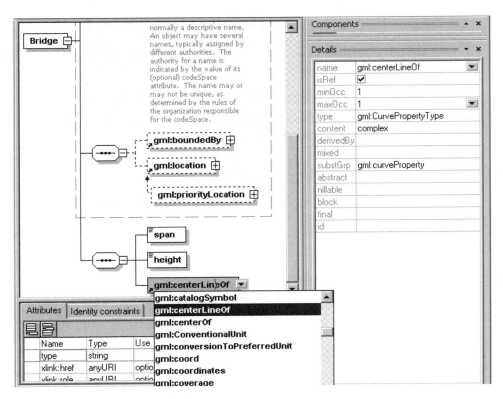

Figure 6 Adding properties to the feature. Reproduced by permission of Altova.

Figure 7 Creating a user-defined simple type. Reproduced by permission of Altova.

To create enumeration values for a simple type, do the following:

1. Select the **Enumerations** tab at the bottom of the **Facets** window.
2. Click on the ▤ (**Append**) icon at the top left corner of the **Facets** window.
3. Enter the enumerated value in the highlighted bar that appears each time you click on the ▤ icon.
4. Repeat for each enumeration value (as shown in Figure 8, in this example, the enumeration values for `MobilityType` are `Fixed`, `DrawBridge`, `HorizontalSwingBridge`, `UniversalSwingBridge`, `TelescopingBridge` and `LiftBridge`).

To associate the `mobility` property with the `app:MobilityType` simple type, do the following:

1. In the **Schema Design** window, select the `mobility` property.
2. In the **Details** window, specify the `type` (for example, `app:MobilityType`), as shown in Figure 9.

D.1.2   Creating a feature with a named type

Note that aside from the `mobility` property, all of the previously created elements in this tutorial do not have named types. As discussed in Chapter 11, an element should have a named type if you plan on deriving from that element's

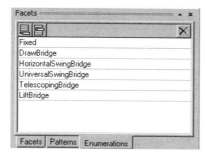

Figure 8 Enumeration values for `MobilityType`. Reproduced by permission of Altova.

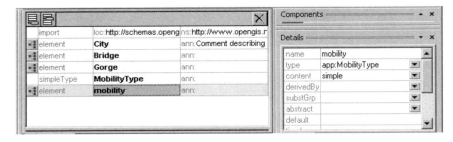

Figure 9 Specifying the simple type for a property. Reproduced by permission of Altova.

content model at a later time. Anonymous types (also known as Venetian blind style) are recommended for schema readability. The following steps show how to use XMLSpy to create a named type instead of an anonymous type:

1. In the **Schema Design** window, add or select an element (for example, `Bridge`). In the **Details** window, set the **type** to the new complex type (for example, `app:BridgeType`) and set the **substGrp** (for example, `gml:_Feature`).
   Note that if you enter a complex type that does not exist yet, the value of the **type** field is displayed in red.
2. In the **Schema Design** window, click the 🗐 (**Append**) icon and select **ComplexType**. Enter the name of the complex type (for example, `BridgeType`) in the new field.
3. In the **Details** window, select the base for the complex type. This is the complex type from which the new complex type derives (for example, `gml:AbstractFeatureType`).
4. To add content to the complex type, click the ▦ schema detail icon to the left of the complex type. Right-click the complex type graphic and select **Add Child > Sequence**.
5. Right-click the sequence and select **Add Child > Element**. Select or enter the name of an element (property) that you want to add to the new complex type. Edit the **Details** window for the element (as discussed in Section D.1.1.4).
6. Repeat step 6 for each element that you want to add (for example, `gml:centerLineOf`, `span`, `height` and `app:mobility`, as shown in Figure 10).
   Note that the order in which you assign elements is the order in which they must be instantiated.

D.1.3   Deriving from a user-defined type

To create a new feature whose content model derives from `BridgeType`, do the following:

1. Click the 🗐 (Append) icon in the upper left corner of the **Schema design** window and select **Element** from the drop-down list.

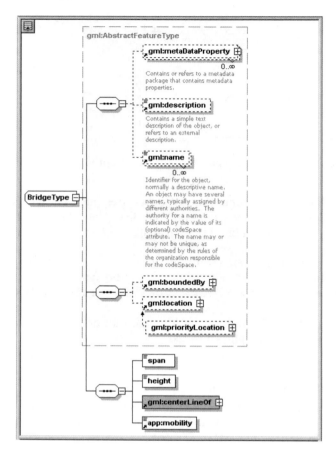

Figure 10 Named complex type with content. Reproduced by permission of Altova.

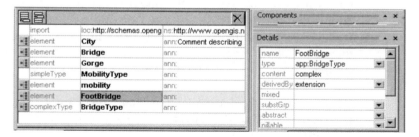

Figure 11 Deriving by extension. Reproduced by permission of Altova.

2. Enter a name for the new element (for example, FootBridge).
3. In the **Details** window on the right side of the **Schema Design** window, select the **type** (for example, app:BridgeType), the **substGrp** (for example, gml:_Feature), and the **derivedBy** option, as shown in Figure 11. Note that in this example the derivation is by extension; however it is also possible to derive by restriction.

4. Click the 昌 schema detail icon to the left of the new element (Foot-Bridge).

5. To add additional elements to the content model, right-click the element graphic (for example, FootBridge) and select **Add Child > Sequence**.

6. Right-click the sequence and select **Add Child > Element.** Select or enter the name of an element (property) that you want to add (for example, maximumWeight). Edit the **Details** window for the element (as discussed in Section D.1.1.4). Figure 12 shows the detailed schema view of the FootBridge feature with an added maximumWeight property.

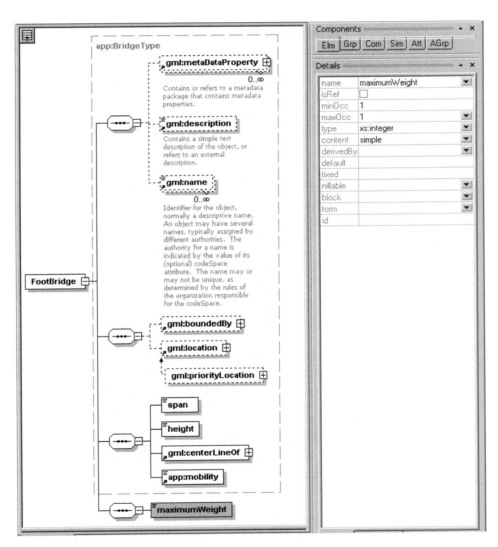

Figure 12 FootBridge with a maximumWeight property. Reproduced by permission of Altova.

D.1.4   Creating feature relationships and associations

As discussed in Chapters 9 and 11, it is possible to use some properties of GML features to express relationships between features. For example, you can have a `Bridge` feature that `spans` a `Gorge`. To illustrate how feature relationships are modelled in XMLSpy, first you need to create the target feature and then create the feature relationship property. Note that in this example, the target feature (`Gorge`) has a named type (`app:GorgeType`), however, this is not mandatory.

D.1.4.1   Creating the target feature

1. Click the ☰ (**Append**) icon in the upper left corner of the **Schema design** window and select **Element** from the drop-down list. Enter the name of the new element (for example, Gorge). In the **Details** window, set the **type** (for example, `app:GorgeType`), the **substGrp** (for example, `gml:_Feature`) and the **derivedBy** value (for example, `extension`).

2. To create the `GorgeType`, perform steps 2–6 from Section D.1.2. Enter `GorgeType` for the `complexType` name and add the appropriate properties (for example, `width`, `depth` and `centerLineOf`), as shown in Figure 13.

D.1.4.2   Creating the feature relationship property

1. To create a user-defined property that expresses the relationship between the features (for example, `Bridge spans Gorge`), click ☰ in the upper left corner of the **Schema design** window and select **Element** from the drop-down list. Enter `spans` as the name of the new element. In the **Details** window, set the **type** (for example, `app:GorgePropertyType`).

2. To create the property's complex type, click the ☰ (**Append**) icon in the upper left corner of the **Schema Design** window and select **ComplexType**. Enter the name of the complex type (for example, `GorgeProperty-Type`).

3. Click the ▦ schema detail icon to the left of the new property complex type. Right-click the complex type graphic and select **Add Child > Sequence**.

4. Right-click the sequence and select **Add Child > Element**. Select the name of an element (in this example, `app:Gorge`) that you want to add to the new complex type.
   Note that in most cases, GML properties can only have one object.

5. In the **Attributes** window (beneath the **Schema Design** window), click the ☰ (**Append**) icon and select **Attribute Group**. Click the arrow at the right side of the new **grp** field and select `AssociationAttributeGroup`, as shown in Figure 14.
   Note that most GML properties have this association group so that `xlink:href` can be used to reference remotely defined values.

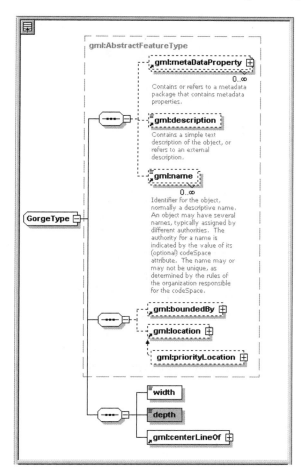

Figure 13 `GorgeType` with properties. Reproduced by permission of Altova.

D.1.4.3   Adding a relationship property to a feature

1. In the **Schema Design** view in XMLSpy, select the element or complex type to which you want to add the relationship property (in this example, `BridgeType`) and click the ▦ schema detail icon.
   Note that in this example, the complex type was selected instead of the element, because the element has a named type. In cases where the elements have anonymous types, select the element instead.

2. Right-click on the sequence (if it has already been created) and select **Add Child > Element**.
   Add a sequence if it has not already been added to the feature. See Section D.1.1.4.

3. In the new element graphic, enter the name of the relationship property (for example, `spans`), as shown in Figure 15.

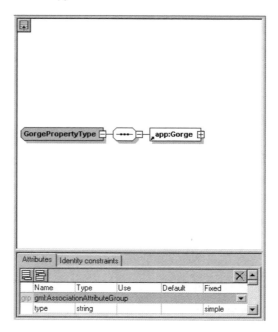

Figure 14 Adding content to a relationship property. Reproduced by permission of Altova.

4. In the **Details** window, select the **type** for the new property (for example, `app:GorgePropertyType`) and set the **minOcc** and **maxOcc** values, as required. In this example, `spans` is mandatory, and both values are set to 1.

D.1.4.4   Creating the relationship for the target feature

In the above example, the `spans` property expresses the relationship of the target feature to the source feature; `Bridge` is the source feature and `Gorge` is the target. When you model a relationship between features, you can also define a property of the target feature that expresses its relationship to the source feature. For example, the `Gorge` feature can be `spannedBy` the `Bridge`, as shown below. To create the `spannedBy` relationship in XMLSpy, you need to create a complex type for the property (in this case `BridgePropertyType`). Follow the steps from Section D.1.4.2 and Section D.1.4.3, except, add a `spannedBy` property to the `Gorge` element.

D.2   Creating GML application instances

This section shows how to use XMLSpy to create a simple GML instance with the elements that you created in the `City.xsd` schema.

D.2.1   Getting started

1. Start **XMLSpy** and select **File > New**. Select **XML Document** as the file type and click **OK**.

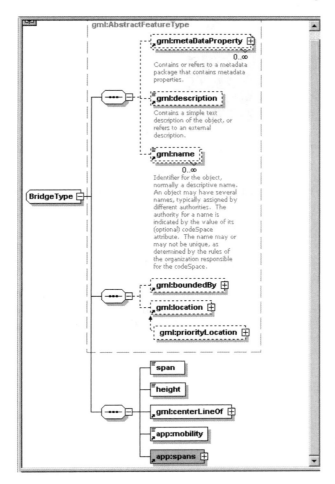

Figure 15 Adding a relationship property. Reproduced by permission of Altova.

2. When you are prompted to decide whether to base the XML document on a DTD or schema, select **Schema** and click **OK**.

3. Enter the GML application schema on which your instance document will be based (for example, `City.xsd`).

4. Once XMLSpy processes your schema, the **Select a root element** box appears. Enter a root element.
This element should be a feature collection (for example, `City` or `FeatureCollection`). After you enter the root element, XMLSpy creates a nearly empty instance document, in which `City` is the root element, as shown in Figure 16.

D.2.2 Entering properties and property values

1. To enter the properties of the root element, do the following:
   a) Select and right-click on the root element, for example `app:City`. Select **Add Child > Element**.

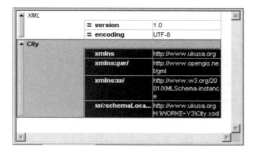

Figure 16 Empty instance document.
Reproduced by permission of Altova.

A list of valid child elements is displayed. These are all properties.

   b) Select one of the elements (for example, gml:featureMember).

2. To add a child element to the property (for example, gml:featureMember), right-click on the element, select **Add Child > Element**.

3. Select or enter the value of interest (for example, Bridge), as shown in Figure 17.
   Be sure not to select any abstract elements (for example, gml:_Feature), because they cannot appear in your GML instance document. Note that your instance document is not valid at this point, since you have not specified the values of the properties of the element (for example, Bridge) that you have entered.

D.2.3   Adding properties to the property value

1. Select Bridge and, if the feature is not already expanded, click the down arrow to the left of the feature.
   This displays all of the mandatory properties of the feature. It does not display any of the optional properties of the feature. Note that there is a box to the right of each property of the Bridge feature. Nothing appears in this box if you select a complex-valued property (for example, centerLineOf) or a text-valued property (for example, span or height).

Figure 17 Selecting a property value.
Reproduced by permission of Altova.

A drop-down list appears if you select properties with enumerative data types (such as `mobility`). This is shown in Figure 19.

2. Enter an integer value in the boxes beside the text-valued properties (for example, `span` and `height`).

3. For the complex-valued properties (for example, `centerLineOf`), do the following:

   a) Right-click on the property, then select **Add Child > Element**.
      A drop-down list appears and displays the possible values (for example, `_Curve` and `LineString`), as shown in Figure 18. Note that if you want to select one of the geometries from the `_Curve` substitution group (for example, `Curve`), you need to enter the value, because abstract elements cannot be instantiated.

   b) Select the appropriate value (for example, `gml:LineString`).

   c) Enter values of the `gml:pos` elements (add more `gml:pos` elements if necessary) by typing the values in the entry box to the right of the appropriate element.

4. For enumerative-valued properties (for example, `mobility`), select the desired value from the drop-down list, as shown in Figure 19.
   This adds values to the enumerative property.

5. To add an optional property to the feature, do the following:

   a) Right-click on the feature (for example, `Bridge`), then select **Add Child > Element**.
      The optional properties are displayed in a drop-down list.

   b) Select the property (for example, `Spans`) and add the value as appropriate (for example, if you select the `spans` property, the value (`Gorge`) automatically appears).

Figure 20 shows the `Bridge` feature with values for all of the properties.

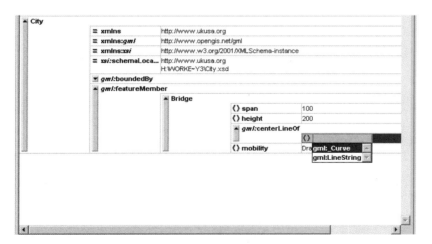

Figure 18 Select a value for a complex-valued property. Reproduced by permission of Altova.

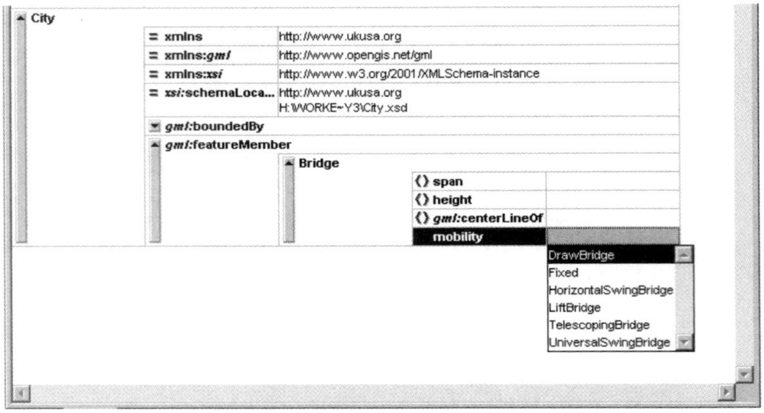

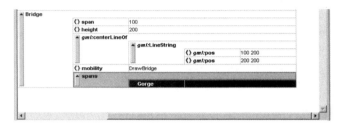

Figure 19 Selecting an enumerative property. Reproduced by permission of Altova.

Figure 20 `Bridge` feature with properties and property values. Reproduced by permission of Altova.

Note: You can check if your instance document is well formed at any time by selecting **XML > Check well-formedness** or by pressing F7. You can also check if the instance document is valid relative to your GML application schema (for example, `City.xsd`) by selecting **XML > Validate** or by pressing F8.

D.2.3.1   Text view of a sample instance document

The following example shows what the sample instance document should look like:

```
<City xmlns="http://www.ukusa.org"
    ⌐xmlns:gml="http://www.opengis.net/gml"
    ⌐xmlns:xsi="http://www.w3.org/2001/XMLSchema-instance"
    ⌐xsi:schemaLocation="http://www.ukusa.org City.xsd">
    <gml:boundedBy>
        <gml:Envelope>
            <gml:pos/>
            <gml:pos/>
        </gml:Envelope>
    </gml:boundedBy>
    <gml:featureMember>
        <Bridge>
```

```
            <span>100</span>
            <height>200</height>
            <gml:centerLineOf>
                <gml:LineString>
                    <gml:pos>100 200</gml:pos>
                    <gml:pos>200 200</gml:pos>
                </gml:LineString>
            </gml:centerLineOf>
            <mobility>DrawBridge</mobility>
            <spans>
                <Gorge/>
            </spans>
        </Bridge>
    </gml:featureMember>
</City>
```

Note that many of the elements are empty, and the attributes are missing. These can be completed in the **View > Enhanced-Grid View** by repeating the procedures for entering features and properties. You can also select **View > Text View** to work directly in the **Text View** if you are familiar with XML. Check for well-formedness and validate the documents at periodic intervals. This is much easier than debugging a complex instance document at a later stage.

Index

about attribute, 122–3, 266
AbstractCoordinateOperationType, 209
AbstractCoordinateReferenceSystemType,
    203, 204, 209–10
AbstractCoordinateSystemType, 205, 211
AbstractCoverageType, 256
AbstractCRSType, 203, 209
AbstractCurveType, 152, 312
AbstractDatumType, 206, 209, 211
Abstract elements, 79, 98, 99, 103–5, 209, 228
AbstractFeatureCollectionType, 107, 111,
    237, 238–9, 341, 356
AbstractFeatureType, 25, 79, 90, 93, 107, 147,
    311, 334, 341
    bridge example, 109–10, 357
    coverages, 257, 259
    default styling, 270
    examples, 103, 109–10, 122, 357
    observations, 236
AbstractGeneralDerivedCRSType, 203
AbstractGeometryType, 151, 152
AbstractGMLType, 96, 98–9, 335, 345, 350
AbstractPointOfInterest, 25
AbstractReferenceSystemBaseType, 202–3
AbstractStyleType, 265
AbstractTimeSliceType, 186–7
AbstractTopologyType, 171
Abstract types, 3, 25, 98, 99, 103–5, 334
    applications schemas, 107, 109
    CRS, 202–3, 209
    underscoring, 79
    WSDL, 17–18, 72, 73, 288, 289–91, 335
Active Server Page (ASP), 282
Adobe, SVG processing, 17, 72
Aerial photos, 237–9
Aggregate geometries, 145, 146
Aggregate values, 231–4, 250
Aggregation relationship, 334
Agriculture, 32
Air transportation, 32
Algebraic topology, 155, 334, 350
Altova, 108, 332
anchorPoint, 206
Angle brackets, 12–13

angle value, 207
Animation, 278, 283
<annotation> element, 115–16
Anonymous types, 110–11, 361
anyURI, 74, 79, 193, 266
app (prefix), 25–6, 82, 84, 88, 108
app:inlineCenterLineOf, 114
app:remoteCenterLineOf, 113–14
app:spans, 118
Application Programming Interfaces (APIs), 37,
    334
Application schemas, 3, 31–5, 73, 76, 334
    catalog services, 49
    coordinate reference systems, 201–2
    core schema import, 107, 123–6, 124, 256, 260
    coverages, 256–60
    creation, 108–20, 256–60
    default styling, 267
    development
        and deployment, 36, 96
        and management, 107–28
    dictionaries, 74, 196, 199
    dynamic features, 179–80
    existing, 126
    features, 90, 108–20
        import schema, 108–9, 201–2
    geometry elements, 150–2
    geospatial web services, 193, 291
    G-XML, 9
    history properties, 186–8
    metadata schemas, 120–3
    networks, 34
    objects, 107
    observations, 237
    range parameters, 260
    registries, 49–50
    target namespaces, 33–4, 107, 108, 345, 354
    time slices, 186–8
    topology properties, 170–1
    types, 110–11
    value objects, 230, 232–4
    values, 228
    WSDL, 51, 52, 291
    XMLSpy tutorial, 354–67

Geography Mark-up Language (GML).   R. Lake, D. S. Burggraf, M. Trninić, L. Rae © 2004 Galdos Systems Inc.
Published by John Wiley & Sons, Ltd   ISBNs: 0-470-87153-9 (HB); 0-470-87154-7 (PB)